T0186903

From Instrumentation to Nanotechnology

From Instrumentation to Nanotechnology

Based on a series of lectures presented at the Advanced Science and Engineering Research Council Vacation School on Instrumentation and Nanotechnology held at the University of Warwick, 16–21 September 1990

Edited by

Julian W. Gardner *and* **Harry T. Hingle**
University of Warwick, UK

GORDON AND BREACH SCIENCE PUBLISHERS

Australia • Canada • China • France • Germany • India • Japan • Luxembourg • Malaysia •
The Netherlands • Russia • Singapore • Switzerland • Thailand • United Kingdom

First published 1991
Second printing 1997

Amsteldijk 166
1st Floor
1079 LH Amsterdam
The Netherlands

Library of Congress Cataloging-in-Publication Data

From instrumentation to nanotechnology / edited by J. W. Gardner and
    H. T. Hingle.
        p.   cm. -- (Developments in nanotechnology, ISSN 1053-7465 :
    v. 1)
    "Lectures presented at an Advanced Science and Engineering
Research Council Vacation School on Instrumentation and
Nanotechnology held at the University of Warwick, 16–21 September,
1990"--
    Includes index.
    ISBN 2-88124-794-6 : $85.00 (U.S.)
    1. Engineering instruments.   2. Nanotechnology.   I. Gardner, J.
W. (Julian W.), 1958–   II. Hingle, H. T. (Harry T.), 1925–   .
III. Advanced Science and Engineering Research Council Vacation
School on Instrumentation and Nanotechnology (1990 : University of
Warwick)   IV. Series.
TA165.F76   1991
681'.2--dc20
                                                    91-28752
                                                        CIP

Contents

Introduction to the Series

The nanotechnological revolution will be as pervasive as the first industrial revolution, for it also introduces concepts and techniques that will influence virtually all aspects of technology and manufacture. Nanotechnology is concerned with the design and manufacture of components with submicrometre dimensions, of larger components with submicrometre tolerances or surface finishes, and of machines with submicrometre precision of positioning or motion. The achievement of ultra-high precision in machining and the manufacture of extremely small devices opens up prospects as diverse and futuristic as massive computing power, medical diagnostic and therapeutic devices inside the bloodstream, global personal communications, still smaller and higher resolution optical devices, and high-economy, low-cost automobiles. The economy of automobile engines has been steadily improving as the precision of component manufacture has increased. The high quality of videocassette recorders is ensured by the reliable, low-cost production of nanometric surface finishes on the recording heads and precise drive spindles. We are seeing in mechanical engineering the same movement towards miniaturization that has occurred over the last three decades in electronic engineering, and we should learn from the analogy; the nanotechnologist must draw from many disciplines to design with confidence in this field. The compartmentalization and consequent simplification of knowledge that works adequately on the large scale make no sense when applied to, for example, submicrometre "swarf" from an ultra-high precision machining process.

It takes a long time to establish a novel enabling technology. It is necessary that the active workers in the field take time to communicate the technology and the excitement they feel in being part of this revolution, whether they address established engineers or recent graduates. In this series of monographs and books on nanotechnology, we endeavour to stimulate, educate and motivate engineers and scientists at all levels in this new field. It will be cross-disciplinary, dealing with instrumentation and measurement, large-scale manufacture, novel manufacturing methods, applications in medicine, and many other relevant topics. The series is designed to encourage fruitful interactions between those who come from different traditional disciplines.

It is most appropriate that the first volume in this new series should arise from a short course intended to open students' eyes to nanotechnology and to give them skills appropriate to it. Many eminent authors have contributed, and the editors – who were responsible for the course programme – have taken pains to integrate the chapters into a logically constructed whole, in which measurement, the cornerstone of any technology, is related to the challenge of manufacturing components and operating machines at the near atomic level.

D. Keith Bowen

Preface

Measurement science is a fundamental forerunner to technology because the design of new systems depends upon data and because signals are needed for control. The usefulness of measurements in turn depends upon our ability to design and manufacture instrumentation of adequate performance. The increasing demand for improved technical performance of instruments requires new levels of both inventiveness in the physical principles used and precision in manufacture. Technological developments in turn feed off the techniques used in instrumentation, and so the cycle of increased performance continues.

This book traces the developments of precision measurement systems from relatively traditional instrumentation, through miniature devices, to the realm of nano-technology: the reader is led gently *From Instrumentation to Nanotechnology*. The book is aimed at practising engineers and physical and chemical scientists who need up-to-date information on the capabilities of modern instrumentation and on developments in nanotechnology. It will present something new to virtually every reader. However, as it does not depend on a detailed knowledge of a specialist subject, it will also be of interest to final-year undergraduate and postgraduate students who want to study these challenging and rapidly developing technologies.

The book is divided informally into sections though the grouping of chapters. Chapter 1 is an overview and describes some recent trends in the development of instrumentation that have led to this new field of science called *nanotechnology*.

Next, mathematical tools are revised. Chapters 2 and 3 provide the reader with a review of some of the basic methods used in signal processing and instrumentation. Readers who are already familiar with topics such as summary statistics, correlation and convolution theory may omit these. Chapter 4 explores some of the principles of the mathematical modelling of instruments and subsequently applies them to a piezoresistive pressure sensor.

Chapter 5 bridges from analysis to instrumentation by considering the design of software. It uses a few simple examples to demonstrate how algorithm design interacts with the hardware design in high-performance systems. This topic is often overlooked due to the more immediate and obvious need for good hardware design, but it will almost certainly be increasingly important in the next few years.

Two important fields of sensor design are then considered in detail in chapters 6 and 7. They review state-of-the-art ultrasonic sensors and silicon microsensors. Silicon microsensors are a recent development from the use of conventional microelectronic technology to fabricate miniature, low-cost devices. They are used in the measurement of either physical paramerers, such as displacement, acceleration and pressure, or chemical parameters, such as pH concentration and odour intensity.

Chapter 8 marks a change in the emphasis of the book from precision instrumentation and microengineering to nanotechnology. It reviews the field of nanotechnology, in its widest definition, covering a range of disciplines from biology to mechanical engineering. We then look briefly at some special manufacturing techniques in chapters 9 and 10 which address the problem of machining materials to an ultra-high precision through the use of energy beam lithography and ductile grinding.

Chapters 11 to 13 describe the use of optical techniques for measuring displacement, surface roughness and particle size at the nanometre level (10^{-9} m). Various optical techniques are exploited in nanotechnology, principally those of diffraction and interferometry. Readers who are unfamiliar with the principles of these techniques may need to refer to a general book on optics. The authors particulary recommend *Fundamentals of Optics* by Professors Jenkins and White.

One of the most exciting recent developments is that of scanning microscopes that can probe the nature of surfaces at the atomic level. Chapters 14 and 15 discuss the principles and mechanisms of such instruments that are being developed to measure surface profiles, such as the scanning tunnelling microscope, and associated nanoactuators required to achieve resolution at the nanometre scale.

The last chapter describes the development of the X-ray interferometer, an instrument which is capable of measuring displacements at the subnanometre level. In fact, the resolution of the instrument is so good that it might be reasonably quoted in picometres.

This book is based on lecture material that was presented at the Advanced Science and Engineering Research Council Vacation School on Instrumentation and Nanotechnology in September 1990 at the University of Warwick. The editors express their gratitude to the Science and Engineering Council for their financial support of the vacation school and the preparation of this book. We would like to thank Jean Reece for typing the manuscript and Frances Jill Linfoot for drawing the figures. We are also indebted to Dr Derek Chetwynd for excellent advice and useful comments on earlier drafts.

Julian W. Gardner
Harry T. Hingle

Contributors

D. Keith Bowen
Centre for Microengineering and Metrology
Department of Engineering, University of Warwick
Coventry CV4 7AL, UK

Peter J. Bryanston-Cross
Centre for Microengineering and Metrology
Department of Engineering, University of Warwick
Coventry CV4 7AL, UK

Derek G. Chetwynd
Centre for Microengineering and Metrology
Department of Engineering, University of Warwick
Coventry CV4 7AL, UK

Sam T. Davies
Centre for Microengineering and Metrology
Department of Engineering, University of Warwick
Coventry CV4 7AL, UK

Michael J. Downs
Division of Mechanical and Optical Metrology
National Physical Laboratory, Teddington
Middlesex TW11 0LW, UK

Ludwig Finkelstein
Measurement and Instrumentation Centre
City University, Northampton Square
London EC1V 0HB, UK

Albert Franks
Division of Mechanical and Optical Metrology
National Physical Laboratory, Teddington
Middlesex TW11 0LW, UK

Julian W. Gardner
Centre for Microengineering and Metrology
Department of Engineering, University of Warwick
Coventry CV4 7AL, UK

David A. Hutchins
Centre for Microengineering and Metrology
Department of Engineering, University of Warwick
Coventry CV4 7AL, UK

R. Peter Jones
Department of Engineering, University of Warwick
Coventry CV4 7AL, UK

Mohsin K. Mirza
Measurement and Instrumentation Centre
City University, Northampton Square
London EC1V 0HB, UK

M. Miyashita
Ashikaga Institute of Technology
Ohmae, Ashikaga City
Tochigi 326, Japan

Peter A. Payne
Department of Instrumentation and Analytical Science
UMIST, Manchester M60 1QD, UK

Jan H. Rakels
Centre for Microengineering and Metrology
Department of Engineering, University of Warwick
Coventry CV4 7AL, UK

Stuart T. Smith
Centre for Microengineering and Metrology
Department of Engineering, University of Warwick
Coventry CV4 7AL, UK

David J. Whitehouse
Centre for Microengineering and Metrology
Department of Engineering, University of Warwick
Coventry CV4 7AL, UK

Chapter 1

TRENDS IN INSTRUMENTATION AND NANOTECHNOLOGY

DAVID J. WHITEHOUSE

The field of instrumentation science is undergoing a period of rapid development. In this chapter the nature of this development is reviewed and in effect the trends from the use of conventional instrumentation to nanotechnology are mapped out.

1.1 INTRODUCTION

Nanotechnology started as a name in 1974. It was introduced by Taniguchi[1] to describe manufacture to finishes and tolerances in the nanometre region. He extrapolated the specifications from existing and past machine tools, such as lathes and grinders, to a new generation of machine tools. He concluded quite correctly that in the late 80's and 90's accuracies of between 0.1 μm and 1 nm would be needed to cater for industries' needs. This has turned out to be true, see Figure 1.1.

It soon emerged that the only way to achieve such results was to incorporate very sophisticated instrumentation and metrology into the design.[2]

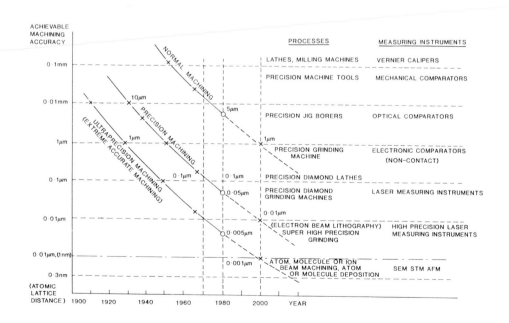

Figure 1.1 The development of achievable accuracy

1.2 NANOTECHNOLOGY SCOPE

This trend in instrumentation and metrology was already developing in the electronics industry where the drive was towards miniaturization for higher packing densities and faster switching. As a result, highly controllable and stable processes such as lithography were introduced. This meant a need arose for the very accurate positioning of work specimens. In turn this resulted in an interest in miniature actuators, motors and accurate slideways for which new technologies have required development. In particular new materials and thin film research were pre-eminent. As well as in electronics and manufacture, new developments on the nanoscale are taking place in the fields of biology and chemistry (see Chapter 8).

In terms of disciplines, therefore, nanotechnology encompasses more than engineering. It is the meeting point at the atomic scale of biology, chemistry, physics and engineering. This overall position is shown in Figure 1.2. Moreover, not only has the science base increased, but also the disciplines within each science, as shown in Figure 1.3.

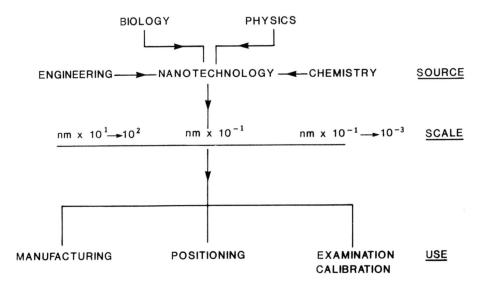

Figure 1.2 The interdisciplinary nature of nanotechnology

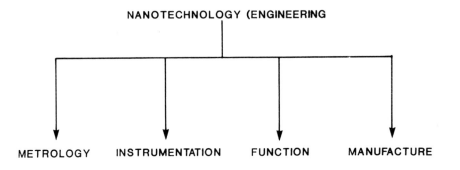

Figure 1.3 Disciplines within a science

2

1.2.1 Nanotechnology and Engineering

Because nanotechnology in name at least started in engineering, it is probably most informative to follow and investigate the growth of the subject in this discipline. Developing from the need to machine more accurately as demands grew for example in the fields of compact discs, gyroscopes etc., came new methods of fabrication with different materials. Together with these applications came the need to make smaller sensors and actuators to enable the non-intrusive control of instruments and machines.

In engineering applications, ninety percent of transducers are concerned with the measurement of displacement, position or their derivatives such as strain, pressure and acceleration. This has resulted in a mechanical micro-world which has emerged from the technology developed for integrated circuits. Already many small mechanisms are being made, including miniature motors of micrometre dimensions. Highly reliable accelerometers are already in use. These devices are fabricated on silicon substrates using extensions of such integrated circuit manufacturing processes as photolithography, thin film deposition and the like (see Chapter 7). Micromotors and articulated microstructures are more ambitious but are now being developed in many parts of the world. Using innovative processes involving bulk micromachining (sculpting silicon with chemical etchants) and surface micromachining (etching layers of thin films deposited on the substrates or bombarding with ion beams) have changed the concept of machining well beyond conventional methods. Also, the new use of parts so machined has created new names in the technical literature. Micro Electro Mechanical systems [MEMs], microdynamics, micromechanics, etc., are technical words that have emerged. They are perhaps best put into perspective by means of a typology mapping shown in Figure 1.4.

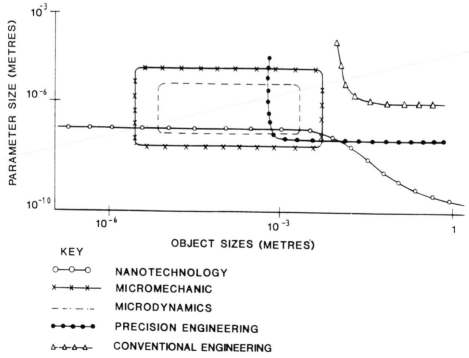

Figure 1.4 Typology of disciplines based on dimensional characteristics

3

In Figure 1.4 conventional engineering includes the process of turning etc., while precision engineering comprises of diamond turning and grinding. The important point is that nanotechnology also utilizes advances in the conventional processes such as ductile (as opposed to brittle) grinding of materials, such as silicon, glass and ceramics (see Chapter 10). Because of the ever increasing tendency to form the essential shape of a component, bulk material removal methods are becoming used less often. The finishing processes, old and modern, are coming into their own. Hence it can be seen that nanotechnology as seen in Figure 1.4 cuts across the boundaries of the other disciplines. For this reason nanotechnology can be regarded as an "enabling technology". Hence its importance and relevance for the future. Nanotechnology will provide the techniques for measuring, investigating and fabricating material at the atomic and molecular level. The growth of nanotechnology as a subject can be seen from Figure 1.5.

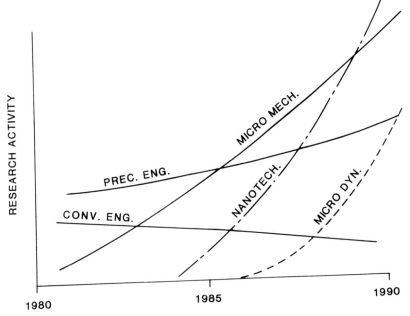

Figure 1.5 Growth of nanotechnology

1.3 NANOTECHNOLOGY INSTRUMENTS

Instrumentation has been developed to explore and measure surface properties down at the atomic level. In engineering terms they have usually been involved with looking at surface topography and boundaries of one sort or another. This requirement goes far beyond the original concept of the dimensional tolerance performance of a machine tool or the attainable surface texture on a component.

1.3.1 Specific Application Areas include:

1. Semiconductor and other electric surface properties in respect of charge injection and charge store;
2. Surface reagents and catalysts in respect of chemical reactions and processes;

3. Surfaces of biological molecules in both liquids and membranes and their changes in real time;
4. Surfaces of magnetic heads, compact discs etc., for storage capacity;
5. Coatings and surfaces of tools for wear properties;
6. Surfaces effects produced by non-conventional mechanisms;
7. Damage monitoring of pure and hybrid materials on the atomic scale;
8. Tribological investigation including bearings, adhesion adsorption etc.

1.3.2 Specific Measurement Requirements:

1. Flaw or defect detection;
2. Structural characterization - lattice parameters;
3. Position and relative position of features;
4. Height or topographic features;
5. Shapes and edge sharpness;
6. Volumetric analysis;
7. Movement of atoms;
8. Time changes of atomic or molecular structures.

A number of instruments are now in use that meet these requirements. Some of these are extensions of earlier instruments used in conventional engineering, such as stylus instrument or optical methods but other methods are new. These include the Scanning Electron Microscope (SEM), Scanning Tunnelling Microscope (STM), scanning Atomic Force Microscope (AFM). To see how recent this technology is, compare the dates of invention of instruments capable to nanometre resolution.

1.3.3 Instrument Development

1. Field ion microscope - Muller (1956) - measures field emission.[3]
2. Scanning electron microscope - Thornton (1968) - measures secondary electrons.[4]
3. Scanning tunnelling microscope - Binnig et al. (1982) - measures charge density.[5]
4. Atomic force microscope - Binnig et al. (1986) - measures atomic and ionic forces.[6]
5. Scanning capacitance microscope - Bugg, King (1988) - surface capacitance.[7]
6. Differential interference microscope - Shonenberger, Alvarado (1989) - optical reflectance.[8]
7. Nano stylus instruments - Garratt, Bottomley (1990) - topography.[9]

Other variants on these instruments which usually involve specially prepared probes are thermal conductance, near field optical evanescence, magnetic force and electrostatic force. See for example Wickramasinghe,[10] and Hansman, Drake, Marti, Gould, Prater.[11] Before attempting to characterize the performance of these instruments, it is useful to describe briefly the basic elements in a few of them.

Stylus contact instruments

In the followings sections, optical, stylus, STM and AFM techniques are briefly outlined. More detailed discussions of these topics are presented in later chapters. The first will be the nano-capability stylus instrument.

In the simplest form the basic instrument is shown in Figure 1.6. It consists of a probe, a specimen and scanning system. This incorporates a "pick-up" which can measure in the z direction over a wide range to a high resolution. To achieve this a miniature interferometer has been developed which caters for the large angular swings of the pick-up. With suitable digital processing 20 bit resolution is possible i.e. 1 part in a million. Nanometre roughness can be detected using this type of instrument. However an even newer generation of stylus contact instruments has been developed which is capable of even finer measurements using zero thermal expansion materials.[12]

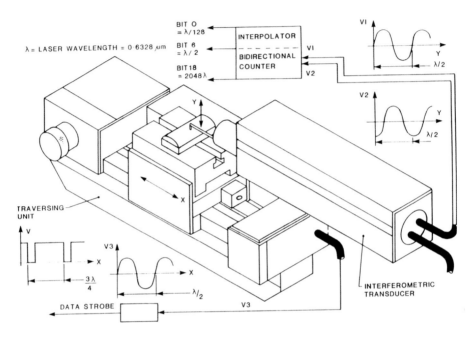

Figure 1.6 The nano capable stylus instrument

Optical methods

Optical methods fall basically into five categories:

    1. Follower methods;
    2. Heterodyne methods;
    3. Scanning methods;
    4. Holographic methods/interferometer;
    5. Diffraction methods.

Basic schematic diagrams of the required optical instrumentation are shown in Figures 1.7a to 1.7e Each instrument is capable of assessing surfaces and films to nanometre resolution but in different ways. Basically there is a compromise between the fidelity of measurement - taken here to be their correlation with stylus methods - and speed of result. This compromise crudely resolves itself into a measure of the numerical aperture of the optical system. Large Numerical Aperture (NA) systems have highly focussed spots which lend themselves well to linear profiling techniques. Low NA systems cover an area, are faster and cost less but as a rule they are less accurate. This is largely due to the difficulty in calibration. Another important point is that at the atomic level there is no fundamental correspondence between optical and stylus methods.

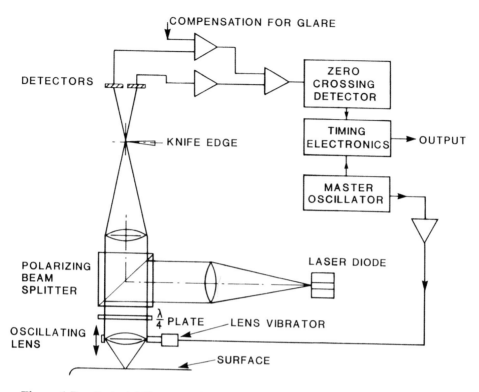

Figure 1.7a Optical follower method

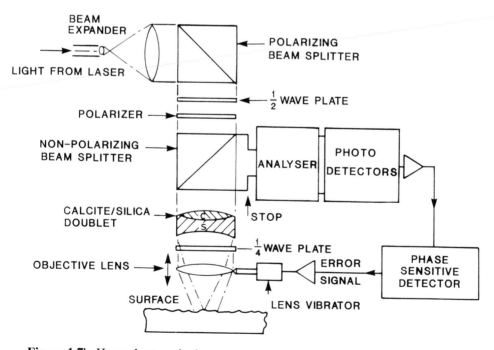

Figure 1.7b Heterodyne method

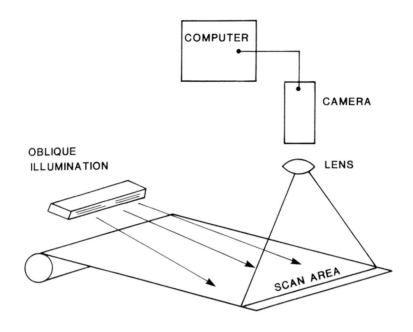

SYSTEM OF SURFACE INSPECTION FOR FLAWS

Figure 1.7c Optical flying scan methods

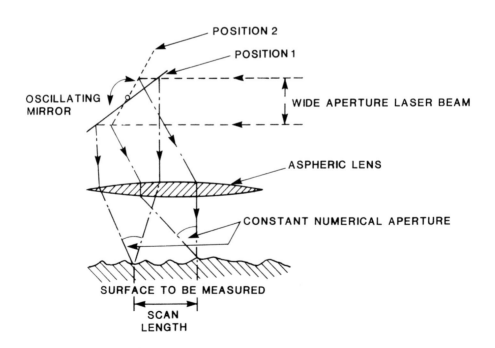

Figure 1.7d Laser scanner showing constant numerical aperature system

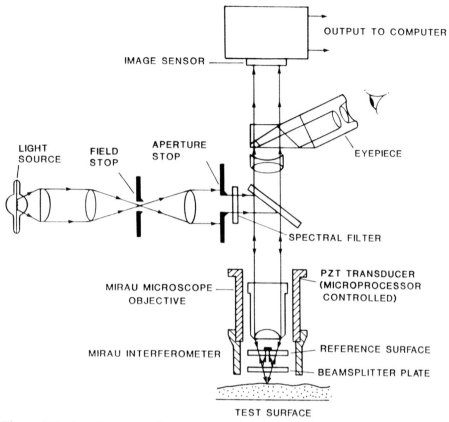

Figure 1.7e Interference methods

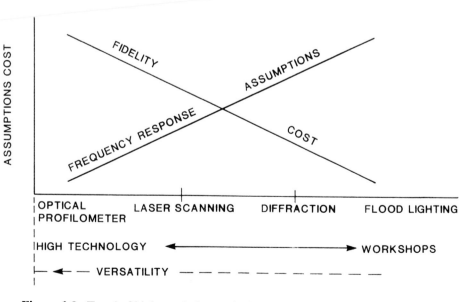

Figure 1.8 Trend of high resolution optical methods

Figure 1.8 shows the general trend of optical, high resolution methods based on focussed systems, diffraction, interference and gloss techniques. These methods have the considerable advantage over other optical methods in that they are non-contacting but they are:

1. Susceptible to a phase change on the surface;
2. Tilt sensitive - difficult to measure form;
3. Limited range/resolution at nanometre level;
4. Difficult to calibrate;
5. Wrong bandwidth/resolution range for nanotechnology.

Inherent in these methods based upon the scattering of light is their limited spatial bandwidth. This is due to the finite spot size at the surface. Diffraction methods do not suffer from this but do not give the opportunity to give topographic profile information. Height resolutions of better than 1 nm are common with these instrumentsbut spatial resolutions of only about a μm are to be expected. For a more detailed discussion of the use of optical methods in nanotechnology see Chapters 11, 12 and 13.

1.3.4 Stylus Methods - Conventional Methods

Early stylus methods employed a sharp diamond stylus to contact the surface and have been discussed. The performance of these suffer to some extent because of the possibility of causing surface damage. This contact problem was thought to be a severe limitation, especially when the stylus method is compared to an optical one. However they have been given a lease of life with the advent of new manufacturing techniques, such as ion beam milling, (see Chapter 9) which enable very fine tips to be made. Such is the modern capability of modern tip generation that it is claimed that single atom tips can now be made! This advance has enabled the probe to be on the same scale of size as the feature being investigated, resulting in good metrology. Together with the use of a sensing phenomenon such as charge density or molecular forces this has enabled a new generation of microscopes to be developed: The Scanning Tunnelling Microscope (STM) and the Atomic Force Microscope (AFM). However, before these new instruments are considered their forerunner, the Scanning Electron Microscope (SEM) should be introduced. This tool effectively bridges the gap between the traditional methods of surface analysis i.e., optical and stylus and the new generation of STM methods.

The principle of the SEM is shown in Figure 1.9. Electrons from a filament are accelerated by a voltage commonly in the range of 1 - 30 kV and directed down the centre of an electron optical column consisting of two or three magnetic lenses. These lenses cause a fine electron beam to be focussed onto the specimen surface. Scanning coils placed before the final lens cause the electron spot to be scanned across the surface of the specimen in the form of a rectilinear raster, similar to that on a television screen. The currents passing through the scanning coils are made to pass through the corresponding deflection coils of a Cathode Ray Tube (CRT) in order to produce a similar but larger raster on the viewing screen in a synchronous fashion. The electron beam incident on the specimen surface causes various phenomena, of which the emission of secondary electrons is the most commonly used. The emitted electrons strike the collector and the resulting current is amplified and used to modulate the brightness of the CRT. The times associated with the emission and collection of the secondary electrons are negligibly small compared with the time taken to scan the incident electron beam across the specimen surface. Hence, there is a one-to-one correspondence between the number of secondary electrons collected from any particular point on the specimen surface and the brightness of the analogous point on the CRT screen. Consequently, an image of the surface is progressively built up on the screen. The SEM has no imaging lenses in the optical sense of the word. The image magnification is determined solely by the ratio of the sizes of the rasters on the CRT screen and on the specimen surface. For example, in order to increase the magnification, on a CRT screen 10 cm across, magnifications of ×100, ×1000, and ×10,000 are obtained by scanning the specimen 1 mm, 0.1 mm and, 0.01 mm across, respectively.

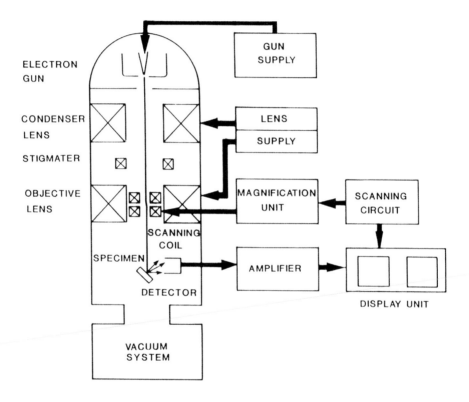

Figure 1.9 SEM Block diagram

One consequence of this is that high magnifications are easy to obtain with the SEM, while very low magnifications are difficult. Thus, for a magnification of ×10 it would be necessary to span a specimen approximately 10 mm across, and this presents difficulties because of the large deflection angles that would be required. For instance, the electron beam may strike the lens pole pieces or aperture at the extremes of the scan, so linearity is not be maintained. The very different operation of the SEM, when compared to most other microscopes, arises from the absence of optical imaging lenses, and so any signal that arises from the action of the incident electron beam (reflected electron, transmitted electron, emitted light, etc.) can be used to form an image on the CRT screen.

Despite the various options on display phenomena and recent advances in stereoscopic detection it is still true to say that the weakness of the SEM is its rather large lateral resolution and its relatively poor vertical resolution (or even interpretation). Even today 30% of SEM pictures use an accompanying profile obtained by other means to give credibility to heights. In contrast the superior spatial and vertical resolutions of the STM and AFM make them such existing new instruments.

Scanning tunnelling and atomic force microscopes

Although much of the interest in STM and AFM is in their ability to "see" atomic and molecular detail they are also unique in their capability of matching the lateral or spatial resolution to the vertical resolution. Typical ranges are 10 μm to 0.01 μm laterally and 0.001 μm vertically. This is an advance over existing optical, stylus, capacitance and SEM methods and it is necessary for nanomeasurement as can be seen in Figure 1.10.

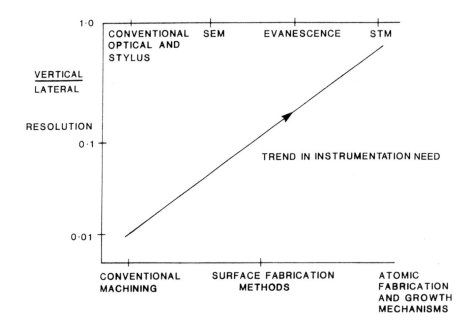

Figure 1.10 Change in aspect ratio of graphical regime as a function of scale of size

At the atomic scale the detail has an aspect ratio of unity that is dictated by the shape of cells, atoms and molecules, whereas in conventional engineering surfaces the aspect ratio is determined by the shape of the unit machining event. This in effect is the depth to width ratio of the effective cutting grain.

Some simple details of the AFM and STM will now be given. Basically they can be used in the modes shown in Figure 1.11 in (a) general mode showing vibration paths, in (b) open loop mode and in (c) closed loop mode.

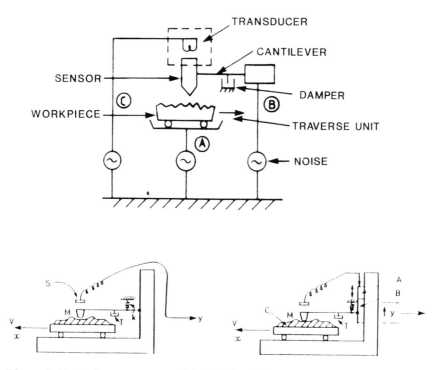

Figure 1.11 Basic arrangement of the STM and AFM

In STM the tip is mounted on a piezoelectric x y z scanner and scanned over a sample to reveal the contours of the surface down to the atomic level. So an STM image is made up of a series of line scans, each displaced in y from the previous one, and displays the path the tip followed over the surface. The principles and operation of an STM are surprisingly simple. An extremely sharp conductive tip (ideally terminating in a single atom) traces the contours of a surface with atomic resolution. The tip can be moved in three dimensions with an x y z piezoelectric translator. An STM actually traces contours of constant electron density at a particular energy determined by the bias voltage. Thus, when there are different atoms in a surface layer, atomic resolution images will depend on bias voltage. In AFM the essential features are a tip, a spring, and some device to measure the deflection of the spring. In practice, the spring deflection sensor can be either based on electron tunnelling to the back of the spring, on optical interference between the back of the spring and a reference plate, or by deflection of a laser-light beam reflected off the back of the spring. In each case, the tip follows a path that is an accurate topograph of the surface. No voltage is applied between the tip and the sample, and no current needs to flow between the tip and the sample. Thus the AFM can image samples with a low electrical conductivity. A review of the development of probe microscopy is presented in Chapters 8 & 14.

1.4 CHARACTERIZING INSTRUMENT PERFORMANCE

In order to examine recent trends in instrumention, it is necessary to consider their application area. There are three important operational parameters to consider:

1. Spatial band width - lateral range/resolution;
2. Vertical bandwidth - height range/resolution;
3. Frequency response or response time.

All of these parameters are important because they are either directly in (1) or (2), or indirectly, (3), related to fidelity of measurement. Usually fidelity or the ability to reproduce faithfully the surface parameter is universely proportional to measurement speed. Take for example the optical systems. The way in which these can be grouped as a function of speed and fidelity is shown in Figure 1.12. This is one way of characterizing surface instruments. The possible deviant is diffraction which under certain circumstances can have high fidelity and speed. Which of these parameters, speed or fidelity, lends most weight depends on whether the application is primarily in manufacture or in research: research applications can be slower but must be better.

Comparing several instruments used in conventional engineering on a frequency/bandwidth criterion shows some definite trends as seen in Figure 1.12. This shows that new stylus instruments which are capable of measuring to nanometer resolution can do so over a range of millimetres albeit rather slowly.

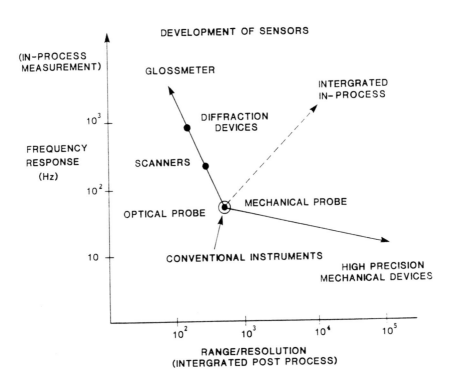

Figure 1.12 Characterization and trends of high resolution instruments

Another change in emphasis as usage develops from conventional engineering to nanotechnology is in the relative weighting of height and spacing information. Traditionally it has been surface heights, steps etc., which have been regarded as most important, probably because of their relevance to the tribological parameters of contact, friction, lubrication and wear.

As usage moves away from basic technology towards chemistry, physics and biology it is spatial information which is often sought. This explains the great emphasis now being placed on nano-mechanism movements. The general development in this area is shown in Figure 1.13.

The information being sought is various but essentially it can be grouped together in the following categories:

1. **Height information**: this involves detecting flaws or omissions. Typical features would include dislocations, distortions, scratches pits, track breakages, scars, blemishes, debris, dust, boundaries, pin holes etc. In any case two options arise; detecting the presence or absence of flaws or measuring and categorizing them. In categorizing flaws and defects, the following information is of interest:
   (i) The shape, position and size of the flaw or defect;
   (ii) The number and distribution.

2. **Aerial information**: this involves searching for and identifying the spatial correlation of features. Interests might be crystalline structure, the order of molecules, isotropy, and corrugation, i.e.
   (i) Unusual event recognition (electronics, technology);
   (ii) Usual event recognition (chemistry, biology).

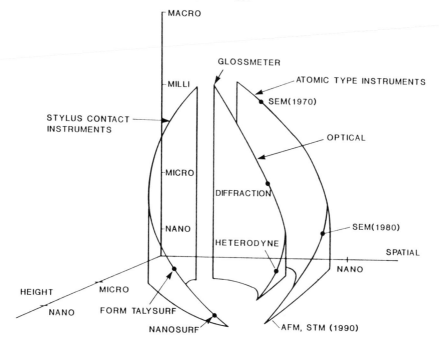

Figure 1.13 Instrument trends in resolution

The crucial feature of instruments is that they readily enable the visualization of the detailed structure of the workpiece or surface. A second more stringent design requirement of the instrument is that it gives quantitative information which is calibrated and traceable.

Figure 1.14 shows that there has been good progress in reducing the scale of size of the feature that can be visualized, although what is missing from this figure is the response time of the instrument.

There has been less progress in obtaining quantitative information from the instruments, mainly because of the difficulties in calibration at the atomic level. This highlights the metrology gap.

Progress is being made in calibration techniques, the key to ultimate calibration standards is exploitation of lattice parameters of pure crystals. Laterally the extremely regular structure of the surface corrugations of a graphite crystal can be used. In the vertical direction, it was once thought that cleavage steps on crystal surfaces could be used as step height standards, for example some were made in topaz.[14] In principle, steps relating directly to the lattice parameter are possible, but in practice it is difficult to determine what is actually present. There are may be cleavage planes of similar spacing, also multiple steps are themselves difficult to characterize. Ultimate calibration now depends on X-ray interferometry to determine small displacements directly in terms of the lattice parameter (see Chapter 16).

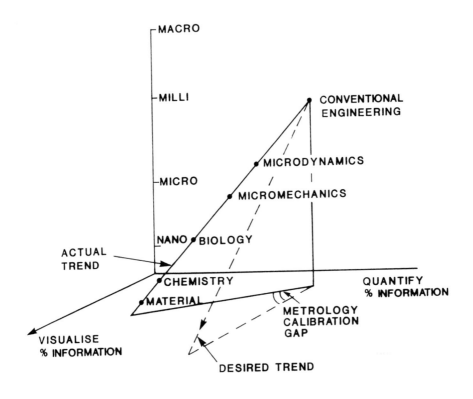

Figure 1.14 Discipline trend as scale of size changes

1.5 FUTURE INSTRUMENTS AND CONCLUDING REMARKS

As mentioned previously, nanotechnology is an enabling technology rather than one just confined to fabrication in the nanometre region. It seems obvious that a first requirement for future instruments is that they are multipurpose and flexible. They should be fast also so that environmental effects such as thermal drift and vibration are minimized. One approach is to make an instrument which is capable of performing many different aspects of nanotechnology with the same piece of apparatus. For example, it is possible for one instrument to machine using ion beam milling, utilize the secondary electrons (SEM) to visualize the surface, profile the topography using an STM adaption, and possibly even to measure nano-friction and hardness. All these operations would be carried out in a continuous vacuum and would probably be able to utilize the same software and computer controller for specimen positioning.[14]

Scaling theory will become more important because of the way the relative importance of physical phenomena change with scale.[15,16] Inertias, for example, reduce dramatically in importance relative to damping factors and especially stiffness.

For a sensor it would metrologically sensible to use single atomic particles to explore individual atomic detail on a surface probably by deflection techniques. It should be possible to achieve this goal. Already individual atoms are being moved about.[17]

At the root of nanotechnology is the smallness of scale of size and the interdisciplinary nature of the subject as a whole. Although some current instruments have resolutions and even probes on the atomic scale, much of the instrumentation has not yet reached this state. It seems that the imbalance between the scale of size of the features being investigated and the human scale ($\sim 10^9$) is very large. This causes problems in nanotechnological instrumentation; future instruments will have to be designed so that more of the instrument itself is on a small scale to reduce this metrology imbalance, to increase speed, hence fidelity and lastly to make advantage of other benefits that arise from the use of submicron dimensions.

ACKNOWLEDGEMENTS

I would like to thank the Science and Engineering Research Council (SERC) for my Senior Fellowship and the staff of the Technical Department of Rank Taylor Hobson Ltd, UK, for useful discussions.

REFERENCES

1. Taniguchi, N. (1983) Current status in, and future trends of, ultraprecision machining and ultrafine materials processing. *Annals CIRP*, **32/2**, 573-582
2. Bryan, J.B. (1979) Design and construction of an ultraprecision 84" diamond turning machine. *Precision Engineering*, **1**, 13-17
3. Muller, E.W. (1956) Field emission microscopy. In Physical Methods in Chemical Analysis, pp.135-182. New York: Academic Press
4. Thornton, P.R. (1968) *Scanning Electron Microscopy*, London: Chapman and Hale
5. Binnig G., Rohrer H., Gerber, C. and Weibul, E. (1982) Surface studies by scanning tunnelling microscopy. *Phys. Rev. Lett.*, **49**, 57-61
6. Binnig, G., Quate, C.F. and Gerber, C. (1986) Atomic force microscope. *Phys. Rev. Lett.*, **56**, 930-933
7. Bugg, C.D. and King, P.J. (1988) Scanning capacitance microscopy. *J. Phys. E.: Sci. Instrum.*, **21**, 147-151

8. Schonenberger, C. and Alvarado, S.F. (1989) A differential interferometer for force microscopy. *Rev. Sci. Instrum.,* **60**, 3131-3134

9. Garratt, J.D. and Bottomley, S.C. (1990) Technology transfer in the development of a nanotopographic instrument. *Nanotechnology,* **1**, 38-43

10. Wickramasinghe, H.K. (1989) Scanned probe microscopes. *Scientific American,* **261**, October, 74-81

11. Hansma, P.K., Drake, R., Marti, O., Gould, S.A.C. and Prater, C.B. (1989) The scanning ion-conductance microscope. *Science,* **243**, 641-643

12. Lindsey, K., Smith, S.T. and Robbie, C.J. (1988) Sub-nanometre surface texture and profile measurement with nanosurf 2. *Annals CIRP,* **37**, 519-522

13. Whitehouse, D.J., Bowen, D.K., Chetwynd, D.G. and Davies, S.T. (1988) Nano-calibration for stylus-based surface measurement. *J. Phys. E.: Sci. Instrum.,* **21**, 46-51

14. Ichinokawa, T., Miyazaki, Y. and Koya, Y. (1987) Scanning tunnelling microscope combined with scanning electron microscope. *Ultramicroscopy,* **23**, 115-118

15. Nakajima, N., Ogawa, K. and Fujimasa, I. (1989) Study on microengines. *Sensors & Actuators,* **20**, 75-82

16. Trimmer, W.S.N. (1980) Micro-robots and micro-mechanical systems. *Sensors & Actuators,* **19**, 267-287

17. Eigler, D.M. and Schweitzer, E.K. (1990) Positioning single atoms with a scanning tunneling microscope. *Nature,* **344**, 524-526

Chapter 2

SIGNAL PROCESSING

R. PETER JONES

This chapter presents a basic overview of signal processing. Its scope embraces 1-dimensional and multi-channel discrete signals (sequences). Techniques for processing static and dynamic signals are covered involving both time and frequency domain approaches. The signal processing techniques presented include methods based on summary statistics, regression and curve fitting, convolution and correlation, spectral analysis, system identification, and digital filtering.

2.1 INTRODUCTION

Signal processing essentially involves mathematical operations carried out on one, or more, source sequences (signals) to generate one, or more, derivative sequences. The general situation is described schematically in Figure 2.1. The source sequence is assumed to be a sequence $\{s_1, s_2, \ldots, s_j\}$, representing either a signal which is inherently discrete, or a sampled version of a continuous signal. The sequence $\{r_1, r_2, \ldots, r_k\}$ results from application of a signal processing algorithm to the source sequence. The source signal usually has a noise component, resulting from measurement error, or random disturbances acting on the system generating the signal.

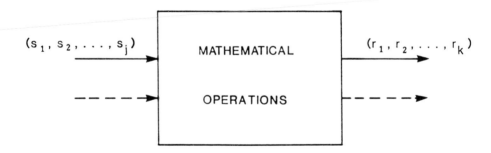

$(s_1, s_2, \ldots, s_j)$ MATHEMATICAL $(r_1, r_2, \ldots, r_k)$

OPERATIONS

Figure 2.1 Signal processing

The signal processing techniques presented here can be grouped into two categories. The first category relates to the situation where the signals can be considered as independent variables. The techniques grouped under this heading include summary statistics, and regression and curve fitting. The second category relates to the situation where the signals are dependent on a further variable, usually time. The techniques grouped under this heading include convolution and correlation, spectral analysis, system identification, and digital filtering.

This chapter on the topic of signal processing has been prepared from an instrumentation and control perspective, with emphasis on the use of signal processing techniques. It embraces 1-dimensional and multi-channel signals, but specifically excludes 2-dimensional signals. Consideration is restricted to discrete signals, or sequences, only. Techniques for processing static and dynamic signals are presented involving both time and frequency domain approaches.

2.2 SUMMARY STATISTICS

The uncertainty associated with random signals prevents consideration of individual values within the sequence describing the signal. In this situation, statistical concepts can be used to generate summary values which characterize the signal.

Consider the situation where a signal s is represented by a sequence $\{s_1, s_2, \ldots, s_n\}$ with values in a range defined by the interval $[s_{min}, s_{max}]$. The frequency histogram for this sequence is obtained by dividing $[s_{min}, s_{max}]$ into N subintervals and counting the number f_i, $i = 1, \ldots, $ N, of sequence values contained in each interval. The graph of f_i against s_i then defines the frequency histogram. This situation is illustrated in Figure 2.2(a), which displays a random sequence consisting of 1000 values, and Figure 2.2(b) which presents the corresponding frequency histogram.

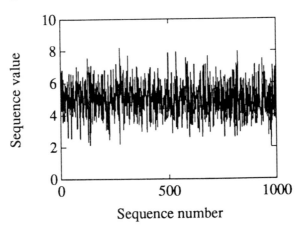

Figure 2.2(a) Random sequence

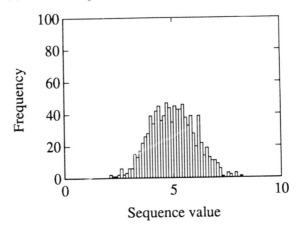

Figure 2.2(b) Frequency histogram

Normalizing the frequency values such that

$$\sum_{i=1}^{N} f_i = 1 \tag{2.1}$$

gives the probability distribution of the random sequence.

The mean or expected value μ of a sequence $\{s_1, s_2, \ldots, s_n\}$ is defined as

$$\mu = \frac{1}{n} \sum_{i=1}^{n} s_i \tag{2.2}$$

A measure of the spread of a random sequence about its mean is provided by the variance, which is defined as

$$\sigma^2 = \frac{1}{n} \sum_{i=1}^{n} (s_i - \mu)^2 \tag{2.3}$$

where σ is the standard deviation. Application of relationships (2.2) and (2.3) to the random sequence presented in Figure 2.2(a) resulted in values of $\mu = 4.998$ and $\sigma = 1.005$.

For a pair of sequences $\{s_1, s_2, \ldots, s_n\}$ and $\{q_1, q_2, \ldots, q_n\}$ the correlation, defined as

$$R_{qs} = \frac{1}{n} \sum_{i=1}^{n} q_i s_i \tag{2.4}$$

provides a measure of the alignment between the sequences, and is analogous to the dot, or inner, product in vector algebra. When $R_{qs}=0$, the sequences are said to be orthogonal. When considering pairs of sequences, the concept of variance generalizes to that of covariance, defined as

$$C_{qs} = \frac{1}{n} \sum_{i=1}^{n} (q_i - \mu_q)(s_i - \mu_s) \tag{2.5}$$

and is the correlation between sequences with their mean levels μ_q and μ_s removed. The covariance, when normalized with respect to the sequence standard deviations, defines the correlation coefficient

$$\rho_{qs} = \frac{c_{qs}}{\sigma_q \sigma_s} \tag{2.6}$$

where, $-1 \le \rho_{qs} \le 1$. When $\rho_{qs}=0$, the sequences are said to be (statistically) independent.

2.3 REGRESSION AND CURVE FITTING

The statistical concepts presented in the previous section provide a useful means of summarizing the characteristics of one or more sequences. Often the situation arises where trend is present in the sequences and it is desired to identify and model this trend. In this case, methods based on regression and curve fitting are appropriate.

Consider the situation where data in the form of sequences $\{s_1, s_2, \ldots, s_n\}$ and $\{q_1, q_2, \ldots, q_n\}$ is available, and it is required that this data be modelled by a function $g(s)$, such that

$$q = g(s) \tag{2.7}$$

matches the data in an appropriate manner. One of two approaches can be employed to obtain a function $g(s)$, viz. interpolation and least squares approximation.

With interpolation $g(s)$ is chosen to be a m-th degree polynomial

$$g(s) = a_0 + a_1 s + \ldots + a_m s^m \tag{2.8}$$

such that $g(s)$ matches the data exactly, i.e.

$$q_i = g(s_i), \quad i = 1, \ldots, n \tag{2.9}$$

If $m=n-1$ then a unique (Lagrange) polynomial can always be obtained. However, a large value of n implies that the approximating polynomial will be high order, and possibly oscillate widely between data points, even when the data varies smoothly. This is often overcome by using piecewise low order interpolating polynomials on subsets of $\{s_1, s_2, \ldots, s_n\}$, e.g. cubic splines $(n=3)$ which result in a function, which when pieced together, is smooth.

With least squares approximation $g(s)$ is chosen to minimize

$$J = \sum_{i=1}^{n} \{ q_i - g(s_i) \}^2 \tag{2.10}$$

This approach implicitly takes account of noise on the sequence $\{q_1, q_2, \ldots, q_n\}$ by attempting to minimize the influence of the equation residual errors ε_i, where

$$\varepsilon_i = q_i - g(s_i), \quad i = 1, \ldots, n \tag{2.11}$$

Often $g(s)$ is chosen to be a polynomial of degree $m < n-1$. In this case, the polynomial coefficients $a = [a_0, a_1, \ldots, a_m]^T$ are the unique solution of the normal equation

$$S^T S \, a = S^T q \qquad (2.12)$$

where $q = [q_1, q_2, \ldots, q_n]^T$ and the matrix S is given by

$$S = \begin{bmatrix} 1 & s_1 & s_1^2 & \cdot & \cdot & \cdot & s_1^m \\ 1 & s_2 & s_2^2 & \cdot & \cdot & \cdot & s_2^m \\ \cdot & \cdot & \cdot & \cdot & \cdot & \cdot & \cdot \\ \cdot & \cdot & \cdot & \cdot & \cdot & \cdot & \cdot \\ 1 & s_n & s_n^2 & \cdot & \cdot & \cdot & s_n^m \end{bmatrix} \qquad (2.13)$$

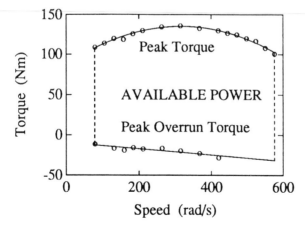

Figure 2.3 Engine torque regression

Linear least squares approximation has a statistical interpretation as a maximum likelihood estimation technique, in the case where the equation residual errors are normally distributed.

Figure 2.3 illustrates the application of least squares curve fitting techniques to engine torque data. The figure demonstrates the fit of a quadratic to the peak torque versus engine speed data, and a straight line to the corresponding peak overrun data.

Any choice of function $g(s)$ which is linear in the parameters a will result in a least squares approximation defined by a linear algebraic equation in the form of equation (2.12). When $g(s)$ is nonlinear in the parameters, the least squares approximation is defined by a nonlinear algebraic equation which is usually solved via iterative optimization techniques.

Least squares approximation is a special case of minimum norm approximation methods. Other minimum norm approximation methods are sometimes more appropriate, e.g. minimax error techniques.

2.4 CONVOLUTION AND CORRELATION

So far the signals have been considered as independent variables. The remainder of this chapter considers the situation where the signals are dependent on time t. Given a sequence $\{t_0, t_1, \ldots, t_n\}$ of equally spaced time instances, then it is assumed that the sequences $\{h_0, h_1, \ldots, h_n\}$ and $\{u_0, u_1, \ldots, u_n\}$ define values of the signals h and u at these time instants, i.e. $h_i = h(t_i)$, $u_i = u(t_i)$.

The convolution operation, defined as

$$y_i = \sum_{k=0}^{i} h_{i-k} u_k \tag{2.14}$$

provides a useful tool for combining the action of dynamic signals. In particular, if h_i represents the response of a linear dynamical system to a unit pulse, the response y_i to any other signal u_i is defined by the convolution operation (2.14).

The autocorrelation function, defined as

$$R_{uu}(k) = \frac{1}{n+1} \sum_{i=0}^{n} u_i u_{i+k} \tag{2.15}$$

represents the mean product of a sequence with a time shifted replica, and essentially extends the statistical concept of expected value to random sequences which are time dependent. Figure 2.4(a) presents a 10 s record, sampled every 0.01 s, of the signal x_k $= \sin(2\pi kT + \phi)$ where ϕ is a random variable which is uniformly distributed on the interval $[0, 3\pi/2]$. The corresponding autocorrelation function is displayed in Figure 2.4(b) which clearly reproduces the 1 Hz component in x_k.

The cross correlation function between two signals u and y, defined as

$$R_{uy}(k) = \frac{1}{n+1} \sum_{i=0}^{n} u_i y_{i+k} \tag{2.16}$$

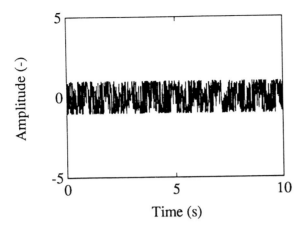

Figure 2.4(a) Random signal

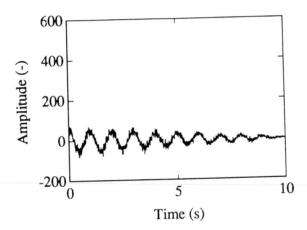

Figure 2.4(b) Autocorrelation of random signal

represents the mean product of one sequence with a time shifted replica of a second sequence, and essentially extends the statistical concept of correlation to random sequences which are time dependent.

Similarly, removal of the mean levels from the sequences before calculation results in the autocovariance function $C_{uu}(k)$, and cross covariance function $C_{uy}(k)$, which are extensions of the statistical concepts of variance and covariance.

2.5 SPECTRAL ANALYSIS

So far signal processing techniques have been considered which operate in the time domain. Spectral analysis establishes a link between the time and frequency domains, and provides techniques for analyzing data in the frequency domain. This link is established through the mathematical concept of the Fourier transform, and its computational manifestation as the FFT (Fast Fourier Transform) algorithm.

Given a time varying sequence $\{u_0, u_1, \ldots, u_n\}$, with time spacing T, then the Fourier transform of this sequence is defined as

$$U_i = \sum_{k=0}^{n} u_k \, \exp\{-j\,(i\Omega)\,(kT)\} \qquad (2.17)$$

where $j = \sqrt{-1}$. U_i is complex valued and is a function of frequency, with frequency spacing

$$\Omega = \frac{2\pi}{(n+1)T} \qquad (2.18)$$

and is termed the Fourier spectrum of u_i.

The corresponding inverse Fourier transform is defined as

$$u_k = \frac{1}{n+1} \sum_{i=0}^{n} U_i \, \exp\{j\,(i\Omega)\,(kT)\} \qquad (2.19)$$

The (discrete) Fourier transform is a periodic function with period $2\pi/T$. Consequently, care must taken with sequences resulting from sampling of a continuous signal, since aliasing errors can be introduced into the Fourier spectrum.

Spectral analysis techniques centre on the power spectral density function

$$S_{uu}(i) = \sum_{k=0}^{n} R_{uu}(k) \, \exp\{-j\,(i\Omega)\,(kT)\} \qquad (2.20)$$

which is the Fourier transform of the autocorrelation function. It is interpreted as a density function for the signal power, expressed as a function of frequency. Figure 2.5(a) displays a 5 s record of the signal $x_k = \sin(2\pi(3)kT) + \sin(2\pi(7)kT) + \sin(2\pi(13)kT) + 1.5\phi$, sampled every $T = 0.01$ s, where ϕ is a normally distributed random variable with zero mean and unit variance.

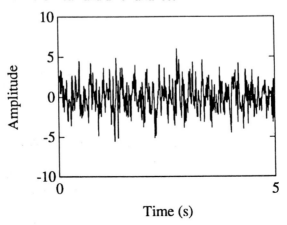

Time (s)

Figure 2.5(a) Random temporal signal

The corresponding power spectral density function is presented in Figure 5(b) which clearly reproduces the 3, 7 and 13 Hz components in the signal.

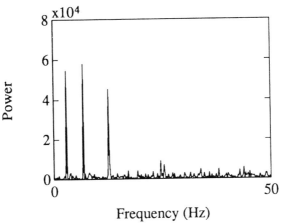

Figure 2.5(b) Spectral density of random signal

The cross power spectral density function

$$S_{uy}(i) = \sum_{k=0}^{n} R_{uy}(k) \exp\{-j(i\Omega)(kT)\} \qquad (2.21)$$

is the Fourier transform of the crosscorrelation function. It is interpreted as the joint power density function, over frequency, of two random signals.

The relation between two signals is quantified by the squared coherency function

$$\Upsilon_{uy}^{2}(i) = \frac{\left| S_{uy}(i) \right|^{2}}{S_{uu}(i)\, S_{yy}(i)} \qquad (2.22)$$

which can be interpreted as a frequency dependent correlation coefficient.

Related to the power spectrum is the power cepstrum

$$C_{uu}(i) = \sum_{k=0}^{n} \log\{S_{uu}(k)\} \exp\{-j(k\Omega)(iT)\} \qquad (2.23)$$

which is the Fourier transform of the log power spectrum function. The power cepstrum is a function of a variable with units of time, but called quefrency to provide a

27

distinction with real time. The cepstrum is essentially a pseudo-correlation function which is aimed at discriminating echo and delay phenomena.

When applying spectral analysis techniques, Smoothing Windows are commonly employed to minimize the effects of errors introduced through truncation of the signals. This involves the application of weights to each of the terms in the summations (2.17) and (2.19). These weights are chosen to emphasize, or de-emphasize, the contribution of individual sequence values in the computation of the Fourier, or inverse Fourier, spectrum.

2.6 SYSTEM IDENTIFICATION

As with the time invariant case discussed in section 2.3, the situation arises where it is desired to model the relationship between two dynamic signals. Models of linear dynamic systems are usually formulated as an Auto-Regressive Moving Average (ARMA) model or a state variable model.

ARMA models characterize the relationship between input and output sequences, $\{u_0, u_1, \ldots, u_n\}$ and $\{y_0, y_1, \ldots, y_n\}$, respectively, via a difference equation in the form

$$y_i + a_1 y_{i-1} + \ldots + a_n y_{i-n}$$

$$= b_1 u_{i-1-k} + \ldots + b_m u_{i-m-k} \tag{2.24}$$

The time shift from t_i to t_{i-1} is usually represented by the backward shift operator, z^{-1}, i.e.

$$t_{i-k} = z^{-k} t_i \tag{2.25}$$

With this notation the difference equation (2.24) is

$$\{1 + a_1 z^{-1} + \ldots + a_n z^{-n}\} y_i$$

$$= z^{-k} \{b_1 z^{-1} + \ldots + b_m z^{-m}\} u_i \tag{2.26}$$

which can be written as

$$y_i = g(z) u_i \tag{2.27}$$

where

$$g(z) = \frac{z^{-k}\{b_1 z^{-1} + \ldots + b_m z^{-m}\}}{1 + a_1 z^{-1} + \ldots + a_n z^{-n}} \qquad (2.28)$$

is the z-transfer function. This input/output z-transfer function model is described schematically in Figure 2.6(a).

Given data sets $\{u_0, u_1, \ldots, u_n\}$ and $\{y_0, y_1, \ldots, y_n\}$, system identification is concerned with finding a z-transfer function $g(z)$ that matches the data in an appropriate manner.

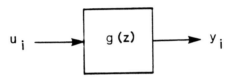

Figure 2.6(a) Transfer function model

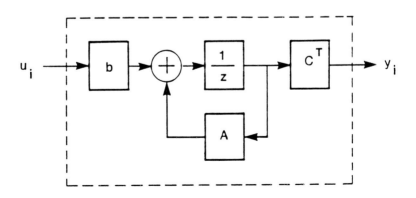

Figure 2.6(b) State variable model

29

One approach involves recasting the ARMA model defined in equation (2.24) as a matrix vector equation. The parameters in the ARMA model are then estimated using least squares techniques, either directly or in a recursive manner suitable for on-line application. An alternative approach is to transform the signals into the frequency domain and to estimate the frequency response via the power and cross power spectral density functions.

State variable models relate the input variable to the output variable via an intermediate variable known as the system state. Linear state variable models take the form

$$x(i+1) = A \, x(i) + b \, u_i$$

$$(2.29)$$

$$y_i = c^T x(i)$$

where $x(i) = [x_1(i), \ldots, x_n(i)]^T$ is a n-dimensional state vector. The structure of a state variable model is shown schematically in Figure 2.6(b).

Given a state variable model of a system, then an equivalent ARMA model can be uniquely obtained, with z-transfer function

$$g(z) = c^T [z \, I_n - A]^{-1} b \qquad (2.30)$$

where I_n is the $n \times n$ identity matrix. However, given a z-transfer function model of a system, it is not generally possible to generate a unique state variable model which is equivalent. This lack of uniqueness gives rise to problems of identifiability in the identification of state variable models from input output data.

2.7 DIGITAL FILTERING

Filtering is a process where unwanted components in a signal are rejected.

Consider a noise corrupted measurement process

$$y_i^o = y_i + n_i \qquad (2.31)$$

where y_i represents the variable under observation, n_i measurement noise and y_i^o the resulting measurement. An estimate $\overline{y}_i$ of the signal y_i can be obtained by filtering the measurement y_i^o via a algorithm in the form

$$\overline{y}_i = g(z) \, y_i^o \qquad (2.32)$$

This filtering process, which is shown schematically in Figure 2.7, can be implemented either as a Finite Impulse Response (FIR) filter, or a Infinite Impulse Response (IIR) filter.

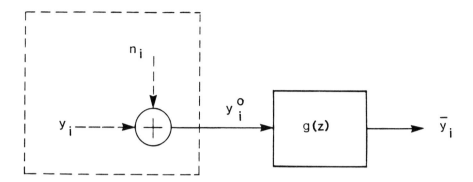

Figure 2.7 FIR and IIR filters

FIR filters have a z-transfer function in the form

$$g(z) = b_0 + b_1 z^{-1} + \ldots + b_m z^{-m} \qquad (2.33)$$

and are characterised by a finite duration impulse response. The design of FIR filters involves the calculation of the filter weights b_0, $b_1, \ldots,$ b_m. These are usually computed to give an acceptable approximation to a desired frequency response using minimax approximation methods.

IIR filters have a z-transfer function in the form

$$g(z) = \frac{b_0 + b_1 z^{-1} + \ldots + b_m z^{-m}}{1 + a_1 z^{-1} + \ldots + a_n z^{-n}} \qquad (2.34)$$

$$= b_0 \frac{\displaystyle\prod_{i=1}^{m} (1 - \beta_i z^{-1})}{\displaystyle\prod_{k=1}^{n} (1 - \alpha_k z^{-1})}$$

and are characterized by a infinite duration impulse response. The design of IIR filters usually starts from an analogue prototype in the form of a Laplace transfer function with acceptable frequency response characteristics. This transfer function is then transformed to an equivalent digital form via the bilinear transformation

$$S = \frac{2\,(z - 1)}{T\,(z + 1)} \tag{2.35}$$

Consider a noise corrupted measurement process

$$y^o_i = g(z)\,v_i + w_i \tag{2.36}$$

where w_i and v_i represent noise processes, y^o_i is the resulting measurement and $g(z)$ is the z-transfer function of a known state variable model. In addition to the estimate $\overline{y}_i$ of the signal y_i, estimates $\overline{x}(i)$ of the unmeasured states $x(i)$ can be obtained by filtering the measurement y^o_i via a algorithm in the form

$$\overline{x}(i+1) = A\,\overline{x}(i) + h\,[y^o_i - c^T\overline{x}(i)]$$

$$\overline{y}_i = c^T\overline{x}(i) \tag{2.37}$$

This process which is known as state estimation is shown schematically in Figure 2.8. The design of a state estimator involves calculation of the gain h. This can be chosen, using stability considerations, to ensure that the eigenvalues of $A - hc^T$ are contained within a unit circle centred at the origin of the complex plane. Alternatively, h can be chosen to optimize a quadratic performance function based on covariance information on the noise signals w_i and v_i. The former technique forms the basis of the Luenberger observer, whereas the latter technique results in the Kalman filter. Figure 2.9 illustrates the operation of a state estimator in tracking the output of a noise process in the form (2.36). The figure demonstrates the ability to converge on the measurement from a poor guess of the initial value of the states.

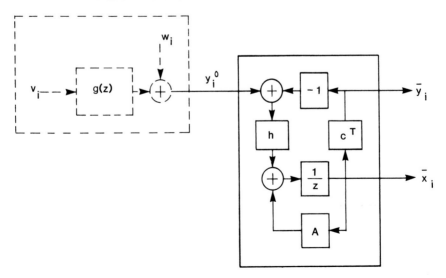

Figure 2.8 State estimator

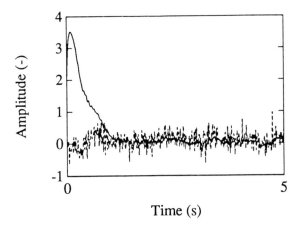

Figure 2.9 Kalman filter

2.8 CONCLUSIONS

This chapter has presented a basic overview of the broad field of signal processing, from the perspective of instrumentation and control. The scope has embraced 1-dimensional and multi-channel signals, with consideration restricted to discrete signals (sequences) only. Techniques for processing static and dynamic signals have been covered involving both time and frequency domain approaches. The signal processing techniques presented include methods based on summary statistics, regression and curve fitting, convolution and correlation, spectral analysis, system identification, and digital filtering.

Signal processing plays an important role in the field of instrumentation, in general, and the subject of nanotechnology, in particular. The next chapter discusses the application of convolution and correlation methods to instrumentation, complementing the discrete approach adopted here through discourse based on continuous analysis. The work on the mathematical modelling of instruments, which is described in chapter 4, embraces the use of system identification techniques. Chapter 5 demonstrates how regression and curve fitting procedures, and digital filtering techniques can provide tools for computer aided precision metrology. Convolution, correlation, and spectral analysis concepts form the basis of optical diffraction techniques for surface roughness measurement, presented in chapter 12. Finally, the X-ray interferometry methods for instrument calibration, presented in chapter 16, involve the use of various summary statistics.

REFERENCES

1. Anderson, B.D.O. and Moore, J.B. (1979) *Optimal Filtering*. Englewood Cliffs: Prentice-Hall
2. Astrom, K.J. (1970) *Introduction to Stochastic Control Theory*. New York: Academic Press
3. Bendat, J.S. and Piersol, A.G. (1980) *Engineering Applications of Correlation and Spectral Analysis*. New York: Wiley
4. Bendat, J.S. and Piersol, A.G. (1971) *Random Data: Analysis and Measurement Procedures*. New York: Wiley

5. Eykhoff, P. (1974) *System Identification - Parameter and State Estimation.* London: Wiley

6. Gabel, R.A. and Roberts, R.A. (1980) *Signals and Linear Systems.* New York: Wiley.

7. Godfrey, K. and Jones, R.P. (Eds.) (1986) *Signal Processing for Control*, (Lecture Notes in Control and Information Sciences, Vol. 79). Berlin: Springer-Verlag

8. Jones, N.B. (Ed.) (1982) *Digital Signal Processing.* Hitchin: Peter Peregrinus

9. Kailath, T. (1980). *Linear Systems.* Englewood Cliffs: Prentice-Hall

10. Ljung, L. (1987) *System Identification - Theory for the User.* Englewood Cliffs: Prentice-Hall

11. McGillem, C.D. and Cooper, G.R. (1974) *Continuous and Discrete Signal and System Analysis.* New York: Holt, Rinehart & Winston

12. Norton, J.P. (1986) *An Introduction to Identification.* London: Academic Press.

13. Oppenheim, A.V. and Schafer, R.W. (1975) *Digital Signal Processing.* Englewood Cliffs: Prentice-Hall

14. Papoulis, A. (1974) *Circuits and Systems: A Modern Approach.* New York: Holt, Rinehart & Winston

15. Parks, T.W. and Burrus, C.S. (1987) *Digital Filter Design.* New York: Wiley

16. Rabiner, L.R. and Gold. B. (1975) *Theory and Application of Digital Signal Processing.* Englewood Cliffs: Prentice-Hall

17. Sage, A.P. and Melsa, J.L. (1971) *Estimation Theory with Applications to Communications and Control.* New York: McGraw-Hill

18. Tretter, S.A. (1986) *Introduction to Discrete-time Signal Processing.* New York: Wiley

19. Young, P.C. (1984) *Recursive Estimation and Time-Series Analysis.* Berlin: Springer-Verlag

20. Ziemer, R.E., Tranter, W.H. and Fannin, D.R. (1983) *Signals and Systems.* New York: Macmillan

Chapter 3

CORRELATION METHODS APPLIED TO INSTRUMENTATION

PETER A. PAYNE

Wiener's Generalized Harmonic Analysis as described by Y. W. Lee, is used to introduce the concepts of correlation-based measurement and instrumentation. Particular attention is given to the noise rejection properties of such an approach and serial and parallel forms of mechanization are covered. Applications areas are briefly described.

3.1 INTRODUCTION

Instrumentation is essentially concerned with the extraction, modification and subsequent use of information. Such a definition includes instrumentation used for process plant, or for the control of complex machines such as aircraft, and also the measuring instrument in which the information may eventually be displayed for the benefit of the human operator. The feature common throughout is that the information that we wish to extract and make use of is almost always corrupted by some form of noise signal. Thus, methods for the detection of signals buried in noise are an important part of the instrument designer's armoury.

Wiener[1] laid the foundations for what has now become known as the statistical theory of communication in work that he conducted between 1925 and 1942 and one of his students of the 1930s published the book *Statistical Theory of Communication*[2] in 1960. Out of this work has come what we now know as the theory of correlation and this has been applied to numerous problems such as signal detection, statistical filtering, and prediction. Since the early 1960s the applications of correlation techniques to problems of a measurement and control nature have increased enormously and correlation methods are now the basis of a host of commercial instruments and instrumentation.

Here we will consider periodic, transient and random signals and relate these to the properties of their correlation functions. The related operation of convolution will also be addressed.

3.2 GENERALIZED HARMONIC ANALYSIS

3.2.1 Periodic Signals

Prior to the work of Wiener, the basis for analysis of information was that of the Fourier series. For a periodic function of time $f(t)$, there is a corresponding series of sinusoids which, when added together, will give rise to the original function of time $f(t)$. Mathematically this is expressed as:

$$f(t) = \frac{a_0}{2} + \sum_{n=1}^{\infty} [a_n \cos(n\omega_1 t) + b_n \sin(n\omega_1 t)] \qquad (3.1)$$

where ω_1 is the fundamental angular frequency which is related to the period T_1 of the function $f(t)$ by the expression $T_1 = 2\pi/\omega_1$. The coefficients a_n and b_n are given

$$a_n = \frac{2}{T_1} \int_{-T_1/2}^{T_1/2} f(t) \cos(n\omega_1 t) dt \qquad (3.2)$$

$$b_n = \frac{2}{T_1} \int_{-T_1/2}^{T_1/2} f(t) \sin n\omega_1 t \, dt \qquad \text{for } n = 1, 2, 3, ... \qquad (3.3)$$

An alternative expression for this relationship (equation 3.1) is given in equation (3.4), making use of the exponential function:

$$f(t) = \sum_{n=-\infty}^{\infty} F(n) e^{jn\omega_1 t} \qquad (3.4)$$

and in equation (3.4):

$$F(n) = \frac{1}{2}(a_n - jb_n) \qquad \text{for } n = 0, \pm 1 .. \pm 3 \qquad (3.5)$$

We may rewrite the expression for $F(n)$ to give:

$$F(n) = \frac{2}{T_1} \int_{-T_1/2}^{T_1/2} f(t) e^{-jn\omega_1 t} \, dt \qquad \text{for } n = 0, \pm 1 .. \pm 3 \qquad (3.6)$$

Equation (3.6) is the Fourier transform of the periodic function $f(t)$. The process of multiplying and obtaining the average value for this product over the period of the signal $f(t)$ and for every value of n, provides us with a means of analysing $f(t)$ to give the amplitudes and phase angles of the individual sinusoids into which $f(t)$ has been resolved. Because we have assumed that our signal $f(t)$ is a periodic signal, then we obtain a complex spectrum $F(n)$, which is a line spectrum. Equations (3.4) and

(3.6) are Fourier transforms of each other and we may separate the amplitude and phase characteristics in the complex spectrum as:

$$|F(n)| = \frac{1}{2}\sqrt{a_n^2 + b_n^2}$$ (3.7)

giving us the amplitude spectrum of $f(t)$ and

$$\theta_n = \tan^{-1}(-b_n/a_n)$$ (3.8)

giving the phase spectrum of $f(t)$.

Correlation

The expression for the cross-correlation function ϕ_{12} between two signals $f_1(t)$ and $f_2(t)$, in this case both periodic, is given by:

$$\phi_{12}(\tau) = \frac{1}{T_1}\int_{-T_1/2}^{T_1/2} f_1(t)f_2(t+\tau)dt$$ (3.9)

Both $f_1(t)$ and $f_2(t)$ have the same fundamental angular frequency ω_1 and τ is a continuous time shift or delay which is independent of t.

If $f_1(t)$ has a Fourier transform of $F_1(n)$ and $f_2(t)$ a Fourier transform of $F_2(n)$, then the Fourier transform of equation (3.9) is given by:

$$F_1^*(n).F_2(n)$$ (3.10)

where * denotes the complex conjugate. This gives the power spectrum of ϕ_{12} and is usually written as $\Phi_{12}(n)$.

Equations (3.9) and (3.10) constitute what is known as the correlation theorem for periodic functions.

If $f_1(t)$ is equal to $f_2(t)$, then equation (3.9) gives rise to the autocorrelation function of $f_1(t)$, $\phi_{11}(\tau)$ which has a corresponding power spectrum $\Phi_{11}(n)$.

It is useful to note that the mean square value of the function $f_1(t)$ is equal to the sum over the range of n of the square of the absolute value of its spectrum. Mathematically this is expressed as:

$$\frac{1}{T_1}\int_{-T_1/2}^{T_1/2} f_1^2(t)dt = \sum_{n=-\infty}^{\infty} |F_1(n)|^2$$ (3.11)

Using the symbols ϕ_{11} and Φ_{11} allows us to write equations (3.9) and (3.10) as:

$$\phi_{11}(\tau) = \sum_{n=-\infty}^{\infty} \Phi_{11}(n) e^{jn\omega_1 \tau} \tag{3.12}$$

and

$$\Phi_{11}(n) = \frac{1}{T_1} \int_{-T_1/2}^{T_1/2} \phi_{11}(\tau) e^{-jn\omega_1 \tau} dt \tag{3.13}$$

these being the autocorrelation theorem for periodic functions. The properties of this function are briefly:

1. Its phase spectrum is zero for all harmonics. This is equivalent to saying that taking an autocorrelation of a signal discards all phase information.

2. It is an even function of τ, thus a cosine series with zero initial phase angles can be used to rewrite equations (3.12) and (3.13) as:

$$\phi_{11}(\tau) = \sum_{n=-\infty}^{\infty} \Phi_{11}(n) \cos(n\omega_1 \tau) \tag{3.14}$$

and

$$\Phi_{11}(n) = \frac{1}{T_1} \int_{-T_1/2}^{T_1/2} \phi_{11}(\tau) \cos(n\omega_1 \tau) dt \tag{3.15}$$

More detailed descriptions of the properties of autocorrelation functions can be found in Lee[2] and Bracewell.[3] The reason that autocorrelation functions are so important is their key role in the analysis of signals in which a periodic function is buried in random noise.

The cross-correlation function $\phi_{12}(\tau)$ and its Fourier transform $\Phi_{12}(n)$ which is the cross-power spectrum of $\phi_{12}(\tau)$ are defined in equations (3.9) and (3.10). It is important to note that the subscript 12 is used to denote the order in which the signals $f_1(t)$ and $f_2(t)$ are dealt with in equation (3.9) and that it means that $f_2(t)$ is displaced by τ. $\phi_{21}(\tau)$ is in general not equal to $\phi_{12}(\tau)$. However, by changing τ to $-\tau$ and making $x = t - \tau$, we have:

$$\phi_{12}(-\tau) = \frac{1}{T_1} \int_{-T_1/2}^{T_1/2} f_2(x) f_1(x+\tau) dx \tag{3.16}$$

Now the right hand side of equation (3.16) is $\phi_{21}(\tau)$ by definition and thus:

$$\phi_{12}(n) = \phi_{21}(\tau) \tag{3.17}$$

The corresponding relationship in the frequency domain is:

$$\Phi_{12}(n) = \Phi_{21}{}^*(n) \tag{3.18}$$

Convolution

If we apply a signal $f_1(t)$ to a system or network whose impulse response is given by $f_2(t)$, then we know that the output from the system or network is given by the convolution integral:

$$\frac{1}{T_1} \int_{-T_1/2}^{T_1/2} f_1(t) f_2(\tau - t) dt \tag{3.19}$$

This expression (equation 3.19) has an appearance of similarity to the cross-correlation function (equation 3.9) but with the very important difference that the shift operation on $f_2(t)$ is backwards. It can be shown that:

$$\frac{1}{T_1} \int_{-T_1/2}^{T_1/2} f_1(t) f_2(\tau - t) dt = \sum_{n=-\infty}^{n=+\infty} F_1(n) F_2(n) e^{j n \omega_1 \tau} \tag{3.20}$$

and that

$$F_1(n) F_2(n) = \frac{1}{T_1} \int_{-T_1/2}^{T_1/2} e^{j n \omega_1 \tau} dt \cdot \frac{1}{T_1} \int_{-T_1/2}^{T_1/2} f_1(t) f_2(\tau - t) dt \tag{3.21}$$

Equations (3.20) and (3.21) represent the convolution theorem for periodic functions.

If we consider the two operations, correlation and convolution, several useful points emerge:

1. The process of obtaining the two functions is precisely the same: however, convolution requires an additional operation of "folding back". (Hence, the use of the word "faltung" in the German language).

2. If the operation of convolution and correlation is carried out on the same two periodic functions of the same fundamental frequency, f_1 and f_2, then the result in both cases gives rise to periodic functions of the same fundamental frequency and harmonics.

39

3. A difference between them is that the spectrum in the case of convolution is the product of the two spectra, but for correlation it is the conjugated product. Symbolically this is:

$$\frac{1}{T_1} \int_{-T_1/2}^{T_1/2} f_1(t) f_2(\tau-t) dt \quad \& \quad F_1(n) F_2(n) \tag{3.22}$$

these being a Fourier transform pair and:

$$\frac{1}{T_1} \int_{-T_1/2}^{T_1/2} f_1(t) f_2(t+\tau) dt \;=\; F_1^*(n) F_2(n) \tag{3.23}$$

which again form a Fourier transform pair.

4. The need to preserve the order of operation for the operation of cross-correlation has been noted. For convolution this is not a requirement since we may show that irrespective of which function, f_1 or f_2, is folded back and shifted, the result is always $F_1(n) F_2(n)$.

3.2.2 Transient Signals

Whereas for periodic signal analysis the Fourier series is employed, for transient (or aperiodic) signals we must employ the Fourier integral. This integral is derived from the Fourier series by allowing the period of the series to approach infinity. This action gives rise to:

$$f(t) = \frac{1}{2\pi} \int_{-\infty}^{\infty} e^{j\omega t} d\omega \int_{-\infty}^{\infty} f(\sigma) e^{-j\omega\sigma} d\sigma \tag{3.24}$$

for an aperiodic signal $f(t)$.

In applying such analysis, by implication we require that the complete integral of $f(t)$ is finite. We may rewrite equation (3.24) to give two expressions for the time and frequency domains:

$$f(t) = \int_{-\infty}^{\infty} F(\omega) e^{j\omega t} d\omega \tag{3.25}$$

and

$$F(\omega) = \frac{1}{2\pi} \int_{-\infty}^{\infty} f(t) e^{-j\omega t} dt \qquad (3.26)$$

In general $F(\omega)$ is a complex function of the angular frequency ω and is a continuous spectrum of the transient signal $f(t)$. Since $F(\omega)$ is complex we may write it as:

$$F(\omega) = P(\omega) + jQ(\omega) \qquad (3.27)$$

where $P(\omega)$ and $Q(\omega)$ are the real and imaginary parts of $F(\omega)$. Thus we have the amplitude density spectrum of $F(\omega)$ as:

$$|F(\omega)| = \sqrt{P^2(\omega) + Q^2(\omega)} \qquad (3.28)$$

and the phase density spectrum is:

$$\theta(\omega) = \tan^{-1}[Q(\omega)/P(\omega)] \qquad (3.29)$$

Correlation

Just as equation (3.8) applied to periodic signals, so for transients we have:

$$\int_{-\infty}^{\infty} f_1(t) f_2(t+\tau) dt \qquad (3.30)$$

which assumes that the complete integrals of both $f_1(t)$ and $f_2(t)$ exist and are finite. If $F_1(\omega)$ and $F_2(\omega)$ are the spectra of $f_1(t)$ and $f_2(t)$ the Fourier transform of equation (3.30) is:

$$2\pi F_1{}^*(\omega) F_2(\omega) \qquad (3.31)$$

Equations (3.30) and (3.31) are a Fourier transform pair forming the basis for the correlation theorem for aperiodic or transient signals. Clearly, if the two functions are identical the Fourier transform pair (equations (3.30) and (3.31)) give rise to the autocorrelation theorem and if they are not identical it is the basis of the cross-correlation theorem. If, in equation (3.30), τ takes the value zero then we may show that the relationship between equations (3.30) and (3.31) becomes:

$$\int_{-\infty}^{\infty} f_1(t) f_2(t) dt = 2\pi \int_{-\infty}^{\infty} F_1{}^*(\omega) F_2(\omega) d\omega \qquad (3.32)$$

This relationship is of immense importance in Fourier analysis of aperiodic functions and is known as a Parseval theorem.

41

Just as before in connection with periodic functions, there exists a great similarity between the correlation and convolution integrals for transient signals which may be deduced from the arguments presented previously.

3.2.3 Random Signals

Periodic and transient signals are reproducible with great precision each time they are generated. Random signals do not display this property. They may arise from numerous sources and are generally referred to as "noise" in the context in which we attempt to extract signals (wanted data) from noise (unwanted data). In the general case no one single source is responsible for generating the noise signal and what we observe is an ensemble or infinite aggregate of individual noise signals.

Ultimately, our concern is the extraction of signals from noise and the application of this to measuring instruments. First though we must examine the properties of random signals in the way we have analysed periodic and transient signals.

Autocorrelation

In the context of random signals we define the autocorrelation function as:

$$\phi_{11}(\tau) = \lim_{T \to \infty} \frac{1}{2T} \int_{-T}^{T} f_1(t) f_1(t+\tau) dt \qquad (3.33)$$

It may be shown that this is an even function of τ. If we set $\tau = 0$ then:

$$\phi_{11}(0) = \lim_{T \to \infty} \frac{1}{2T} \int_{-T}^{T} f_1^2(t) dt \qquad (3.34)$$

This is the mean square value of $f_1(t)$ or the mean power if $f_1(t)$ represents current or voltage in a $1\,\Omega$ load.

We may also deduce that when τ tends to infinity then $\phi_{11}(\tau)$ tends to zero assuming $f_1(t)$ contains no d.c. offset or periodic components. A further property of $\phi_{11}(\tau)$ is that is is continuous for all values of τ, provided that it is continuous at $\tau = 0$. $\phi_{11}(\tau)$ is also of maximum value at the origin.

If we write the autocorrelation function $\phi_{11}(\tau)$ and its Fourier transform $\Phi_{11}(\omega)$ as a pair we have:

$$\phi_{11}(\tau) = \int_{-\infty}^{\infty} \Phi_{11}(\omega) e^{j\omega t} d\omega \qquad (3.35)$$

and

$$\Phi_{11}(\omega) = \frac{1}{2\pi} \int_{-\infty}^{\infty} \phi_{11}(\tau) e^{-j\omega\tau} d\tau \qquad (3.36)$$

$\Phi_{11}(\omega)$ is the power density spectrum of $f_1(t)$ and if we let $\tau = 0$ in equation (3.35) we have:

$$\phi_{11}(0) = \int_{-\infty}^{\infty} \Phi_{11}(\omega) d\omega \qquad (3.37)$$

giving rise to $\Phi_{11}(\omega)$ in terms of the mean square value of $f_1(t)$. The power density spectrum is a real even non-negative function of ω.

Cross-correlation

In terms of random functions cross-correlation is very important. In particular, the cross-correlation function between input and output of linear systems forms one basis for characterizing such systems.

In considering the cross-correlation of two random functions we deal with two ensembles of which they are members and it is essential that these member functions are paired. Thus, if $f_1(t)$ is a member function of the first ensemble and $f_2(t)$ a member function of the second ensemble, then by definition the cross-correlation function of these random functions is:

$$\phi_{12}(\tau) = \lim_{T\to\infty} \frac{1}{2T} \int_{-T}^{T} f_1(t) f_2(t+\tau) dt \qquad (3.38)$$

The order of operation denoted by the subscripts is important and by $\phi_{21}(\tau)$ we mean:

$$\phi_{21}(\tau) = \lim_{T\to\infty} \frac{1}{2T} \int_{-T}^{T} f_2(t) f_1(t+\tau) dt \qquad (3.39)$$

As before:

$$\phi_{12}(\tau) = \phi_{21}(-\tau) \qquad (3.40)$$

In general, $\phi_{12}(\tau)$ is neither even nor odd and the maximum value is not necessarily at $\tau = 0$. However, just as with the autocorrelation function $\phi_{12}(\tau)$ tends to zero as τ tends to $\pm\infty$, provided neither component has a d.c. or periodic component.

As before, we may write a Fourier transform pair involving the cross-correlation function $\phi_{12}(\tau)$ and the cross-power density spectrum $\Phi_{12}(\omega)$:

$$\phi_{12}(\tau) = \int_{-\infty}^{\infty} \Phi_{12}(\omega) e^{j\omega\tau} d\omega \qquad (3.41)$$

$$\Phi_{12}(\omega) = \frac{1}{2\pi} \int_{-\infty}^{\infty} \phi_{12}(\tau) e^{-j\omega\tau} d\tau \qquad (3.42)$$

So far we have briefly reviewed the methods of analysing periodic, transient and random signals in preparation for discussing instruments and instrumentation used for dealing with them or, more usually, with combinations of them. The discussion has been based on the second chapter of Lee[2] to which the reader is directed for a more complete and lucid treatment of this important subject.

3.3 MEASUREMENT OF FREQUENCY AND IMPULSE RESPONSE

As an illustration of the use of Generalized Harmonic Analysis we will consider the measurement of the impulse and frequency response of linear systems and the instrumentation requirements implicit in this. A brief examination of a modification to these methods will also be described aimed at extending some of those techniques so that non-linear systems may also be dealt with. Throughout the following sections the major emphasis will be on how we can measure system impulse or frequency response in the presence of noise and both random and periodic noise signals will be considered.

The traditional method of measuring a single input, single output linear system's frequency response consists of injecting a sinusoidal signal at the input and observing the resultant signal at the output. The action of applying the input signal will of course generate a transient disturbance in the System Under Test (SUT) which will in time die away leaving only a phase shifted and attenuated (or amplified) version of the input sinusoid. The measurement consists of establishing the amplitude and phase of the output signal for all values of frequency of interest. Frequency response methods are often applied to the "open loop" analysis of control systems as part of the design process and they can also be of use in the analysis of closed loop performance for the purpose of performance assessment.[4]

Although the analysis of amplitude and phase can be performed on an oscilloscope, if the signals are corrupted by noise this becomes more difficult and various filtering methods have been employed to assist in this. By far the most popular is that based on the correlation approach outlined above (section 3.2.1 on Periodic Signals). The following analysis is based on that presented in Lamb and Payne.[5]

First we choose a zero time reference such that any transient due to application of the test signal $f_1(t) = a\sin(\omega t)$ has sensibly fallen to zero. Then at the output from the SUT the signal will be $f_2(t) = ka\sin(\omega t + \theta) + n(t)$ where k is the amplification constant and θ is the phase angle, both of which are functions of the angular frequency ω (radians per second). $n(t)$ is a noise signal. The manner in which we then proceed is illustrated in Figure 3.1 which is the basis for a class of instruments known as a Transfer Function Analyser (TFA) or more correctly (and recently more often) as a Frequency Response Analyser (FRA).

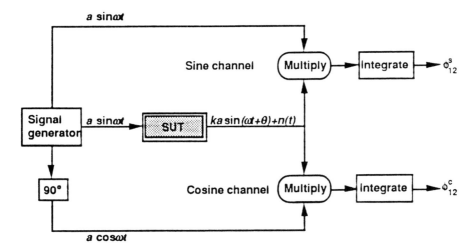

Figure 3.1 Simplified block diagram of a frequency response analyser

As shown, we proceed by generation of a second quadrature signal, $a\cos\omega t$ and both the sine and cosine channels of the instrument follow identical mechanisms. Looking first at the sine channel we see that the output of the integrator is given by:

$$\phi_{12}^{s}(\tau) = \frac{1}{T}\int_{0}^{T} a\sin\omega t\,[\,ka\sin(\omega(t+\tau)+\theta)+n(t+\tau)\,]\,dt \qquad (3.43)$$

We use equation (3.9) since we are dealing with periodic signals. $n(t)$ is assumed to be a noise signal and its effect will be analysed later. We note also that T is an integer multiple of the period of the input sine wave, ie:

$$T = \frac{2\pi N}{\omega} \qquad\qquad N \text{ integer} \quad (3.44)$$

Examination of Figure 3.1 reveals that no time shift is employed, i.e. $\tau=0$, in which case equation (3.43) reduces to:

$$\phi_{12}^{s}(0) = \frac{ka^2}{2}\cos\theta + \frac{a}{T}\int_{0}^{T} n(t)\sin(\omega t)\,dt \qquad (3.45)$$

For the cosine channel we may obtain a similar result, ie:

$$\phi_{12}^{c}(0) = \frac{ka^2}{2} \sin\theta + \frac{a}{T}\int_0^T n(t)\sin(\omega t)dt \qquad (3.46)$$

If, the for the moment, we assume that the noise signal is zero, then we may combine equations (3.45) and (3.46) to yield:

$$k = \frac{2}{a^2} \sqrt{[\phi_{12}^{s}(0)]^2 + [\phi_{12}^{c}(0)]^2} \qquad (3.47)$$

and

$$\theta = \tan^{-1}\left[\frac{\phi_{12}^{c}(0)}{\phi_{12}^{s}(0)}\right] \qquad (3.48)$$

giving us estimates of the system amplitude ratio k and phase angle θ for a given value of ω.

Returning to equations (3.45) and (3.46) we see that these estimates of k and θ are affected by the presence of $n(t)$ and it is essential that we establish the error caused by finite values of $n(t)$. If we first assume $n(t)$ to be a wide bandwidth random signal of spectral density ρ, then an estimate of the variance σ^2 caused by the presence of $n(t)$ is given by Cummins[6]:

$$\sigma^2 = \frac{1}{T^2}\int_0^T dt \int_0^T \overline{s(t)s(t')n(t)n(t')}dt' \qquad (3.49)$$

where $s(t)$ denotes either $\sin\omega t$ or $\cos\omega t$ and $\overline{}$ denotes an ensemble averaging process. If $s(t)$ and $n(t)$ are uncorrelated, then:

$$\sigma^2 = \frac{1}{T^2}\int_0^T dt \int_0^T \phi_{ss}(t'-t)\phi_{nn}(t'-t)dt' \qquad (3.50)$$

where ϕ_{ss} and ϕ_{nn} are the autocorrelation functions of signals ($\sin\omega t$ and $\cos\omega t$) and noise, respectively.

Now for both signals:

$$\phi_{ss}(t) = \frac{a^2}{2}\cos\omega t \qquad (3.51)$$

46

and for the noise:

$$\phi_{nn}(t) = \rho\delta(t) \tag{3.52}$$

where $\delta(t)$ is the Dirac delta function.

Substituting equations (3.51) and (3.52) into equation (3.50) gives, after some manipulation:

$$\sigma^2 = \frac{a^2\rho}{2T} \tag{3.53}$$

or in terms of $T = 2\pi N/\omega$:

$$\sigma = \frac{a}{2}\sqrt{\frac{\rho\omega}{\pi N}} \tag{3.54}$$

In terms of the signal-to-noise ratio S/N, of this measurement we have:

$$\frac{S}{N} = ka \begin{bmatrix} \cos(\theta) \\ or \\ \sin(\theta) \end{bmatrix} \begin{bmatrix} \frac{\pi N}{\rho\omega} \end{bmatrix}^{1/2} \tag{3.55}$$

From this we see that the use of cross-correlation provides us with a powerful method of controlling the signal-to-noise ratio of the measurements. An improvement in S/N can be achieved by increasing a, reducing ρ or increasing N. Usually we wish to keep a to as small a value as can be tolerated since this reduces disturbance to the SUT and often ρ is not under our control. Thus, only adjustment of N is possible. Note also from equation (3.53) that the standard deviation σ is proportional to $\sqrt{T}$, the square root of the measurement period, as we might expect.

In some circumstances we encounter "noise" signals that are of a deterministic nature or perhaps more often, noise signals comprising a mixture of random and deterministic signals. It is important therefore to derive expressions for σ and S/N for such cases.

Let us assume that we have a sinusoidal noise signal. Going back to equations (3.45) and (3.46) and assuming $n(t) = 0$ then we have:

$$\phi^s_{12}(0) = \frac{ka^2}{2}\cos\theta = R \text{ (the real part)} \tag{3.56}$$

and

$$\phi^c_{12}(0) = \frac{ka^2}{2}\sin\theta = Q \text{ (quadrature part)} \tag{3.57}$$

47

We may denote the errors in R and Q due to the noise signal $n(t)$ by δR and δQ, respectively.

For a sinusoidal noise signal of the form:

$$n(t) = a_n \sin(\omega_n t + \theta_n) \tag{3.58}$$

we may determine its effect by evaluating δR and δQ by a root mean square averaging procedure, ie:

$$\delta R^2 = \left[\frac{1}{T} \int_0^T a_n \sin(\omega_n t + \theta_n) a \sin(\omega t) dt \right]^2 \tag{3.59}$$

and

$$\delta Q^2 = \left[\frac{1}{T} \int_0^T a_n \sin(\omega_n t + \theta_n) a \cos(\omega t) dt \right]^2 \tag{3.60}$$

If we now assume that the measurements of R and Q made to establish δR and δQ are of a serial nature, then we must average equations (3.59) and (3.60) over all phase angles θ_n between 0 and 2π, ie:

$$\overline{\delta R^2} = \frac{1}{2\pi} \int_0^{2\pi} \delta R^2 \, d\theta \tag{3.61}$$

and

$$\overline{\delta Q^2} = \frac{1}{2\pi} \int_0^{2\pi} \delta Q^2 \, d\theta \tag{3.62}$$

Evaluation of equations (3.61) and (3.62) gives:

$$\overline{\delta R} = \frac{a_n \sqrt{2}}{\pi N} \frac{1}{(\omega_n/\omega)^2 - 1} \sin(\pi N \omega_n/\omega) \tag{3.63}$$

and

$$\overline{\delta Q} = \frac{a_n \sqrt{2}}{\pi N} \frac{(\omega_n/\omega)}{(\omega_n/\omega)^2 - 1} \sin(\pi N \omega_n/\omega) \tag{3.64}$$

Again, we see that the reduction of the error in measuring R or Q may be brought about by reducing the amplitude of the noise signal a_n or increasing N, thus increasing the integration time T.

3.3.2 Impulse Response Measurement

It is well known that the impulse response function for a linear system can be closely approximated by the cross-correlation function between a Pseudo-Random Binary Sequence (PRBS) as an input and the resultant system output. To do this we must mechanize the process described by equation (3.9), although the integration limits will be modified since we start such an experiment at time $t=0$ and the upper limit will be T, defined by the length of the PRBS employed. We should also note that such an experiment may be conducted by either:

1. Setting τ to some value, say τ_1, and evaluating the cross-correlation function - a serial experiment;

2. Using M channels simultaneously giving rise to an experiment which uses M values of delay and gives M points on the cross-correlation function - a parallel experiment.

In the presence of a deterministic noise signal the noise rejection properties of the two approaches will differ as explained below.

Assuming that we apply a PRBS $f_1(t)$ to a linear system whose impulse response is $h(t)$ and the output from the system is $f_2(t)$ then from equation (3.9), with modified integration limits, we have:

$$f_2(\tau) = \int_0^\infty h(t) f_1(\tau - t) dt + nt \qquad (3.65)$$

If we now evaluate the cross-correlation function between $f_1(t)$, the input and $f_2(t)$, the output, using equation (3.9) we obtain:

$$\phi_{12}(\tau) = \int_0^\infty h(z) \phi_{ss}(\tau - z) dz + \phi_{ns}(\tau) \qquad (3.66)$$

where $\phi_{ss}(\tau)$ is the autocorrelation function of the input PRBS and $\phi_{ns}(\tau)$ the cross-correlation function between signal and noise.

In the absence of noise:

$$\phi_{12}(\tau) = \int_0^\infty h(z) \phi_{ss}(\tau - z) dz \qquad (3.67)$$

and if $h(z)$ is effectively zero for $\tau > T$, then the integration may be halted at T.

Also we may choose the characteristics of the PRBS such that for a given system is the Dirac delta function. Thus:

$$\phi_{12} \approx kh(t)$$ (3.68)

where the factor k is the equivalent impulse factor for the chosen PRBS.

We now consider the noise term in equation (3.66) and the variance of the error in $\phi_{12}(\tau)$ due to this term may be written as:

$$\sigma^2 = \frac{1}{T^2} \int_0^T dt \int_0^T \overline{n(t)n(t')} s(t+\tau)s(t'+\tau)dt'$$ (3.69)

using the same approach as for equation (3.49). If the noise and signal are uncorrelated then:

$$\sigma^2 = \frac{1}{T_2} \int_0^T dt \int_0^T \phi_{nn}(t'-t)\phi_{ss}(t'-t)dt'$$ (3.70)

In the case of wide bandwidth random noise of spectral density ρ, then as before:

$$\sigma^2 = \frac{\rho}{T_2} \int_0^T dt \int_0^T \delta(t'-t)\phi_{ss}(t'-t)dt'$$ (3.71)

which reduces to

$$\sigma^2 = a^2 \frac{\rho}{T}$$ (3.72)

where a is the amplitude of the PRBS. Assuming that this signal has a clock period of λ then the equivalent impulse factor k is given by:

$$k = a^2\lambda$$ (3.73)

and we may eventually show that the signal-to-noise ratio is given by:

$$\frac{S}{N} = h(t)a\lambda \sqrt{\frac{T}{\rho}}$$ (3.74)

50

Once again, we see that for a given PRBS length and $h(t)$, S/N can be increased by increasing a the amplitude of the signal (not usually desirable) or by increasing the integration time T. A reduction in ρ is usually a trivial solution of any real problem.

As before, we now consider deterministic forms of noise signal and we will use a sinusoidal "noise" as an example. As pointed out earlier, in this case the method of mechanizing the cross-correlation instrument will be of importance.

Serial mode cross-correlation experiments

If the noise signal is $n(t) = b\sin(\omega t)$, then:

$$\phi_{nn}(t) = \frac{b^2}{2} \cos\omega t \tag{3.75}$$

and

$$\phi_{ss}(t) \approx k\delta(t) \tag{3.76}$$

and inserting equations (3.75) and (3.76) into equation (3.70), we obtain:

$$\sigma^2 = \frac{kb^2}{2T^2} \int_0^T dt \int_0^T \delta(t'-t)\cos\omega(t'-t)dt' \tag{3.77}$$

which reduces to

$$\sigma^2 = \frac{kb^2}{2T} \tag{3.78}$$

giving

$$\frac{S}{N} = \frac{h(t)}{b}\sqrt{2T} \tag{3.79}$$

showing again how increasing T increases S/N.

Parallel mode cross-correlation experiments

Here the problem is that once the experiment is started, a fixed phase relationship is established between the start of the PRBS at $t=0$ and the "noise" signal. We therefore express $n(t)$ as:

$$n(t) = a_n \sin(\omega t + \theta_n) \tag{3.80}$$

in order to preserve this phase relationship in our analysis.

To proceed we also need to expand $f_1(t)$, the PRBS input signal in terms of its Fourier series. As pointed out above (see section headed Periodic signals) this will give rise to a complex line spectrum in accordance with equation (3.6). A solution to this has been given by Lamb.[7] Thus:

51

$$f_1(t) = \sum_{r=1}^{\infty} A_r \sin(\omega_r t + \theta_r) \tag{3.81}$$

and

$$\phi_{12}(t) = \frac{1}{T} \int_0^T \left[\sum_{r=1}^{\infty} A_r \sin(\omega_r t + \theta_r) \right] a_n \sin(\omega_n (t-\tau) + \theta_n) dt \tag{3.82}$$

To proceed we assume that $\omega_r = n\omega_0$ and $\omega_n = m\omega_0$ where ω_r is the PRBS fundamental frequency and m and n are integers. Considerable further analysis eventually leads to:

$$\phi_{12}(\tau) = A_r a_n C \sin(\omega_r t + \theta) \tag{3.83}$$

where $C^2 = A^2 + B^2$, $A = \cos(\theta_r + \theta_n)$, $B = \sin(\theta_n - \theta_r)$ and $\theta = \tan^{-1}(B/A)$

This gives a noise contribution of sinusoidal form at the noise frequency, the amplitude of which will vary with the relative phasing of the noise sinusoid and the spectral line A_r.

3.3.3 Speeding up the Measurement

Mention has already been made of the parallel mode of cross-correlation applied to PRBS signals as a means of obtaining an estimate of a system impulse response. Clearly, this brings about a speed-up of the overall measurement assuming that each point on the impulse response requires the same integration time. In effect we can trade-off measurement time for complexity of the instrument. However, since the entire computation will, today, be carried out in the digital domain there is little argument over the choice to be made and a modern cross-correlator will comprise an analogue-to-digital converter followed by a computational unit plus some form of display or alternative output.

Parallel methods of measurement have also been applied to frequency response analysers and, again, much of the instrument is of a digital nature.[8] Since that time (1979) digital signal processing has advanced phenomenally and parallel frequency response analysis is entirely possible by means of software on a personal computer[9], as is PRBS-based impulse response determination.

Parallel processing in either the time or frequency domain brings with it a major problem if we attempt to deal with non-linear systems. The reason for this is that in either case (impulse or frequency response) the signals employed normally comprise line spectra with simple harmonic relationships with respect to the fundamental. Hence, in the presence of non-linearities, harmonic generation feeds "error signals" into incorrect spectral lines, causing large differences to be apparent when comparisons are made between serial and parallel modes of measurement.

Some reduction in this harmonic error generation is possible using an optimization algorithm in the frequency domain.[10] This places lines in selected regions of the spectrum in order to build up a specific test signal which ensures that each line generates the minimum number of interfering harmonics. Experiments show[11] that such a signal can reduce the error (difference between results of single and multiple frequency response data) by significant factors.

3.4 APPLICATIONS OF CORRELATION-BASED INSTRUMENT TECHNIQUES

In many cases the dynamic performance of a system needs assessing at various stages in its lifetime, such as the research, development, production, maintenance and post-repair stages. The methods discussed earlier provide a means of doing this in the presence of noise and many examples exist of the use of impulse or frequency response data as a means of discriminating between "good" and "bad" systems and then further diagnosing faults to a subsystem level. Another major use of these methods is associated with the need to identify the transfer function of plant around which a controller is to be placed.

Auto- and cross-correlation techniques have also been applied to areas as diverse as flow measurement[12] and the improvement of the performance of ultrasonic pulse echo techniques and this latter example is described below. Many other examples exist of the use of these techniques which represent one of the most powerful measurement methods available.

3.4.1 Golay Code Correlation system for Improved Ultrasonic Pulse Echo Instruments

It is well known that the conventional ultrasonic pulse echo system is limited both in detection range, by the average power (energy) radiated, and in resolution, by the acoustic pulse width. Hence, there are conflicting requirements for deep penetration range and resolution, particularly in the testing of highly absorbent materials such as plastics and rubber. However, the same limitation is not present in systems using correlation detection of a large time-bandwidth transmitted signal. Correlation systems transmit a long duration broad bandwidth ultrasound signal and reconstitute the returned echoes to give high resolution short duration pulse. The resolution of the correlation system is determined entirely by the bandwidth of the received echo signal.

Several types of transmit signals have been used in ultrasonic correlation detection systems. These include random noise, m-sequences, m-sequence modulated Radio Frequency (RF) signals and Barker-code modulated RF signals. All these transmit signals enable the correlation system to obtain high signal-to-noise ratio enhancement, which helps to retrieve echo signals buried in receiver noise. However, all these systems, except those which use continuously transmitted m-sequences, suffer from an inherent limitation which arises from so-called range sidelobes produced during the correlation process. This problem relates to the imperfect autocorrelation function that most codes produce for sensible periods of integration. However a type of paired pseudo-random code exists (Golay code) which can theoretically reduce range sidelobes to zero after only two code lengths. A system based on this has been designed and built and as is shown in Ding and Payne[13] improvements in signal-to-noise ratio of more than seven-fold are achieved when compared with a simple pulse echo system using only signal averaging.

3.5 CONCLUSIONS

The concepts associated with auto- and cross-correlation based measurements have been developed beginning with the Generalized Harmonic Analysis approach of Wiener[1] and

Lee.[2] Modern dynamic analysis instruments have been considered and the problems of non-linear systems addressed briefly. Finally, some indication of the broad range of application areas has been indicated.

REFERENCES

1. Wiener, N. (1948) *The Extrapolation Interpolation and Smoothing of Stationary Time Series*. New York: John Wiley
2. Lee, Y.W. (1960) *Statistical Theory of Communication*. New York: John Wiley
3. Bracewell, R.M. (1963) *The Fourier Transform and its Applications*. New York: McGraw-Hill
4. Brown, J.M., Towill, D.R. and Payne, P.A. (1972) *Radio & Electronic Engineer*, **42**(1), 7-20
5. Lamb, J.D. and Payne, P.A. (1971) In *Proceedings of the fifth IEEE Conference on Circuits and Systems* (Asilomar, 1971), pp.548-554
6. Cummins, J.D. (1964) *A note on errors and signal to noise ratio of binary cross-correlation measurements of system impulse response*, UKAE Report AEEW No. R329
7. Lamb, J.D. (1970) *IEEE Trans. Automatic Control*, **AC-15**(4), 478-480
8. Jawad, S.M. and Payne, P.A. (1979) In *Proceedings of IEEE Conference Autotestcon*, (Minneapolis, 1979), pp.192-196
9. Vakilian, H. (1987) Application of Multi-frequency Signals to the Determination of the Dynamics of Chemical Plant. *PhD Thesis*, University of Manchester Institute of Science & Technology, UK
10. Lawrence, P.J. and Payne, P.A. (1975) Optimum Multifrequency Testing, *DAG Tech. Note*, 77, UWIST, Cardiff
11. Lawrence, P.J (1980) Multifrequency Testing of Nonlinear Systems, *PhD Thesis*, University of Wales Institute of Science & Technology, UK
12. Beck, M.S. and Plaskowski, A. (1987) *Cross Correlation Flowmeters - Their Design and Application*, Bristol: Adam Hilger
13. Ding, Z.X. and Payne, P.A. (1990) *Meas. Sci. & Technol.*, **1**, 158-165

Chapter 4

MATHEMATICAL MODELLING OF INSTRUMENTS - APPLICATION AND DESIGN

MOHSIN K. MIRZA & LUDWIK FINKELSTEIN

Mathematical modelling can dramatically reduce the lead time and cost of designing instruments. This chapter presents a general method for developing mathematical models for multi-energy systems and gives an account of a popular and powerful numerical technique (Finite Element Method) used for solving continuum problems. Applications of modelling techniques are illustrated with practical examples. Development of a simple design methodology through the efficient use of mathematical models is also discussed.

4.1 INTRODUCTION

The number of applications for instruments has dramatically increased in the last decade particularly in the area of automatic control systems. For example, new cars are being designed with engine management systems, active suspension, anti-locking breaking and four wheel steering systems. It is envisaged that high performance cars now being designed may have as many as sixty sensors. With these new applications the specifications for the design of instruments have become more stringent. Modern sensors are required to be low cost, robust, accurate and compact. Conventional semi-empirical methods for the design of instruments to meet these requirements are not feasible. The wide availability of cheap and powerful computers coupled with powerful numerical techniques has led to the possibility of developing accurate mathematical models to study the static and dynamic behaviour of instruments.

The Measurement and Instrumentation Centre (MIC) at the City University has been involved in research on mathematical modelling and computer aided design of instruments for more than 15 years. A general review of the expertise has been published previously.[1,2,3] The work demonstrates that mathematical modelling and Computer Aided Design (CAD) techniques can dramatically reduce the cost and lead times for design.

This chapter gives a brief description of the different types of modelling techniques with applications for solving practical problems. It also presents an efficient use of mathematical models to streamline the design process. This is demonstrated in the area of snap-action diaphragms where a numerical model is used to generate non-dimensional performance curves which could be used to develop a simple design methodology.

4.2 MATHEMATICAL MODELS

An instrument system is essentially an information machine which transforms an input variable (measurand) into a convenient output (usually electrical) such that the output together with the transformation law can be used to deduce the input. For example in an electrical linear pressure sensor the output is voltage v and the transformation law is a straight line of gradient k such that $v = k p$ then p is determined from $p = v / k$.

Confronted with a requirement for a new design, the designer may adopt one of two approaches[2]:

1. Design the system using subsystems of known performance;
2. Design the subsystems from first principles.

To illustrate matters suppose a pressure sensor is to be designed using bellows. Then a typical system would be that the displacement in bellows under a pressure load is detected by some form of displacement sensor (a Linear Variable Differential Transformer, LVDT, perhaps) to produce an electrical output. A system maybe configured from available subsystems (bellows and LVDTs). At this level the designer may benefit from a mathematical model of the total instrument in terms of the subsystem characteristics. This type of modelling has been called functional modelling.[4]

The other requirement is that the instrument designer may have to design the sensing subsystems of an instrument. Suppose no suitable bellows can be found for the pressure sensor illustration. In this case it will be necessary to design the bellows in terms of its geometrical form and dimensions and the materials of which it is made. The models which relate the shape and materials to the behaviour of the subsystems are called physical models. Such models require the calculation of "field" quantities from the solution of governing partial differential equations. For elastic elements this would be the displacement field; for electromagnetic devices - electric and/or magnetic fields; for flow velocity and pressure fields and for temperature sensors - heat flux and temperature fields. From such "fields" other quantities of interest such as strains, stresses, stiffnesses, masses, resistances, inductances, discharge coefficients can be obtained by integration, differentiation or other simple operations.

4.3 FUNCTIONAL MODELLING

4.3.1 System Elements

Functional models are used for static or dynamic analysis of a physical system. In essence functional modelling is a generalization of the electrical circuit modelling problem: given a basic set of elements connected in a prescribed manner, how does the total system respond to given inputs? The difference from conventional electrical circuit modelling is that it must be able to handle different energy domains. The generalization is achieved through identification of "effort" and "flow" variables (also known as across and through variables) for the different energy domains.[4,8] The flow variable f acts at a single point in space while the effort variable e acts between two points. They are termed as energy rate variables and their product $e . f$ represents power. The variables for different energy domains are shown in Table 4.1.

The elements are classified according to the manner in which power is handled (either dissipated, stored, converted or transmitted). This leads to the following classification of elements:

One port elements. Energy flow for one port elements can be described by a pair of system variables. Resistive or dissipative elements R and two types of storage elements, L (e.g. inductive) and C (capacitative), are one port elements. Examples of the one port elements for different energy domains can be found in Table 4.2.

Variables / Energy System	effort accumulation $ea = \int e\, dt$	effort e	flow f	flow accumulation $fa = \int f\, dt$
Electrical	flux linkage λ	voltage $v = \lambda$	charge $i = \dot{q}$	charge q
Mechanical Translation	displacement x	velocity $v = \dot{x}$	force $F = \dot{p}$	momentum p
Mechanical Rotation	ang. disp. θ	ang. velocity $\omega = \dot{\theta}$	torque τ	ang. momentum h
Fluid	pressure momentum Γ	fluid pressure p	volume flow rate $Q = \dot{V}$	volume V
Thermal pseudo *	—	temperature T	heat flow rate $q = \dot{H}$	heat content H

Table 4.1 System modelling variables
* In thermal systems the variables are termed pseudo because heat flow rate q is itself power. e $\times$ f is not power.

Elements / Energy System	Dissipator R $e = R.f$	Flow Store C $fa = C\ e$	Effort Store L $ea = L.f$
Electrical	$v = R.i$	$q = C.v$	$\lambda = L.i$
Mechanical Translation	$v = 1/b\ .\ F$	$p = m\ .\ v$	$x = 1/k\ .\ F$
Mechanical Rotation	$\omega = 1/B\ .\ \tau$	$h = I\ .\ \omega$	$\theta = 1/\kappa.\ \tau$
Fluid	—	$V = Cf\ .\ P$	$\Gamma = Lf\ .\ Q$
Thermal	—	$H = Ct\ T$	—

Table 4.2 System elements and their linear constitutive relations.

Converters. Converters are elements with two or more ports which transfer, without storage or dissipation, energy supplied at one or more ports to energy output at other ports. Distinction is made between transforming converters also called transformers (TF elements) and gyrating converters also called gyrators (GY elements). Linear relationships for converts having two ports indexed 1 and 2 are:

$$\text{Transformer: } e_1 = h.e_2 \quad \& \quad f_2 = h.f_1 \tag{4.1}$$

$$\text{Gyrator: } e_1 = r.f_2 \quad \& \quad e_2 = r.f_1 \tag{4.2}$$

where h and r are the transduction coefficients in equations (4.1) and (4.2).

Controlled elements. The constitutive relationships for physical elements depend on its geometric and material properties. An input at a port may alter one or both of these features of an element without any energy conversion taking place. A multiport element in which an input at one port affects the terminal relations at other port or ports, without any energy transformation between them, is called a controlled element. A wire strain gauge is a typical example of a two port controlled element. An input at the mechanical port (displacement) alters the relationship between the effort and flow variables (voltage, current) at the electrical port without energy transformation taking place between the two ports.

These together with energy source and some purely signal transforming elements allow modelling of a wide variety of engineering systems and are particularly applicable to instrumentation.

The discussion so far has dealt with the components of multi-energy system. To develop functional models of such systems, interconnection of the subsystems must be considered. The flow and effort variables in physical systems obey generalized Kirchhoff's laws.[4,8]

1. *Compatibility Constraint:* if a set of ports are connected to form a closed loop then the sum of all the effort variables is zero.
2. *Continuity Constraint:* if a set of energy ports have a common terminal then the sum of flow variables at the port must be zero.

Systematic development of mathematical models using system elements can be carried out in a number of ways.[8,9] We will consider a graphical approach called Bond Graph Analysis.

4.3.2 Bond Graphs

The fundamental concept behind bond graphs is the energy bond used to indicate the interaction of different energy ports. The energy bonds connect the components of the system. The flow of energy from one component to another is indicated by the system variables (flow and effort). Graphical representation of the compatibility and the continuity constraints is carried out using effort and flow junctions. A multi-energy port with a constant flow is called an effort junction. The algebraic sum of all the effort variables at the effort junction is zero. Similarly a multi-energy port with a constant effort is called a flow junction. The algebraic sum of all the flow variables at a flow junction is zero. Graphical representation of the two junctions is shown in Figure 4.1.

The two junctions represent generalized Kirchoff's Laws. The direction of the half arrows indicate the direction of the energy flow associated with the variables.

EFFORT JUNCTION

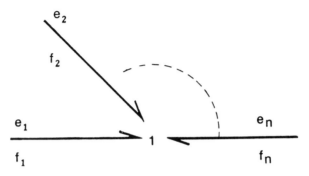

$$f_1 = f_2 = \ldots = f_n$$
$$e_1 + e_2 + e_n = 0$$

FLOW JUNCTION

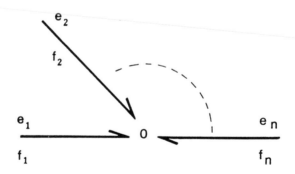

$$e_1 = e_2 = \ldots = e_n$$
$$f_1 + f_2 + \ldots + f_n = 0$$

Figure 4.1 Graphical representation of effort and flow junctions.

4.3.3 Electro-mechanical Force Sensors

To illustrate the development of functional models using bond graphs, an electro-mechanical force transducer will be considered (Figure 4.2).

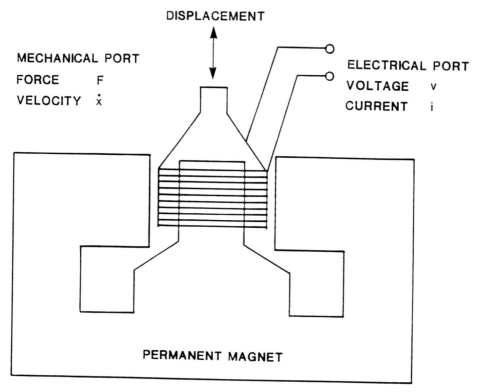

Figure 4.2 Electro-mechanical force transducer.

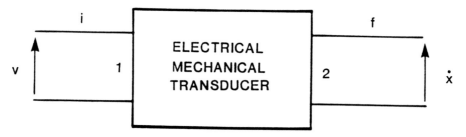

Figure 4.3 Two port network representation of electro-mechanical transducer.

It is a two port device; one port being electrical and the other mechanical. The device works on the principle of an electric motor. The input can be applied at either of the two ports resulting in an output on the other port. Hence the device can be idealized by a bilateral converter (TF) as in Figure 4.3 where:

$$v = h\dot{x} \quad \& \quad f = h_i \tag{4.3}$$

where $h=BL$ is the transduction coefficient, B is the flux density experienced by the conductor of length L.

In reality there are impedances associated with each energy port as shown in Figure 4.4.

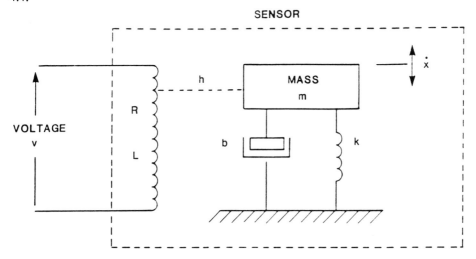

Figure 4.4 Schematic for electro-mechanical transducer.

The electrical port can be represented by an effort junction since there is a constant flow (current). Similarly the mechanical port can be represented by a flow junction because of the shared or constant effort variable (velocity), as is shown by the Bond graph Figure 4.5

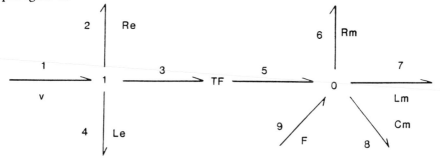

Figure 4.5 Bond graph for electro-mechanical transducer.

The equations for the system can be obtained from the compatibility and the continuity constraints. For the electrical port (effort junction), ignoring the capacitive effects, the equations are:

$$f_1 = f_2 = f_3 = f_4 = i \quad \& \quad (e_1 - e_2 - e_3 - e_4) = 0 \tag{4.4}$$

The relationships between flow and effort variables for inductive and dissipative elements are given in Table 4.2. In Laplace domain equation (4.4) can be re-written as:

$$v = Lsi + Ri + h\dot{x} \tag{4.5}$$

where s is the Laplace parameter.

Similarly for the mechanical port (flow junction) the equations can be written as:

$$e_5 = e_6 = e_8 = e_9 = \dot{x} \tag{4.6}$$

$$\&\quad (f_5 - f_6 - f_7 - f_8 + f_9) = 0 \tag{4.7}$$

or
$$F = ms\dot{x} + (k/s\dot{x}) + b\dot{x} - hi \tag{4.8}$$

In matrix form the equations for the systems can be expressed as:

$$
\begin{bmatrix} v \\ f \end{bmatrix} =
\begin{bmatrix} (Ls + R) & h \\ -h & (ms+k)/(s+b) \end{bmatrix}
\begin{bmatrix} i \\ \dot{x} \end{bmatrix}
\tag{4.9}
$$

The electrical and mechanical parameters (L, R, h, m, k, b) can be obtained experimentally or through detailed physical models which relate these parameters to the dimensions and the materials of the subsystems. Solution of the ordinary differential equations above can be obtained in the usual manner and will yield the dynamic response of the system.

4.3.4 MEDIEM

A computer package, acronymed MEDIEM (Multi Energy Domain Interactive Element Modelling), has been developed at the MIC for functional modelling of engineering systems.[5,6] Representation of a system being modelled is carried out using a graphical language called "structure graphs". The method was developed by combining the advantages of bond graphs and another representation method called linear graphs. [8] The disadvantage of bond graphs is that the topological structure of the system being modelled is lost. Linear graphs are not very orderly and they produce complex representations. The main improvement in structure graphs is that the representation has close resemblance to the schematic diagram of the system being modelled. The package MEDIEM has a comprehensive set predefined elements analogous to the system elements discussed in previous sections. Interaction with the package is performed through combined use of a mouse and keyboard. A fixed menu displays a list of different types of elements and set of operational instructions. Development of a model essentially involves reproduction of a structure graph of the system on the screen. Once the problem has been defined the program generates the differential equations and solves them. The package allows rapid and efficient model development from which time and/or frequency response can be displayed.

A structure graph for the previous example of the electro-mechanical transformer is shown in Figure 4.6.

An integration element "AAI" is introduced to observe the displacement of the coil when a step voltage input is applied. The output from the package for the coil displacement is shown in Figure 4.7.

4.4 PHYSICAL MODELLING

A functional model provides an insight into overall behaviour of an instrument based on input/output characteristics of its constituent subsystems. In order to determine these characteristics it is often essential to develop detailed physical models. A physical model relates the functionality of a subsystem to its shape, dimensions and the material properties. Such models require solution of "field" quantities which are obtained from the solution of the appropriate partial differential equations. For elastic elements this would be displacement field, for electromagnetic devices - electric/magnetic fields, for flow problems - pressure fields and for temperature sensors - heat flux and fields. From these "fields" other quantities of interest such as stresses, strains, masses, resistances, inductances etc. can be obtained by integration, differentiation or other simple operations.

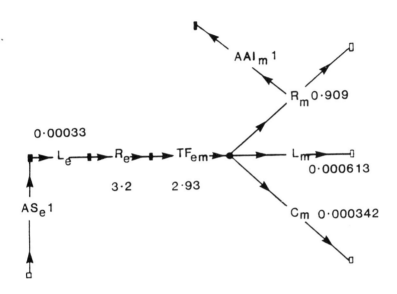

Figure 4.6 Structure graph of electro-mechanical transducer.

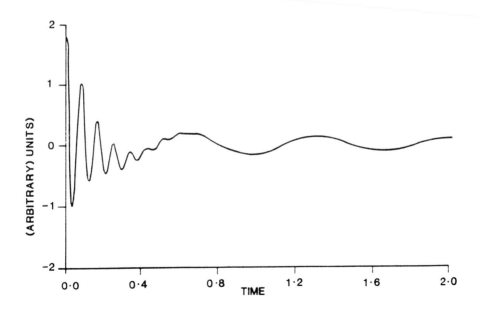

Figure 4.7 Coil displacement due to step voltage input.

4.4.1 Finite Element Method

Practical engineering problems described by partial differential equations cannot be solved accurately by analytical methods. A powerful numerical technique for solving a wide range of continuum problems is called the Finite Element Method (FEM). The method has gained considerable popularity in recent years. There are a number of excellent books available on the subject.[10,11,12]. More recent developments on the subject can be found in such publications as the International Journal of Numerical Methods of Engineering, and Computers and Structures.

Origins of the finite element method stem from the Matrix method used for the analysis of complex frame structures, since then it has extended to solve continuum problems. The method is based on piecewise approximation of the governing differential equations. The steps involved in finite element analysis are:

1. Discretization of the domain. The first stage in the finite element method involves approximating the solution region with an assemblage of elements simple shape (lines, triangles, quadrilaterals and tetrahedrals etc.). Since the elements can be put together in a variety of ways, they can be used to represent extremely complex geometries. There are a variety of elements available for the purpose.

2. Selection of the interpolation functions. In a continuum problem the field variable (displacement, temperature etc.) has infinite values since it is a function of each generic point in the solution domain. In finite elements the unknown field variable is defined at the element nodes. Interpolation functions are used to find its value anywhere within an element. The choice of interpolation functions depends on the order of the equations.

3. Establishment of element equations. Element equations can be derived in a number of ways: (i) direct method, (ii) variational method, (iii) weighted residual method, (iv) energy balance method. The direct method can be applied to very simple cases. Regardless of the method used, the resulting element equations are of the format:

$$(k^e . a^e) - F^e = 0 \qquad (4.10)$$

where k^e is called the element stiffness matrix, a^e unknown variable vector and F^e is the loading vector.

4. System equations. The contributions of individual elements are combined together to form system equations. The matrix form of the system equations is the same as the element equations except that it has many more elements.

5. Solution of the system equations. The set of simultaneous equations are solved, after imposing the boundary conditions, to obtain the values of the unknown field variables. If the equations are linear then standard solutions techniques can be employed. For nonlinear equations the solution procedure is more complicated and an iterative approach such as the Newton-Raphson method has to be used.

6. Further calculations. Other parameters of interest are calculated at this stage i.e. in elasticity problems the solution of the system equations yields the nodal values of displacements from which strains and stresses may be determined. Generally finite element analysis is carried out in three stages:

(i) Pre-processing: At this stage the discretization of the solution domain is carried out and the boundary conditions, i.e. loading data and the constraints, are defined.

(ii) Analysis: Analysis involves formulations of the element equations and hence the system equations. Solutions of the system equations to yield nodal variables.

(iii) Post-processing: At this stage calculation of other parameters of interest and graphical display of the results.

The reason for carrying out finite element modelling in three stages are:

1. The computing requirements for the different stages are different. For stage 1 and stage 3 interactive graphical facilities are essential, whereas at stage 2 large number crunching is involved and can be submitted as a background job.

2. For general purpose analysis, computer programs for each stage are large enough to require separate existence.

Because of the increasing power of the work-stations, self-contained packages such as APPLE and PEARS[13] are becoming more popular. However, the domain of application of these packages is limited e.g. PEARS can be used for linear analysis of plane or axi-symmetric elasticity problems only. Whereas general purpose finite element packages, such as LUCAS,[14] can be used to solve elasticity and thermal problems. They have an extensive library of elements which could be used for linear and nonlinear analysis of problems with complex geometries and boundary conditions. The analysis of a piezo-resistive sensor will be used to illustrate some of the aspects of the finite element modelling techniques.

4.4.2 Piezo-resistive Pressure Sensor

Piezo-resistive sensors are widely used in industry for the measurement of absolute and gauge pressures. Typical applications of the sensor are in the automotive industry. They are used in engine management systems, and for the measurement of air and hydraulic pressures in advanced braking and suspension systems.

The sensing element of a piezo-resistive pressure sensor consists of a semiconductor diaphragm with four stress sensitive gauges located at the diaphragm surface, Figure 4.8.

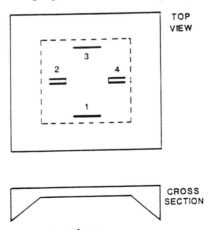

Figure 4.8 Piezo-resistive pressure transducer.

The gauges are formed by the diffusion of impurities of the opposite type. The *p-n* junction thus formed act as an electrical stress sensor. The gauges are connected together to form a bridge network. Application of pressure generates a stress field in the diaphragm which alters the resistance of the gauges and hence the output voltage. Schematically the behaviour of the sensor is shown in Figure 4.9.

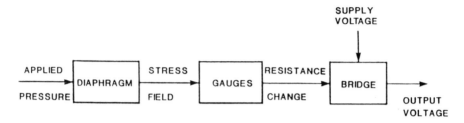

Figure 4.9 Schematic for piezo-resistive sensor.

The device is configured such that the change in resistance of opposite gauges is identical. Therefore the bridge output is given by:

$$\frac{V_o}{V_i} = \frac{\delta R_a / R \; - \; \delta R_b / R}{2 \; + \; \delta R_a / R \; + \; \delta R_b / R} \tag{4.11}$$

where R is the nominal resistance of the strain gauges, δR is the change in the resistance and depends on the stress levels.

$$\frac{\delta R}{R} = \left[\Pi \; + \; \frac{(1 \; + \; \upsilon)}{E} \right] (\sigma_x \; - \; \sigma_y) \tag{4.12}$$

where E, is Young's modulus, υ is Poisson's ratio, Π is a material dependent parameter and σ_x and σ_y are the average stresses experienced by the gauges in x and y directions. In order to find σ_x and σ_y a finite element model was developed using a general purpose package "LUCAS". Since the structure is bisymmetrical only a quarter section of the device needs to be modelled. The centre diaphragm is modelled by an eight noded semi-loof shell element while the tapered boundary is represented by two layers of twenty noded brick elements. Since the shell elements represent the mid-plane of the diaphragm, to transfer bending stresses from across the diaphragm thickness to the brick elements, three noded beam elements of comparatively very high stiffness are incorporated between the shell elements and the brick elements.

The discretization of the sensor and the deformation in the device due to applied pressure is shown in Figure 4.10(a). Contours of constant stresses at the diaphragm surface are shown in Figure 4.10(b).

The package produces results for the stresses at the nodes of the elements. To calculate the average stresses experienced by the gauges σ_x and σ_y, the gauge length is divided into ten equally spaced intervals. Stress values are determined at these points using bi-linear interpolation functions. The average value of these stresses are then used to determine variations in the resistance of the gauges and hence the bridge output. A good agreement was found between the computed results and experiments conducted on a one bar device manufactured by Lucas Automotive Ltd.

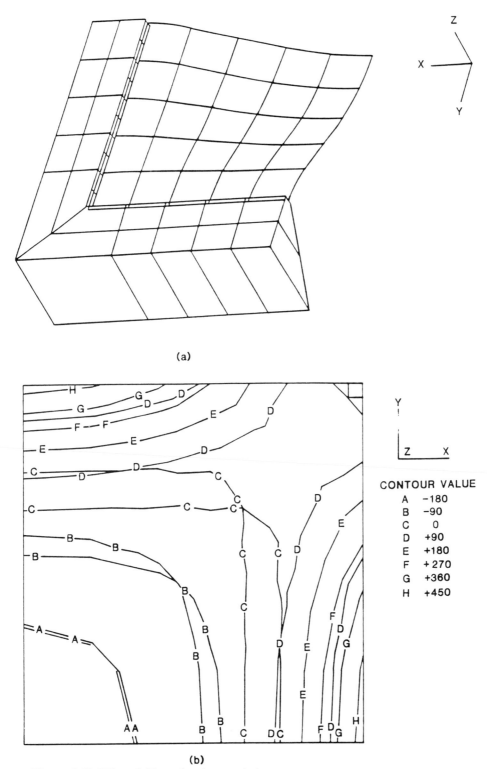

(a)

(b)

Figure 4.10 FE modelling of a piezo-resistive sensor.

CONTOUR VALUE

A	−180
B	−90
C	0
D	+90
E	+180
F	+270
G	+360
H	+450

4.5 DESIGN OF INSTRUMENTS

4.5.1 Design Curves

The process of designing a new range of instruments starts with the establishment of the specifications i.e. input/output requirements, geometric and material constraints, and acceptable error levels etc. The outcome of the design is the identification of the parameters necessary to define completely the instrument. Mathematical models described in the previous sections are best suited for the purpose of analysis. Design of a new instrument through repeated use of these models is tedious and time consuming. A better approach is to use these models to generate "dimensionless" performance curves for a class of instruments. These curves can then be used to develop a simple design methodology.

4.5.2 Snap-action Diaphragms

A snap action diaphragm is a convex shaped thin shell structure, Figure 4.11.
  When the diaphragm is subjected to increasing pressure from the convex side there is a corresponding increment in the centre-displacement (see Figure 4.12).

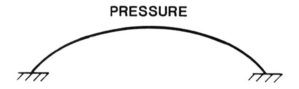

Figure 4.11 Snap action diaphragm.

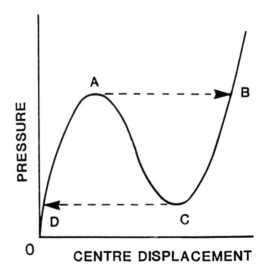

Figure 4.12 Pressure-deflection characteristic.

68

At a critical value of the applied pressure "B", the diaphragm becomes unstable and snaps through to new stable position "C". Similar behaviour is displayed when the pressure is gradually reduced (after the diaphragm has snapped through).

Snap-action diaphragms are generally used as force or pressure operated switches or relays. They are ideally suited for this purpose because of their positive action, large displacements and good reproducibility. The nonlinear equations for snap-action diaphragms cannot be solved analytically. Therefore, a finite element program was developed. Mathematical details of the model and the computer implementation can be found elsewhere.[15]

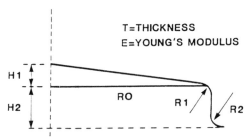

Figure 4.13 Parametric representation.

A systematic procedure for generating non-dimensional performance curves will be illustrated for conical shaped diaphragms. The steps involved in establishing the performance curves are:

1. *Parametric representations.* A diaphragm is said to be conical when its major load carrying section can be defined by a straight line. The parameters required to define such a class of diaphragm are shown in Figure 4.13.

2. *Sensitivity analysis.* Sensitivity tests are carried out to find the parameters (design variables) which have considerable effect on the diaphragm performance. The behaviour of conical diaphragms was found to be critically dependent on R_o, H_1, T and E. When designing a new diaphragm only these parameters need to be considered. Engineering common sense guided by geometric constraints can be used to estimate the values for the other parameters.

3. *Dimensionless groups.* The design variables are combined together to form dimensionless groups. These groups may be obtained through dimensional analysis or based on simple mathematical models. The groups for a conical diaphragm based on Andreeva's model[16] are:

$$(P/E) \cdot (R_o/T)^4, \quad H_1/T, \quad W/T \tag{4.13}$$

where P is the applied pressure and W is the centre displacement.

4. *Generalization of finite element results.* Results obtained from the finite element program for a particular case are generalized by presenting them in non-dimensional performance curves. Results from the finite element program for various H_1/T are shown in Figure 4.14. The concept of dimensionless curves is based on similarity. The curves were obtained for a specific case but are applicable to any conical diaphragm with the same parameter ratios.

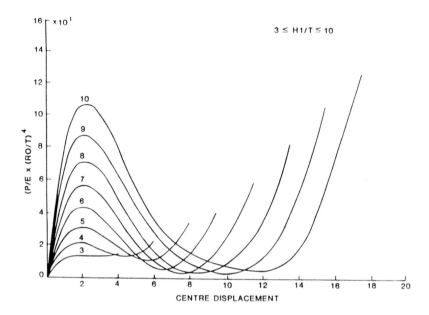

Figure 4.14 Performance curves for snap action diaphragm.

The performance curve contains all the essential information regarding the class of instrument. The results can be rearranged in different ways to facilitate the establishment of a simple design methodology.[1,15] Once the design variables have been determined, values for the other parameters required to completely define a conical diaphragm (see Figure 4.13), can be assigned using geometric constraints guided by engineering common sense. The design based on non-dimensional curves is approximate and the parameters may need fine tuning to obtain accurate results.

4.6 CONCLUSIONS & DISCUSSION

Mathematical modelling plays an important part in the design of instruments. It provides insight into instrument behaviour and can be used to accurately predict instrument performance from the specification of subsystem properties in the case of functional modelling, and geometric and material properties for physical modelling.

A suggested approach for the analysis and design of instruments is that mathematical modelling should be carried out in a hierarchical fashion. During the initial stages simple analytical mathematical models should be developed to identify the instrument transformation law and block diagram decomposition to reveal subsystems and their own transformations. As the need the accuracy increases, more complex models based on the methods discussed in this presentation can be adopted.

Finally, the role of mathematical models in the process of design should not be regarded as a substitute for experimentation with prototypes for designing new instruments. An efficient use of mathematical models can lead to streamlining the entire process. This streamlining is illustrated in the area of snap-action diaphragms, where a finite element model is used generate dimensionless performance curves which could be used to develop a simple design methodology.

ACKNOWLEDGEMENTS

The authors wish to acknowledge the Science and Engineering Research Council for their continued support of the research work on Mathematical Modelling and Computer Aided Design of Instruments.

REFERENCES

1. Abdullah, F., Mirza, M.K. and Rahman, M.M. (1988) Computer Aided Design of Elastic Elements in Instrumentation. ACTA IMEKO XI. In *Proceedings of the 11th Triennial World Congress of the International Measurement Confederation* (Houston, 1988). Plenaries, invited papers, computers, and intelligent systems, pp 223-238

2. Abdullah, F. (1984) Mathematical Models and their use in force Measuring Instruments. In *Measurement and Instrumentation for Control*, edited by M.G. Mylroi & G. Calvert. London: Peter Peregrinus Ltd

3. Abdullah, F. and Finkelstein, L. (1983) Computer aided design of instrument sensors. In *Fortschritte durch digitale Mess- und Automatiesierungstechnik: INTERKAMA-Kongress, (Fachberichte Messen Stevern Regln :20)*, edited by M. Surbe & M. Thoma, pp. 551-562. Berlin: Springer-Verlag

4. Finkelstein, L. and Watts, R.D. (1978) Mathematical Models of Instruments: Fundamental Principles. *J. Phys. E: Sci. Instrum.*, **10**, 566-572

5. Liebner, R.D., Abdullah, F. and Finkelstein, L. (1982) Structure Graphs: A New Approach to Interactive Computer Modelling of Multi-Energy Domain Systems. *ASME Trans. J. Dynamic Systems, Measurement and Control*, **104**, 143-50

6. Bailey, W.N. and Abdullah, F. (1989) Interactive graphics-based computer-aided system dynamic modelling. *Trans. Inst. MC.*, **11**, 127-137

7. Doeblin, E.O. (1990) *Measurement Systems: Application and Design*. New York: McGraw Hill

8. Wellstead, P.E. (1979) *Introduction to Physical System Modelling*. UK: Academic Press Inc. Ltd

9. Nuebert, H.K.P. (1975) *Instrument Transducers*, London: Oxford University Press

10. Hubbner, K.H. (1975) *The Finite Element Method for Engineers*, USA: John Wiley & Sons Inc.

11. Zienkiewicz, O.C. (1977) *The Finite Element Method*, 3rd edn. UK: McGraw-Hill Book Company Ltd

12. Brebbia, C.A. (1985) *Finite Element Systems: A Handbook*. UK: Springer-Verlag

13. APPLE and PEARS, Rutherford and Appleton Laboratory (RAL), Chilton, Oxford.

14. Finite Element Analysis Ltd. (1984) *LUSAS User's Manual*, London

15. Mirza, M.K. (1983) Mathematical modelling and computer aided design of snap-action diaphragms. *PhD Thesis*, Department of System Science, City University, London, UK

16. Andreeva, L.E. (1966) *Elastic elements of instruments*. Israel program for scientific translations, Israel, 1966.

Chapter 5

ALGORITHMS FOR COMPUTER AIDED PRECISION METROLOGY

DEREK G. CHETWYND

Low cost, high quality microcomputing has strongly influenced the design and application of precision metrology instrumentation. Post-processing of data can extract additional information from conventional instruments and provide the basis of new approaches. This article examines the application of some classes of algorithms that are particularly relevant to surface metrology but have also a wider significance. It is argued that a knowledge of the physics of the instrument and of the reason for the measurement is necessary to the selection of good algorithms. Since this is often unclear, real instruments use inherently compromised solutions. Specific areas discussed include the generation of trend lines and reference figures by least squares and minimax methods; digital filters; peak sensitive parameters and sampling regimes; and error compensation.

5.1 INTRODUCTION

Every computer aided metrology system involves the processing of a digitized sequence of data points. The point of this processing might be to better detect a wanted signal from a noisy background; to select a sub-set of the available information that is of direct interest; to summarize the data in terms of a small number of parameters; to correct for the effects of drift or misalignment in a gauging system (detrending, for example); or to compensate for known systematic errors so as to extend the system performance beyond that of electromechanically similar instruments. It is likely that any good quality signal processing suite of software, either a set of subroutines or a stand alone package, will contain algorithms relevant to the majority of these tasks. Thus the challenge of enhancing instrument performance might seem to reduce to the production of a software interface onto the package of one's choice. Of course, things are less simple in practice. There are many cases where changes in the algorithm or the manner in which the data sampling is executed can, separately or in combination, affect the results obtained from nominally identical measurements. The only way of interpreting the significance of these changes is through a knowledge of the physical behaviour of the instrument and an understanding of why the measurement is being carried out. Surface metrology provides some good examples of these effects and so will be the primary vehicle used in this discussion. The implications of the examples have a much wider scope of applicability.

The examples are concerned with the assessment of straightness, roundness or surface finish from data provided either by a specialized instrument or by a coordinate measuring machine. Although some machines provide three-dimensional data, it is common practice to assume that what is provided is a profile, that is a description of the shape of the air/solid interface of a thin, predefined section through the workpiece. Only profile descriptions will be considered here since they are simpler to visualize.

A coordinate measuring machine provides measurement data by using a mechanical probe, usually spherical and of 1 to 10 mm radius, which is brought into contact with the surface at reasonable pressure. The position of the probe holder is monitored, typically by gratings, relative to a set of axes fixed to the machine and defined by its slideways. This position is reported as (x, y, z) coordinates which are taken to represent the position of a point on the surface since the offset from the holder to the probe tip is fixed. This is not strictly true for a finite tip size since contact can be at different points and sometimes correction terms are included. Typical readout precision is around 1 µm but overall accuracy is more typically 10 µm. Some modern machines use a non-contact optical probe but for present purposes these will be ignored after noting that, since the physical principle of the detection is different, the exact nature of the data compared to the surface it is representing may be somewhat different.

Surface metrology instruments are characterized by the use of a linear gauging system which measures the variation of the distance between the workpiece surface and a datum surface during a relative movement of the gauge and workpiece. In many cases the datum is generated in space by the motion of a precision rotary bearing or linear slideway, although a physical datum surface can be used. The gauge may be sensitive to a few nm or less but its range is only rarely as much as 1 mm. In the majority of instruments the workpiece surface is detected through contact with small stylus, with an effective tip radius of typically a few µm for surface finish up to perhaps 1 mm for straightness or roundness measurements.

Whichever type of instrument is used, we cannot assume that the workpiece can be placed on the instrument sufficiently precisely that the raw profile data represents directly the deviations from form. Thus a major task of computation is to establish from the measured data the position of the ideal surface of which the true profile is taken to be an approximate realization. This is the idea of computing a best fit reference figure in the coordinate frame of the instrument relative to which profile deviations may be assessed as if in a workpiece coordinate frame. For straightness and roundness the reference figures are clearly the straight line and the circle. For surface finish measurements a straight line may be used, in the same way as signals subject to sensor drift might be detrended, but often a "wavy" mean line is generated through the action of a standard filter. Once the profile is re-expressed relative to its reference, a concise description of it is required. Sometimes, often in quality control applications, a simple peak or RMS amplitude might be adequate. However the functional performance of a surface is likely to depend on its mechanical, electrical or thermal contact, or optical scattering properties. To summarize these in a few parameters is extremely problematic since there is currently only a poor understanding of the interaction of geometry and function. The attempt provides rich ground for the use, and mis-use, of more sophisticated computer algorithms.

5.2 COMPUTING REFERENCE FIGURES

Straightness and roundness measurement involve very similar decision processes for reference fitting. The notion of a best fit implies that the deviations of the data from a trial figure will be assessed and some function of them be invoked to measure the quality of the fit. In this context the deviations are generally referred to as "residuals" and the most common criteria of fit are least squares (minimum sum of the squares of the residuals) and minimax (minimum value for the largest residual). During manufacture a workpiece acquires a definite set of deviations from the design size and shape, but the residuals, which provide our estimate of those deviations, vary according with the parameters of the reference figure. The implications of various definitions of residuals and fitting criteria will be explored for straightness profiles. Roundness measurement will then be considered since it well illustrates some further important points.

5.2.1 Least Squares Straight Lines

Most basic texts on statistics include formulae for the least squares line through a set of data. Not always stressed is that this is a linear least squares method that is applicable only when some implicit assumptions can be adequately met. Basically a straight line is fitted to the measurands (ordinates) y_i, which are subject to error or noise, at set points (abscissae) x_i which are assumed error free. Mathematically, for the straight line

$$y = ax + b \tag{5.1}$$

we seek the values of the coefficients a and b such that the residuals ε_i defined from

$$ax_i + b = y_i + \varepsilon_i \qquad : i = 1, .., n \tag{5.2}$$

have the minimum sum of squares value. These values of a and b are taken to provide a best estimate of the line about which the measurands are assumed to be scattered.

Using a vector and matrix notation, which is convenient when the defining equations are linear in the parameters, equation (5.2) becomes

$$H A = Y + \varepsilon \tag{5.3}$$

where

$$H = \begin{bmatrix} x_1 & 1 \\ \vdots & \vdots \\ x_n & 1 \end{bmatrix} \;\; ; A = \begin{bmatrix} a \\ b \end{bmatrix} \;\; ; Y = \begin{bmatrix} y_1 \\ \vdots \\ y_n \end{bmatrix} \;\; ; \varepsilon = \begin{bmatrix} \varepsilon_1 \\ \vdots \\ \varepsilon_n \end{bmatrix} \tag{5.4}$$

For the least squares criterion that $\Sigma \varepsilon_i^2$ (or $\varepsilon^T \varepsilon$) is minimum, it can be proved that the best estimates of the parameters are given by

$$A = (H^T H)^{-1} H^T y \tag{5.5}$$

where the superscripts T and -1 denote transposition and inversion, respectively.

This is the general solution and incorporates the versions often quoted in which some terms are pre-evaluated, perhaps by assuming equally spaced abscissae.

A most important point about this solution is that the residuals are defined as measured parallel to the y-axis, otherwise a simple addition could not be used in equations (5.2) and (5.3). It is this which forces the linearity of the formulae and so guarantees that there is a unique solution to the fit. This model is often used unthinkingly but it is only valid if it makes physical sense in terms of the original measurement to consider the residuals acting in this manner. For example, if we use a coordinate measuring machine to take a set of points consisting of coordinate pairs along the length of a nearly straight feature, the deviations are usually considered to be normal to the reference line not parallel to an instrument axis. Thus the residuals should be considered as shown in Figure 5.1(b) not as in 5.1(a).

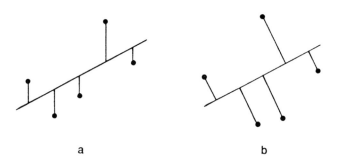

a   b

Figure 5.1 Residuals of profile data points to the best straight line. (a) linear formulation and (b) normal formulation.

Calculating the residuals according to this definition, the form of equation (5.2) is replaced by

$$a x_i + b = y_i - \varepsilon_i \sqrt{(1+a^2)} \qquad (5.6)$$

which is distinctly non-linear. As it happens, this is one of the few non-linear optimizations that can be directly solved, though it cannot be guaranteed that there is only one minimum solution. It is clear merely from looking at the formula that the computational task is greater than with a linear formulation.

Consider now the situation on a stylus based straightness measuring instrument. In practice a linear least squares fit is used. Every few years, it seems, a proposal to use the more accurate form of equation (5.6) is placed before Standards committees. The argument for this is essentially that rehearsed in the paragraph above. The counter-argument for retaining the simpler method often invokes the geometry of the instrument. Its operation depends upon the instrument datum surface being aligned quite closely to the trend of the workpiece surface. Typically the traverse might be 10 mm or longer, while the vertical displacement might be some tens of μm or less. Thus the slope of the reference line relative to the instrument axis (a in equation (5.6)) must be very small and only an insignificant approximation is involved in using equation (5.2) rather than equation (5.6).

There is nothing wrong with this reasoning, the approximation is slight under typical conditions. There is, however, a flaw in the premise on which the reasoning is based. In the idealized instrument the gauge measures distance from the datum to the workpiece unidirectionally and that direction is a function of the construction of the instrument and not of the alignment of the workpiece. The individual point measurements, and their associated uncertainty, are made normal to the instrument axis. The object of fitting the reference is to model the misalignment so that it may be removed, hence the residuals should be defined normal to the instrument axis. Linear least squares is not merely a good approximation. It is in principle a more accurate method than the alternative (more complex) formulations. Treating the data in isolation and applying mathematical reasoning cannot discover this. Only by an appreciation of the instrument physics can the correct definition of residuals be discovered. That best definition is not necessarily the same for all instruments performing nominally the same task.

5.2.2 Minimum Zone Straight Lines

In metrology the term minimum zone is commonly used for the best fit obtained by a minimax (or Chebychev[1]) criterion. The minimum zone straight lines are a pair of parallel lines so placed that they enclose the profile between them and that their separation is a minimum. The minimax line is that placed such that $Max(\varepsilon_i)$ is minimum, which is readily seen to be the line parallel to and midway between the minimum zone lines.

The arguments regarding the method of defining the residuals are exactly the same as for least squares fitting since the instrument conditions are not changed by a different choice of best fit criterion. For the majority of systems the linear formulation is correct and the minimax fit, formally expressed as a constrained optimisation, will be a linear programme

$$
\begin{aligned}
&\text{Minimize} &&z &&(5.7)\\
&\text{Subject to} &&ax_i + b + z \geq y_i\\
&\text{and} &&ax_i + b - z \leq y_i, \quad \forall\, i
\end{aligned}
$$

Solving this requires iterative methods but linear programmes are one of the few classes of constrained optimization that have guaranteed convergence onto a unique solution. Before discussing how this might be done efficiently, it is useful to examine why it is so. Parameter space plotting is a good tool for this purpose.

A parameter space, or solution space, is a conceptual frame in which each parameter is plotted along an orthogonal axis. This provides a geometrical view of the problem which is intuitively useful when there are three or less parameters and mathematically useful in any number of parameters. The total n-dimensional space described by those axes contains one point for each conceivable combination of parameter values. In an optimization, the objective function plots as a family of surfaces when equated to different constants and solution consists of finding the largest or smallest such constant that can exist. In a constrained optimization, only part of the parameter space, the feasible region, contains valid combinations of parameters. Solution methods then concentrate on the intersection of objective function surfaces with the feasible region. The boundaries of the feasible region are formed by the intersection of the equality conditions of the constraints which will also be surfaces in the space.

For the simplest illustration, consider the solution to the linear programme of equation (5.7) for the case when $b = 0$. Each data point will generate a constraint pair which produces boundary lines in an (a, z) space as

$$
\begin{aligned}
z &= -x_i a + y_i\\
z &= x_i a - y_i
\end{aligned}
\qquad (5.8)
$$

that plot as shown in Figure 5.2. Examination of the original inequality directions of the constraints shows that feasible combinations must lie above the lines as plotted. Since we seek to minimize z, the objective surfaces are lines parallel to the a-axis and the best solution is clearly where the constraints cross. With extra constraints overplotted the situation does not change materially because the feasible region will continue to be a polygon. It is a geometric consequence of the constraints being linear in the parameters that the best solution will lie at an intersection and that there will be a single best point (or a continuous region between two equally good intersections if part of the polygon happens to be exactly horizontal).

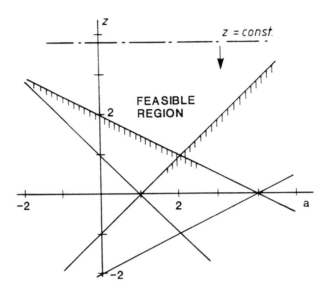

Figure 5.2 Parameter space representation of a simple linear programme for the minimum zone straight line. Scales are arbitrary since units depend on the exact formulation of the geometrical problem.

The guaranteed convergence onto a unique solution is because the feasible region is convex, that is any two points within it or on its boundary may be connected by a straight line lying wholly within the region. It may be solved with high efficiency because the linearity of the constraints and objective function force the solution to lie at an intersection on the boundary so that only a small set of potential solutions need be examined.

By contrast, Figure 5.3 shows typical forms of the constraint boundary

$$z(a) = (xa - y) / \sqrt{(1+a^2)} \tag{5.9}$$

that would occur if the normal formulation of residuals according to equation (5.6) were used in equations (5.7) and (5.8). There is no doubt that a nicely behaved boundary to the feasible region will not be obtained. Since the lines are curved, the minimum may not lie at an intersection so a much more extensive search of the solution space will be required. Worse, the intersection of the lines will rarely, if ever, create a convex region so there will be several local minima and no means of telling which is the global one without testing all the candidates. In practice it is difficult to stop search algorithms getting stuck on local solutions and, as it cannot be determined *a priori* how many of them there are, there is no certainty of finding the global minimum.

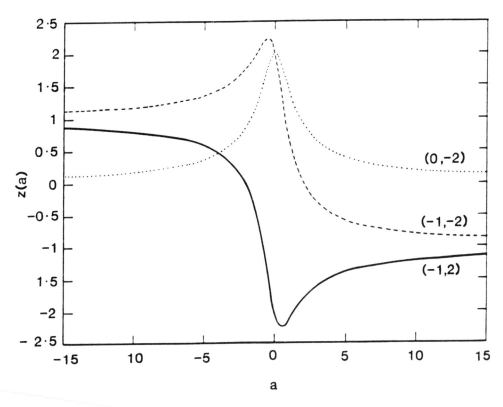

Figure 5.3 Typical constraint lines in a (z, a) parameter space for minimum zone straight lines with residuals defined normal to the reference. Scales are arbitrary.

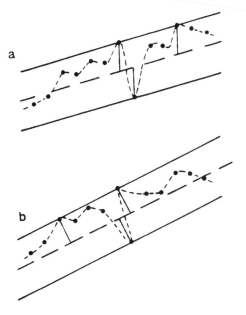

Figure 5.4 Minimax lines to the same data with residuals defined (a) normal to datum axis (b) normal to the line.

Figure 5.4 shows an example of minimax fitting to the same data with the two definitions of residuals. The smaller zone is that of Figure 5.4(b) so in the absence of physical knowledge of the instrument this would seem the better solution. Note however that relative to the horizontal (i.e. the instrument datum) and to the reference line the sequential order of the points is different. Physically this is nonsense for it implies that the underlying profiles are different yet they come from the same measurement. The instrument configuration shows which version is correct.

A small bonus from behaviour of this sort is that if we know the order in which points occur along the real surface then we can select the correct solution. This can be used in reality or as a "thought experiment" to help determine which type of residual definition should be used for a given measurement. It also suggests that a small restriction on the freedom of the user, for example that data must be taken in sequential order, can replace considerable complexity of software in attempting to untangle which of several candidate solutions is the most appropriate one. In most cases data would in any case be so collected for it would be the easiest way to do it. A little common sense in the design of measurement schemes can do much to remove ambiguity.

Perhaps before leaving this topic it should be stressed that common sense must be used together with mathematics in specifying algorithms. A classic example of this is the best fit line to one cycle of cosine and one cycle of sine. With both least squares and minimax the solution for cosine is the independent axis as expected. With sine both tilt the line quite sharply. This is disconcerting only because we know that the mean of a continuous wave would be horizontal. A little thought confirms that the trend must be downwards during one cycle of sine. If the wave were slightly distorted so that it looked like a surface with an S-curve, the momentary confusion would probably not arise.

Returning now to the solution of the linear minimum zone fit, one approach is to use the formal techniques of linear programming. This is a good, rigorous approach but more efficient geometrical methods exist because of the particular nature of the problem to be solved. Iterative procedures known as exchange algorithms can be applied directly to the data. Specifically here the Stiefel Exchange Algorithm may be used. This algorithm can be derived from the theory of linear programming[1] and so is proven to converge to the correct solution. Consider the data in Figure 5.4(a) with residuals defined vertically. A straight line is defined by two points so two parallel lines require three, the extra constraint controlling the freedom on the choice of zone width. These points will contact the zone lines (or lie equidistant from the central line). Clearly two must lie on one boundary and one on the other. In the figure the points lie alternately on each extreme. In this configuration attempting to change the slope by pivoting the line pair about any of the defining points forces the zone to grow. Therefore this is a local minimum in the zone and, since the problem is linear, also the global solution. This alternation provides a criterion for checking whether the optimum solution has been found. In an exchange algorithm a trial set of points is selected and if they do not form a solution one of the set is replaced by a point which previously violated the constraints. The procedure is

1. Choose three points and fit a line pair through them such that the alternation rule is satisfied.

2. Test whether any other points lie outside the zone so defined.

3. If no, then the minimum zone has been found, else choose any point lying outside the lines. (Conventionally the most distant is chosen on the grounds that it may well have a strong influence on the eventual solution)

4. Replace (exchange) one of the original defining points by the newly chosen one such that the alternation rule is still satisfied.

5. Fit a line pair to the new set and return to 2.

The fit of the line pair through three points and the exchange in step 4 are always unique so the iterations progress smoothly. It is simple to implement and very fast to execute. Further discussion of why this is so will be deferred until examples of roundness measurement have been examined.

5.2.3 Least Squares Roundness References

Roundness measurement involves the collection of data representing the profile of a ball, shaft or hole and applying to it a best fit reference circle. The centre position and the radius of the reference may themselves be important parameters (for example in measuring the distance between hole centres) and the deviation from circularity is measured by the residuals of the data with respect to the reference. As in the case of straightness measurement, the correct definition of the residuals influences the algorithmic approach to computing the reference and is related to the instrument geometry.

On a coordinate measuring machine the profile is represented by Cartesian ordinate pairs to which a circle should be fitted with the residuals measured radially from its centre. Any formulation of that geometry is going to require a non-linear function of the reference parameters. However, a linear representation can be found by using polar coordinates from an origin close to the true centre. A circle eccentric by a small distance e centred at (a, b) can be written as

$$r(\theta) = a\cos\theta + b\sin\theta + \sqrt{(R^2 - e^2\sin^2\theta)} \tag{5.10}$$

which can be approximated by

$$r \approx a\cos\theta + b\sin\theta + R \tag{5.11}$$

with an error $\leq e^2/2R$ at any angular position. This is a standard approximation used since the early days of the subject[2] the limits of which have been thoroughly investigated.[3] It is often referred to as the "limacon approximation". An important property of both equations (5.10) and (5.11) is that the radial and angular position are expressed from the assumed origin not from the reference centre. This modifies the way in which results should be interpreted under some conditions.

Specialized roundness measuring instruments are based upon a circular instrument datum, generated by a precision rotation, from which variations in radial distance to the workpiece surface are measured. The gauge is ideally uniaxial and so there is no doubt that the profile ordinates and their associated measurement error are expressed in a polar frame centred on the instrument axis. The residuals should be defined to be radial in this frame. The formulation of equations (5.10) and (5.11) is correct for this situation and the limacon approximation even less approximate than it might seem. Real instruments express only radial variation at the surface skin of the workpiece and absolute radius is lost. The detailed implications of this inherent distortion will not be explored here, but the linear radial parameterisation of the limacon model eases the handling of the suppressed radius. Under most circumstances it models the measured profile of a perfect workpiece more precisely than does a true circle.[4]

Fitting a least squares reference is obviously much simplified if a linear model is used and Standards all specify formulae based on the limacon approximation.[5] The general solution to the least squares limacon (residuals defined radially from the instrument centre) is found from equation (5.5) in which now

$$
A = \begin{bmatrix} a \\ b \\ R \end{bmatrix} \; ; \; H = \begin{bmatrix} \cos\theta_1 & \sin\theta_1 & 1 \\ \vdots & \vdots & \vdots \\ \cos\theta_n & \sin\theta_n & 1 \end{bmatrix} \; ; \; Y = \begin{bmatrix} r_1 \\ \vdots \\ r_n \end{bmatrix} \qquad (5.12)
$$

Omitting limits and subscripts, which are always n and i, for clarity, this gives

$$
\begin{bmatrix} a \\ b \\ R \end{bmatrix} = \begin{bmatrix} \Sigma\cos^2\theta & \Sigma\sin\theta\cos\theta & \Sigma\cos\theta \\ \Sigma\sin\theta\cos\theta & \Sigma\sin^2\theta & \Sigma\sin\theta \\ \Sigma\cos\theta & \Sigma\sin\theta & n \end{bmatrix}^{-1} \begin{bmatrix} \Sigma r\cos\theta \\ \Sigma r\sin\theta \\ \Sigma r \end{bmatrix} \qquad (5.13)
$$

The shape of the terms in this variance-covariance matrix offer plenty of scope for designing sampling schemes which simplify the computation. All the off-diagonal terms can be forced to zero simultaneously by choosing a four-fold symmetry of samples around the circle, that is for a sample at θ there should also be ones at $\theta+\pi/2$, $\theta+\pi$ and $\theta+3\pi/2$. A simple and common way of doing this is to uniformly sample with a multiple of four points. Other non-uniform schemes can also be devised. For any four-fold scheme $\Sigma\cos^2\theta = \Sigma\sin^2\theta = n/2$ where a total of n samples are taken. Equation (5.13) then reduces to the Standard formulae

$$
a = \frac{2\Sigma r_i\cos\theta_i}{n} \; ; \; b = \frac{2\Sigma r_i\sin\theta_i}{n} \; ; \; R = \frac{1\Sigma r_i}{n} \qquad (5.14)
$$

The first derivations of these formulae were through the integration of continuous profiles using analytic results for the integrals of the trigonometric functions over one cycle. This was fine for the first instruments, which used analog computers for centreing. However, when discrete versions were specified for hand evaluation of a graph and later for digital computers problems arose through a poor appreciation of algorithm design. It was assumed that the integral could be replaced by a summation providing a "reasonable" number of samples were used. In most Standards this has appeared as a "reasonably large even number" without further justification. The need for four-fold symmetry was found only when a true discrete least squares solution, as outlined here, was undertaken. As it happens, the error involved in this example is not generally serious, but the potential danger of a less than rigorous approach is clear. Using a continuum approach to develop least squares formulae for incomplete arcs[6] and performing a Simpson rule type of numerical integration is less satisfactory. The errors in the parameter estimates increase considerably more rapidly as the arc length is reduced than if equation (5.13) is used directly.

5.2.4 Boundary Value Roundness References

Standards define three alternatives to least squares as criteria for reference fitting in roundness measurement. These are the minimum radius circumscribing circle (or ring gauge circle), the maximum radius inscribing circle (or plug gauge circle) and the minimum radial zone circles. The first two are defined in the obvious ways and the argument for their use on shafts and holes, respectively, is seen from their common names. The minimum zone is formed by two concentric figures which enclose the profile while having a minimum difference in radius.

The circumscribing circle is unique for any profile, by elementary geometry, but the inscribing circle may not be. Consider, for example, the extreme case of a just open figure of eight. An examination of the parameter space for the fits shows this clearly. The constraints are that the data all lies at smaller, or greater, radial distances than the reference, so the boundaries of the feasible region are given by equation (5.10). They are paraboloids in the three-dimensional parameter space. Figure 5.5 shows the R - a plane through typical constraint surfaces. For the circumscribing case, R is being minimized so the feasible region is convex and a unique solution point exists. The inscribing criterion is searching up the R axis and so meets the non-convex side of the boundary envelope, indicating that multiple local solutions may exist.

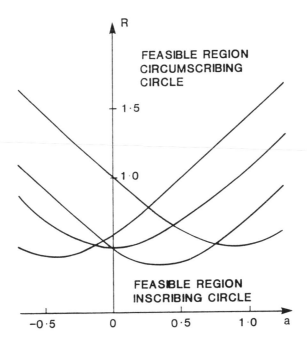

Figure 5.5 Typical constraint boundaries in (R, a) parameter space of circumscribing and inscribing circles. Units are arbitrary but of dimension length.

The minimum zone criterion involves four parameters and so is not illustrated. However it can be shown that the feasible region is not necessarily convex and profiles having two distinct solutions have been invented.[3] The equivalent limacon approximations all generate linear constraints, so all three fits have unique solutions.

The advantages of computation gained by using a linear parameterization and its quality in modelling many instrumental effects both lead to the limacon approximation being used in practice. All three boundary limacons can be expressed and solved as linear programmes, for example the ring gauge is

$$\text{Minimize:} \quad R \qquad\qquad (5.15)$$
$$\text{Subject to:} \quad H\ A \geq Y$$

where the definitions are as in equation (5.12). Given the similarity in the shape of equations (5.15) and (5.7) it is no surprise that it has been formally proved that all three boundary limacons may be solved by exchange algorithms with guaranteed convergence.[7]

The actual exchange algorithms have exactly the same form as that given in the previous section. Only the rule for selecting a set of defining points need be altered. Both the circumscribing and inscribing figures involve three points (constraining freedoms in two centre ordinates and the radius). The geometric condition that the points define a limiting figure is that they form a triangle which encloses the origin of the coordinate frame. An alternative statement of this is that adjacent contacts must subtend an angle at the origin of less than π. The only further difference between the fits is in the interpretation of which points enter the defining set. For the circumscribing case it is one of those outside the reference, for the inscribing case it is an inside one.

The minimum zone limacons have four defining points that, akin to the Stiefel exchange, must lie alternately on the inner and outer boundaries as angle from the origin is scanned.

The geometrical criteria for all three references also hold for true circles provided that the rules are applied relative to the centre of the circle used in the iteration rather than to the origin. It is this apparently slight change of geometry which causes all the complexity and uncertainty in the optimization process.

The circumscribing limacon provides a very clear illustration of how the exchange algorithms work and why they are so efficient. Solving by conventional means a linear programme, such as equation (5.15), that has three parameters and n constraints involves a computational effort related to that needed to invert an n-square matrix. This is because H must be padded out to contain a full set of linear equations, one for each constraint, by introducing extra ("slack") variables. If equation (5.15) is called a primal linear programme then there must exist a corresponding dual programme in which the rôles of A and Y are reversed and the constraints formed from H^T. The two programmes can be related to each other at any iteration in their solution, but the most important point is that their optimal solutions contain exactly the same information. It does not matter which we choose to solve. However in the dual of equation (5.15) there are only three constraints and so the computational effort relates to inverting only a 3-square matrix. In computational terms it is always better to solve "long, thin" programmes in one form or the other and in this case the dual should be chosen.

The exchange algorithms are derived from the dual programme and this is the major reason for their fast operation. The ability to set up an exchange algorithm depends on particular structures of the problem formulation which allow a geometric interpretation of the dual even though it is the primal that describes the real geometrical problem.[7] The details of this will not be repeated here. The effect in geometrical terms is worth exploring. Any limacon which encloses all the profile points represents a feasible solution for the circumscribing figure. It is not the required solution unless it is also optimal, that is it obeys the rule on subtended angles. Any limacon which does not enclose all the points is an infeasible solution. However an infeasible solution may be regarded as optimal if it is the best solution for all the points that it does actually enclose. Thus there are four possible situations as illustrated by Figure 5.6.

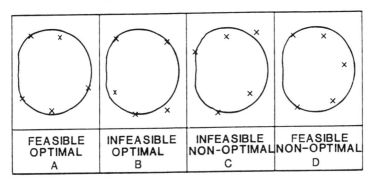

Figure 5.6 Classes of trial solution for the circumscribing limicon.

A conventional solution process would search the feasible solutions for an optimal one. The primal linear programme, solved by the simplex method, offers a good way of doing this since it provides an orderly search through an identifiable sub-set of the solutions one of which must be optimal. The exchange algorithm (and the methods of the dual programme) iterates through optimal solutions until a feasible one is found. In practice a very small set of points tend to dominate the alignment of the fit and by chosing the most violating point at each iteration they are rapidly identified. Thus, although it will not necessarily occur, even with 500 or more profile points three iterations are often sufficient to obtain the solution.

Exchange algorithms are obviously of great importance in metrology and can also be exploited in other fields. The requirements are first to provide a valid linearization since otherwise it is unlikely that convergence can be proven. This needs a knowledge of the instrument principles and the purpose of the measurement. Second the shape of the matrices in the mathematical expression of the optimization must be exploited. This needs knowledge of the theory of the computation. A good implementation will only arise if both needs are met.

5.3 PROFILE FILTERING

5.3.1 Surface Roughness Filters

As with many types of signal, a profile representing the surface roughness of a workpiece usually requires filtering in order to select the information desired. Surfaces have structure on many scales of size and can often be modelled as fractal.[8] Structures beyond a certain size may be irrelevant to certain functions, for example wavelengths much larger than the contact region are unlikely to influence wear rate, and should be filtered out. Sometimes short wavelengths should be removed. For many years analog instruments have used a conventional (and Standardized) filter consisting of two buffered simple Capacitance-Resistance (CR) stages, although, unusually, the cut-off wavelength was defined as that giving about 80% transmission. Latest Standards have incorporated the conventional definition of 50% transmission.

In many applications and often in metrology it is the power spectral density or the RMS amplitude of the signal which is of concern. In such cases the phase characteristic of the filter is unimportant. However, the shape of the profile is critically important when contact or optical properties are being investigated. All conventional filters have phase shift which depends upon wavelength and so tend to distort profile shapes as different frequency components of the signal are shifted relative to each other. A linear phase characteristic is therefore desirable, though not strictly realizable. There is some case for using a Bessel filter characteristic from this point of view but near linearity in phase is only achieved in higher order filters. Generally, we have only a poor feel for the value at which the cut-off should be set since our knowledge of functional significance is poor. It is wise therefore to use a filter with a relatively gentle roll-off in case we choose inadvisedly. Hence no serious attempts to supersede the 2CR filter had been made.

With digital signal processing on instruments the situation changes completely. It is, of course, possible to model the Standard 2CR filter to almost any desired precision by a digital filter (see chapter 2). Finite Impulse Response (FIR) methods convolute a discretized version of the theoretical impulse response, truncated in the time axis, with the digitized profile. However recursive techniques in which the transfer function is simulated by a feedback process with a weighted sum of points from both the input and output sequences is much more efficient. In a typical application there might be several hundred samples in one cut-off length and a one cut-off long FIR does not give great precision. Clearly the convolution involves large numbers of multiply operations for each output point. In contrast, a low order recursive (IIR) filter involves only a handful of multiplies per point. There are a few subtleties which can affect the quality of a recursive simulation of an analog filter. For a high pass section, the implementation usually involves creating a low pass section and taking its complement. With linear electronics at reasonable frequencies this is easy since the propagation delay of a network is small. However the basic shape of the algorithm for a low-pass recursive section is

$$y_0 = Px_0 + Qy_{-1} \tag{5.16}$$

where y_i and x_i are the output and input sequences respectively and P and Q functions of the filter time constant and sampling interval.

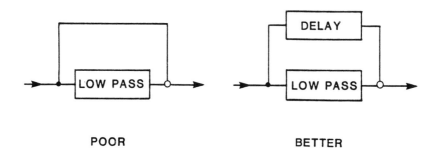

Figure 5.7 Implementing recursive high filters by complementing a low pass section.

This implies that, in any real implementation, the output sequence will lag behind the corresponding input sequence by an amount depending upon the sampling interval. Forming the complement should account for this by incorporating a delay into the direct

Forming the complement should account for this by incorporating a delay into the direct path as illustrated by Figure 5.7. Theoretical arguments[9] show that some common filter types have an extra half interval delay. Alternative implementations can provide an integer delay. Taking this into account and allowing for the unusual definition of cut-off enables high performance filtering on low cost instruments.[10]

Digital computers provide a ready means of temporary storage for a profile that allows the designer to break the formal link with real time processing. The ideal of a zero phase filter cannot be produced in real time because it would have a symmetrical impulse response which implies that it must "know" about the future signal. Once the signal is stored, we do in one sense know about its future and so many new options for filter design arise. A zero phase version of the 2CR filter is readily produced either by a symmetrical FIR filter or by a recursive implementation incorporating "time reversal". This is a trick, long considered well known in some circles, which still has the power to surprise: that it provides some phase compensation is intuitively obvious, that it provides exact phase cancellation is less so. A data sequence is passed normally through a single stage filter and the output stored. This data is now reprocessed through the same filter section in reverse order. The amplitude attenuation from these operations multiply in the same way as for two series sections to give a second order filter. However since the effective direction of time is reversed in the second stage it phase advances where the first retards and vice versa. The phase shifts cancel out at all wavelengths. Although not a real time technique, the first section can run in real time and the computation is fast so results can be available very soon after the input is completed.

Whitehouse[11] first suggested that computers give the freedom to specify an ideal filter characteristic for the surface assessment based on function rather than what could be easily produced. For reasons discussed earlier, a steep roll-off is not desirable and so a zero phase filter was proposed with a frequency domain characteristic which is unity in the pass band and decreases linearly from the cut-off on the log-log Bode diagram with a slope of 3:1 or 2:1. This could be implemented by Fourier transforming the data, applying the characteristic and inverse transforming, but even with FFT algorithms this remains computationally unattractive. Instead it is implemented as a FIR filter, though its impulse response decays relatively slowly so that a long convolution is needed. Consequently, it has rarely been used in practice.

In recent years there has been much discussion in Standards committees, though little in the open literature, about alternative "ideal" filters. A zero phase characteristic is their most important feature. After that an agreed impulse response is being sought with a view to all instrument specifications including a statement of the deviations of their filter implementation from that ideal. As well as being metrologically sensible, it should be easily described mathematically and unambiguously convertible to an algorithm. If implementation of the ideal requires an FIR filter, much care must be taken to specify, for example, how many sample points it contains. Given that there is no strong evidence to suggest that one amplitude characteristic is preferable to other fairly similar ones on functional grounds, this author considers the time reversed recursive 2CR filter to be a good choice. It has the advantages that its characteristic is very close to the analog Standard and its formal specification does not involve factors such as a point of truncation.

On the other hand, a well defined FIR filter could conceivably run very close to the ideal in almost real time if implemented on a modern, fast digital signal processing chip. With this in mind, one strong proposal has been for a triangular weighting function which is truly a finite impulse response characteristic and involves no potential truncation errors. Unfortunately it has a rather poor frequency response shape. Instead a Gaussian filter seems likely to be chosen. In this the sequence representing the impulse response is defined as

$$s_i = (1/\alpha\lambda) \exp[-\pi(h_i/\alpha\lambda)^2]$$ (5.17)

where λ is the cut-off wavelength, h the sample interval, i an index and α a normalization constant. This form has the advantages of an explicit mathematical description and that, although it has to be truncated, it decays to a small value in a relatively small number of sample intervals. Its monotonic decay makes it easy to specify suitable truncation points. However its real attraction for a Standard is probably that it has also a mathematically explicit frequency response

$$G(\omega) = \exp[-\pi(\alpha\omega/\omega_c)^2]$$ (5.18)

where ω_c is the spatial frequency at the cut-off. This combination of a calculable impulse response and a theoretic frequency response curve with which real implementations can be compared should ensure its selection for Standards.

5.3.2 Surface Slope and Differentiation

Since surface slope is important to both the tribological and optical behaviour of surfaces, there is regular use of numerical differentiation algorithms on surface profile data. All such algorithms are varieties of finite difference approaches, the simplest being the back difference

$$y_1' = (y_1 - y_0)/h$$ (5.19)

A disadvantage of this formula is that it is asymmetric. It takes account only of what the profile has been doing and not of where it is going to. Thus the derivative calculated does not really correspond to the slope at y_1 but to the slope at half an interval before y_1. The Lagrangian formulae are usually preferred because they are symmetrical about the point at which the derivative is taken. The formulae can be derived for any odd number of samples by simple rules. For the chosen points a correct order polynomial (for example, a quadratic through three points) is fitted exactly. This polynomial can be differentiated analytically and the value at the mid-point found by substitution. Calculating these procedures explicitly involves considerable effort but need not be done. With equally spaced profile data points the rules for fitting the polynomial coefficients are fixed and so the derivative becomes a simple formula involving points to either side of the position of interest. The three and five point formulae are

$$_3y_0' = (y_1 - y_{-1})/2h$$ (5.20)

$$_5y_0' = (-y_2 + 8y_1 - 8y_{-1} + y_{-2})/12h$$ (5.21)

All the differential formulae are short FIR filters, a fact not always recognized even though electronic differentiators clearly involve filter networks. In particular the Lagrangian formulae are zero phase filters. Since they are filters, their effectiveness can be examined using transfer function and related methods.[12] Intuition might suggest that, as differentiation is notoriously sensitive to noise on the measured profile, a higher order formula involving an averaging over several points would be favourable. On slowly changing signals this is generally the case. As the wavelengths present in the profile approach the effective cut-off of the formula being used an attenuation occurs and all the formulae underestimate the real slope. With the Lagrangian formulae the error in RMS slope at any wavelength decreases as the order of the formula increases, although with diminishing returns for the extra computational work. However, many finishing processes produce profiles which are better described by random signals with moderate point-to-point correlation. None of the formulae seem good at estimating RMS slopes for these forms. The three-point one is poor compared to theoretical methods of slope estimation based for example on zero crossing density.[13] It is not clear whether estimates really improve beyond using around seven points.

From a mathematical point of view it is a nonsense to be attempting to find the slope of a profile which is almost fractal in nature. Differentiation implies a continuous form which would not be present. Nevertheless there remains a physically useful concept of local slope which must be extracted from the data. There is a case, not widely explored, for defining slope in terms relating directly to its functional significance for a particular situation. For optical applications the surface could be modelled directly as a sequence of flat facets. The simplest model takes each facet to lie between adjacent sample points and their slopes are then given by equation (5.19). Estimates of the slope of longer facets might be taken from the slope of the least squares straight line through a suitable number of successive points. This implies step discontinuities between facets if taken as a literal description, but it may be useful as an approximation in an area where little high quality information is available. Again, simple formulae can be derived for equispaced data. Interestingly, with three points this approach yields the same formula as the Lagrangian method. However this is not true for higher orders, the five point form is

$$y_0' = (2y_2 + y_1 - y_{-1} - 2y_{-2}) / 10h \qquad (5.22)$$

This gives much higher weighting to the more distant points than does the Lagrangian formula. In a true facet model the slope y_0, should be forced onto the whole profile length from y_{-2} to y_2. Obviously equation (5.22) could also be used as an FIR filter. There appears to have been no work to investigate the method divergence expected between such approaches.

While no firm conclusion can be drawn about the "best" approach to differentiation, this very statement is an important conclusion. The most appropriate algorithm depends on the nature of the signal and the reason for making the measurement. No general purpose instrument having a fixed algorithm can be expected to perform equally well on all types of surface. That being the case, the five or seven point Lagrangian formulae represent a good compromise selection.

5.4 ALGORITHMS FOR SHAPE AND PEAK SENSITIVE PARAMETERS

The Nyquist sampling criterion is so well known that it may seem surprising that sampling rate (or interval) determination is at all problematic. Either a rate is chosen that is a little over twice the maximum frequency present in the signal or an "economic" rate chosen and an anti-aliasing filter applied to the signal to limit its highest frequency. The occasionally heard claim that sampling should be at, say, ten times the Nyquist rate so as not to miss sharp peaks is a misreading. If there are sharp peaks, then the signal has high frequency components and it is to these that Nyquist relates, not to the dominant frequency. (All the same there is some sympathy for the view of sampling higher than the expected rate providing it is not too expensive since unexpected events do occur.)

There can be genuine difficulties when parameters sensitive to signal shape are evaluated in a literal manner since the Nyquist criterion is based on the preservation of Fourier transform and power spectral content.

The difficulty is illustrated by comparing the sampling rate sensitivity of profile amplitude measurement using RMS and R_a (the arithmetic average height). Aliasing through undersampling has the effect of folding the power spectrum back upon itself but does not in principle reduce the total power recorded. Hence RMS ought to be accurately assessable from undersampled data, although in practice the possibility of pathological cases such as samples synchronized to the zero-crossings of the profile or just the extra uncertainty in short records make such a procedure unwise. The theoretical value of R_a for a unit sinewave is $2/\pi$. Sampling with four points per wavelength, starting at 0 relative angle the value computed by simple summing is 0.5 while starting at $\pi/4$ gives 0.707. Nyquist does not guarantee correct recording of R_a because the algorithm depends on the shape of the profile. The more points taken in the cycle, the better in principle does the computation match the theoretical value. The simple algorithm treats the profile as a series of steps at heights set by the sample points and it is this that causes the parameter sensitivity to sample position. Oversampling is one solution, using a more sophisticated algorithm or a less sensitive parameter are probably better solutions. On real profiles the situation is not as serious as this example suggests since there is nearly always a steady decrease in amplitude with shortening wavelength so that the long wavelengths which tend to dominate the R_a will be strongly oversampled. On the other hand more subtle shape and sampling interactions are possible.

Tribological properties of surfaces are often related to the geometry of surface peaks. The measurement of these is rich with examples of algorithm sensitivity. The simplest algorithm for detecting a peak is to test whether y_0 is greater than both y_1 and y_{-1}. Samples could be taken very close together so that the central point is very close to the true peak, but peaks of gentle curvature may then be missed because quantization causes successive points to take the same value. As the standard stylus on some instruments will ideally have a flat tip 2 μm across, sampling closer than that could lead to difficulty. No real stylus is perfectly flat and electronic noise will also be present so the greater practical difficulty may be that several "peaks" are seen because of jitter imposed onto a smooth feature. In either case the wrong answer is found because of instrument imperfections, particularly the non-linear low pass filtering imposed by the sensing system. The missing of peaks can be overcome by allowing y_0 to be replaced by any number of successive equal valued points. Counting noise on the peak can be reduced by thresholding in which the signal must rise and fall more than some fraction of the instrument range predetermined by its expected noise level.

Thresholding is arbitrary and an obvious source for divergence between different implementations of nominally the same algorithm and yet its use is often unavoidable. A pragmatic approach to setting the threshold is called for. By plotting the number of peaks detected against the threshold level for typical signals it may be possible to select between instrument noise and real features. Figure 5.8 illustrates the best result from such tests. The peak count is expected to fall as the threshold is increased but the presence of a plateau region suggests that the fall to the left has been through the elimination of noise and that little profile information is lost until the right hand drop-off. Selecting a suitable threshold is then easy. If no plateau is found this method is no help, except that it may indicate that the instrument is far from ideal for the task being undertaken.

Thresholding also addresses the question of what is meant by "peak". In some tribological models the mathematical definition used above is in-built and so should be used in parameter assessment. The R_z parameter[14] of roughness amplitude shows a different case. Its evaluation requires the identification of the five largest peaks in the profile and it is clear from the Standard that an English rather than mathematical definition of peak is required.

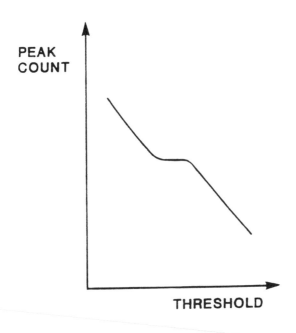

PEAK COUNT

THRESHOLD

Figure 5.8 Ideal form of peak count vs threshold curve for determining set points empirically.

Generally only one of a series of subsidiary peaks on a large feature is counted unless there is a very clear valley between them. However, human selection, unlike some early algorithms, does not use high-spots, that is the profile between successive crossings of the profile mean line. Many tests using profile graphs suggest that the human definition of a peak is the highest point of a feature bounded to either side by valleys lower than it by at least 10% of the profile amplitude. For instrument use this is often translated as a threshold of 5% of range on the simple peak algorithm. Note that this value is much larger than would be needed to discriminate against instrument noise and there is no tribological reason for supposing that close peaks are more or less significant for having a 10% valley between them. The algorithms for R_z are set purely to agree with a human perception of what should be included.

Digitized surface profiles can be regarded as Markov processes and subjected to random process analysis. Much of the early development of this theory was aimed at communications where the purpose is usually to discover weak periodic signals within the random noise. In tribology the noise itself is the main feature and so differences of emphasis are to be expected but many of the results are common.[15] For example, the densities of peaks, defined by the simple three-point model, and of mean line crossings can be predicted from ratios of the zero, second and fourth moments of the power spectral density, see, for example, the articles by Sayles, Gibson and Whitehouse for further details.[16]

One consequence of modelling by Markov chains is that the closer samples are taken the higher the correlation between successive samples if the surface statistics are assumed stationary (and little else is practicable). This implies that parameter computation which involves the relative values of successive ordinates will necessarily be dependent upon sampling interval. The RMS slope based on the three point Lagrangian formula will decrease as a half power law with increasing sample interval while the average peak or valley curvature, also defined by simple three point formulae, decreases as a 3/2 power law. With real profiles the model breaks down at smaller intervals partly because it is clearly inappropriate to have the descriptors tending to infinity, but mainly because of the obscure filtering imposed by the stylus or other sensor. If computer simulated random profiles are subjected to a simulated measurement with a perfect stylus the model holds well until the sampling interval becomes smaller than the stylus tip dimension.[17] Figure 5.9 shows an example using mean peak curvatures in which sampling interval is expressed as normalized to the stylus dimension.

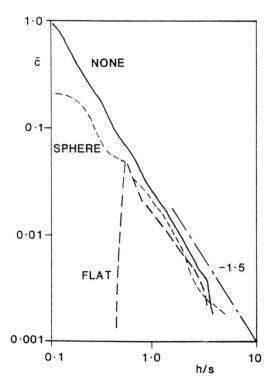

Figure 5.9 Simulated relationship of mean peak curvature $\bar{c}$ to sampling interval h (normalized to stylus dimension s) in the presence of a perfect stylus.

For a flat stylus the model breaks down completely since the three point peak definition cannot work if samples are less than half the stylus width apart. With a circular tip, the breakdown is much less violent but still distinct. Valley curvatures are much less seriously affected since the stylus tends to show them as too narrow and too sharp but less often distorts them as grossly as it does peaks. Attempts by the author to reproduce this behaviour on a real instrument have had little success. There are some experimental difficulties in changing stylus while maintaining closely the same profile track. However the main difference seems due to the instrument operation. Figure 5.10 shows the shape of curve typically obtained (it is not a specific measurement but a composite simplified for clarity).

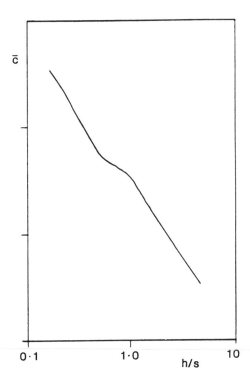

Figure 5.10 General trend of experimental results attempting to reproduce Figure 5.9 with a nominally flat stylus. Scales are logarithmic, units vary with different testpieces.

As the sample interval reduces the model is initially followed closely and there is a tendency to find a falling away in the trend at around the stylus size before the trend recovers to be close to the model prediction. The reason is probably that in the real instrument wide bandwidth random noise is introduced after the stylus truncation has occurred and this eventually dominates the peak detection. Also the stylus itself will have small scale structure which will interact with the peaks. In practice the centre of the flattened region of the curve might be used as an estimate of the effective stylus size: a value of around half the nominal dimension is usually obtained. Since the valley statistics are less seriously affected, a crude on-line measure of stylus size could be found in the changing ratio of mean peak and valley curvatures.

One method of avoiding a problematic interaction between the algorithm and the sampling scheme is to find an alternative algorithm. For the situations just discussed this is difficult because it is the assumed profile shape as defined by reconstructing the sampled data sequence which is the source of the geometric distortion. Algorithms based on continuous functions fitted to the sequence, possibly types of cubic spline, might be constructed but it is not clear whether they would offer any real improvement. Another approach, valid only for determining averaged parameters, is to introduce assumptions about the profile structure such that the required information can be estimated from less sensitive calculations. This is not totally satisfactory for the assumptions will never be exactly correct in practice and so we trade one set of errors for another. It can be an effective tool if it is known *a priori* that the profiles will have statistics not far from those assumed, as is the case with random finishing processes.

Consider the mean peak curvature. Experience has shown that many profiles can be reasonably simulated by a three point Markov process having an exponential autocorrelation function. This cannot be a true model because it is discontinuous in its derivatives at zero lag which implies infinite power spectral moments. Various slight modifications to the assumed correlation function can avoid this difficulty.[18] Many statistics of the profile can be derived directly from the estimated Markov process parameters, that is from the values of the first three points of the profile autocovariance function. The theoretical results given by Whitehouse and Phillips[18] can readily be reformulated as algorithms for these statistics. The benefit of this procedure is that the autocovariance coefficients are estimated from all points of the profile sequence not just those around the peaks. Most of these points are likely to lie on flanks of gentle slope where the finite size of the stylus causes little distortion. Consequently the coefficients are not very sensitive to variation in sampling intervals comparable to the stylus dimension and so neither are parameters such as mean peak curvature derived from them. Simulation results demonstrate that the indirect algorithm does indeed follow the expected 3/2 power law more closely as the sample interval is reduced below the stylus size.[17] This tempts us to accept it as a more accurate approach. It should be stressed that the reasonableness of this belief comes from a feeling of what is intellectually preferable and not from pragmatic evaluation. There is no evidence to indicate which algorithm better describes what is actually present on a real surface traversed by a real stylus. Moreover there is no obvious way of obtaining such evidence.

5.5 SYSTEMATIC ERROR COMPENSATION

Computer error compensation (sometimes and probably over-ambitiously called error correction) has become popular in many surface and dimensional metrology instruments and in larger machine tools. On one level it might be no more than the process of averaging several readings taken in the presence of noise in order to gain slightly greater precision through the dependence of the standard error on the mean on the number of values taken. More usually it is taken to imply that some of the systematic errors of the instrument have been determined and the output modified to account for their presence. Such procedures require that the errors are systematic in the strictest sense and repeat exactly with similar operations of the instrument since, by definition, their size cannot be determined only from the output of a measurement of an unknown workpiece. Sometimes extra sensors are incorporated in order to estimate the error from a more general model, for example temperature sensing to compensate for linear expansion or for the refractive index of air for a laser interferometer. Another important class is the correction of sensor non-linearity. For most of history an assumption of at least local linearity in nature has been unavoidable if useful models and measurements were to be obtained. The arithmetic and presentational power of modern computing is now permitting non-linear systems to be handled with consequent improvements in accuracy. It was once valid, in practice, to regard a non-linear device as "inaccurate"; now almost any device with a known, monotonic input/output characteristic may be regarded as accurate.

These topics are too broad to cover here, but a few comments will be made about a third major class: computer compensation for geometrical errors of a machine. Occasionally an error can be directly modelled, e.g. when a mechanism is used to approximate a straight line motion, but usually they are caused by manufacturing imperfections in the machine. They are systematic on any one machine but are random in form between nominally identical machines. Thus the errors must first be independently determined. On computer numerically controlled machine axes this might be done by using a laser interferometer to make precise displacements which are logged to give calibration both for scale mismatching and axis misalignment. On a coordinate measuring machine a similar procedure could be adopted or a special test-piece with precisely calibrated features could be measured. In either case what is built up is a map of the exact displacements between certain nominal coordinate positions within its measuring volume. Whether the calibration data is stored and used, with interpolation, to correct at any position or whether it is parameterized and used in a vector model, the mathematical manipulation is sufficiently complex to cast doubts upon the degree of accuracy improvement that can be obtained. For example, since a slideway may have independent errors in all kinematic degrees of freedom and there are also axis to axis misalignments, a simple three axis (xyz) machine displays 21 independent error profiles which must be combined into its overall volumetric description. Further discussion can be found in the literature, for example.[19,20,21]

The situation is considerably more tractable in surface metrology where essentially a single cross-sectional profile is taken so that only an equivalent section across the instrument space need be known. Also it may be acceptable to always sample at the same points in that space, for example to take always 512 or 1024 points in the revolution of a roundness instrument. In such cases a complete error profile may be stored and simply subtracted from each measurement in real time. For the very highest attainable precision and if there is confidence that the systematic errors remain constant over the time of a measurement even though they may drift over longer periods, error separation techniques can be used. The formal approach to separation involves making two or more measurements which are identical except in a phase shift introduced between the workpiece and instrument. By comparing the resulting profiles, the contributions from the workpiece and the instrument may be deduced. The concept goes back to the art of craftsmen in making precise flat surfaces. For straightness measurement the profile may be recorded first with one end of the workpiece at the start of the traverse and then with the other end there. Alternatively the workpiece may be left in one orientation but measured while to either side of the instrument datum so that a "bump" on the datum records as a dip in the workpiece one side and a bump the other.[22] Averaging the two profiles then removes most effects of the datum imperfection. In roundness measurement a similar reversal technique can be used.[23] Here the workpiece is rotated by an angle π relative to the datum spindle between the measurements, as is the sensor probe so that again a bump on the datum appears once as a bump and once as a dip in the combined profiles. The reversal process is not easy to automate and other methods have been devised in which the workpiece is stepped sequentially through several angles relative to the datum.[24] A best fitting of harmonics allows the separate estimation of the workpiece and datum errors. Theoretically it is less good than reversal since it always involves estimation from fewer independent data than there are variables.

In practice there is little to choose between them since the physical reversal cannot be perfectly executed.

It is inherent in the methods described that a spindle (or other datum) error is definable. Intuitively it is clear what is meant by this concept: the profile produced by a perfectly aligned measurement of a geometrically perfect workpiece. However its physical form is not fixed but a function of how the instrument is used. In practice this is adequately identified by simple implementations of the algorithms, although care is needed in defining how the errors are actually generated when assessing the accuracy of the compensation.[25]

A stored error profile can be found by any of the methods just described using a good but imperfect test-piece. Simpler but a little less accurate is to use a testpiece which may be regarded as perfect compared to the precision required of the compensated instrument. On modern roundness instruments a compensation corresponding to decreasing the effective spindle error by an order of magnitude is quite possible. It is then difficult to make adequately good masters so reversal is usually used in practice. On machine tools, masters are still commonly applied.

As a final example, consider probe compensation on a CMM. With a ball ended probe, many machines offer a test against a datum sphere from which the probe radius may be calculated. That radius may then be automatically subtracted from, say, a hole diameter assessment. It is implicit in this operation that the contact between the probe and workpiece is in line with the assessment direction, for example, on the line connecting the calculated centre of a hole to the centre of the probe. In practice this will never be exactly so and there are possibilities of gross errors if the probe brushes accidentally against another feature of the workpiece. It seems unlikely that analysis only of the measured data can resolve fully where any surface is relative to the locus of the probe centre. Computer correction can only be really effective in this case if a new generation of probes is developed which provide vector information about the contact.

5.6 CONCLUDING REMARKS

This chapter has explored briefly the classes of algorithms used in computer aided metrology and given more detailed information about some of them. As illustrated by Figure 5.11, a typical measurement system will involve most of the topic areas covered. The measurement process and the subsequent digitization of its output introduce errors and distortions of various types. The software is expected to extract high quality information about the real surface from the data. Thus computed references and filters are major components of modern instruments. However, the algorithms may introduce new distortions unless they correctly reflect the geometrical relationship between the probe and the workpiece. Algorithms to provided parametric descriptions of the surface structure inevitably involve implicit assumptions about the definition of terms such as "peak" and about the interpretation of the true shape from a discretized representation of it. Again it is easy to fall into traps in which a poor choice of algorithm produces plausible but inaccurate information.

As microengineering and nanotechnology grow, there will be increasing demand for instruments of exceptional precision. Purely electromechanical designs would, if possible at all, be prohibitively expensive to produce. Thus computer aided metrology has become essential. However, improving instrument performance is a matter not of adding a computer to fix problems but of high quality system design. It may now be better to design for a reasonably good datum which is highly repeatable, and readily computer correctable, rather than for one having the best average accuracy at the expense of some measurement to measurement variation. For example, elastic mechanisms which only approximate a desired motion but do so in a predefined manner may be preferred on high precision traverses. The availability of computer compensation challenges the instrument designer to produce systems, mechanics, electronics and algorithms, which best exploit what the computer can offer.

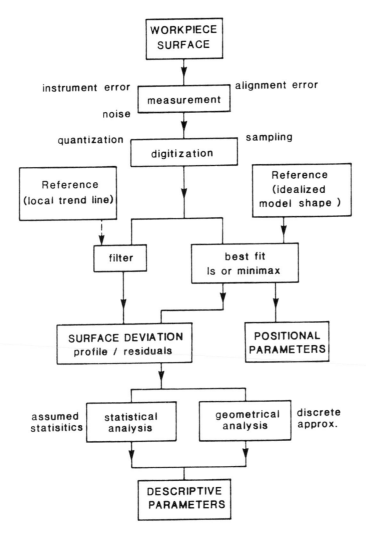

Figure 5.11 Information flows through a typical computer based surface metrology instrument. Upper case labels indicate possible process end points. Lower case labels indicate processes or predefined information, with external qualifiers noting potential error sources.

REFERENCES

1. Osborne, M.R. and Watson, G.A. (1968) On the best linear Chebyshev approximation. *Comput.J.*, **10**, 172-177
2. Reason, R.E. (1966) *Report on the measurement of roundness.* Leicester: Rank Taylor Hobson

3. Chetwynd, D.G. (1979) Roundness measurement using limacons. *Precision Engineering*, 1, 137-141

4. Chetwynd, D.G. and Phillipson, P.H. (1980) An investigation of reference criteria used in roundness measurement. *J.Phys.E:Sci.Instrum.*, 13, 530-538

5. see e.g. British Standards Institute (1982) BS3730 *Assessment of departures from roundness*, London: BSI

6. Whitehouse, D.J. (1973) A best fit reference line for use in partial arcs. *J.Phys.E:Sci.Instrum.*, 6, 921-924

7. Chetwynd, D.G. (1985) Applications of linear programming to engineering metrology. *Proc.Instn.Mech.Engrs.*, 199 B2, 93-100

8. see e.g. several papers from the 4th International Conference on Metrology and Properties of Surfaces, in *Surface Topography*, 1 (1988)

9. Haykin, S.S. and Carnegie, R. (1970) New method of synthesizing linear digital filters based on convolution integral. *Proc. IEE*, 117, 1063-1072

10. Kinsey, D. and Chetwynd, D.G. (1973) Some aspects of the application of digital computers to the on-line measurement of surfaces. *Acta IMEKO*, 601-616

11. Whitehouse, D.J. (1968) Improved type of wavefilter for use in surface finish measurement. *Proc.Instn.Mech.Engrs.*, 182 3K, 306-318

12. Chetwynd, D.G. (1978) Slope measurement in surface texture analysis. *J. Mech. Eng. Sci. (I.Mech.E)*, 20, 115-119

13. Bendat, J.S. (1958) *Principles and applications of random noise theory*. New York: John Wiley and Sons

14. British Standards Institute (1988) BS 1134 Part 1 *Assessment of surface texture methods and instrumentation*, London: BSI

15. Whitehouse, D.J. and Archard, J.F. (1970) The properties of random surfaces of significance in their contact. *Proc. Roy. Soc. Lond.*, A316, 97-121

16. Thomas, T.R., editor, (1982) *Rough surfaces*. London: Longmans

17. Chetwynd, D.G. (1979) The digitization of surface profiles. *Wear*, 57, 137-144

18. Whitehouse, D.J. and Phillips, M.J. (1978) Discrete properties of random surfaces. *Phil. Tran. Roy. Soc. Lond.*, A 290, 267-298

19. Lotze, W. (1986) Precision length measurement by computer-aided coordinate measurement. *J.Phys.E:Sci.Instrum.*, 19, 495-501

20. Burdekin, M., Di Giacomo, B. and Xijing, Z. (1985) Calibration software, an application to coordinate measuring machines. *Software for Coordinate Measuring Machines*, edited by M.G. Cox and G.N. Peggs, pp. 1-7, London: National Physical Laboratory

21. Busch, K., Kunzmann, H. and Waldele, F. (1985) Calibration of coordinate measuring machines. *Precision Engineering*, 7, 139-144

22. Thwaite, E.G. (1973) A method of obtaining an error free reference line for the measurement of straightness. *Messtechnik*, 10, 317-318

23. Donaldson, R.R. (1972) A simple method of separating spindle error from test ball roundness error. *CIRP Ann.*, 21, 125-126

24. Whitehouse, D.J. (1976) Some theoretical aspects of error separation techniques in surface metrology. *J.Phys.E:Sci.Instrum.*, 9, 531-536

25. Chetwynd, D.G. (1987) High-precision measurement of small balls. *J.Phys.E:Sci.Instrum.*, 20, 1179-1187

Chapter 6

ULTRASONIC SENSORS

DAVID A. HUTCHINS

The aim of this chapter is to describe the types of sensor that are used for the generation and detection of ultrasonic motion in liquids and solids. The discussion starts with a description of the types of wave modes that are possible in these media. This is followed by a treatment of their propagation characteristics. Both are required before an understanding of the design and operation of the various types of sensor can be formed. Later sections deal with the sensors themselves and their performance, together with examples of their applications in precision measurements.

6.1 INTRODUCTION

6.1.1 Wave Modes

Ultrasound is concerned with acoustic waves whose frequency is beyond that of normal human hearing, and is usually taken as being frequencies above 20 kHz. In liquids and gases, the principal mode of propagation is as a longitudinal wave, where particle motion is in the same direction as the wave propagation. The velocity of this wave depends on the medium through which it is travelling. In general, the wave velocity increases from gases to liquids, with the velocity being highest in solids.

In solids, the situation becomes complicated by the presence of additional modes, due to the fact that the solid can support shear stresses. Thus, in the bulk solid, an additional shear wave can exist, where the particle motion is perpendicular to the direction of propagation.[1] This wave is transverse in nature, and can hence be polarized. Its velocity of propagation is always slower than that of the longitudinal wave.

Other modes are also present if the solid is considered to have surfaces (i.e. is not treated as an infinite solid). Solution of the relevant equations show that at the surface of a semi-infinite solid, a wave can travel without radiation of energy into the bulk material. This Rayleigh or surface wave[2] has a velocity that is slightly slower than that of the shear mode. The Rayleigh mode has a complicated wave motion, which can be illustrated schematically as in Figure 6.1. Note that the amplitude decays with depth, and has decayed to $1/e$ of its value at the surface at a depth of one wavelength.

An additional set of wave modes are present in layered materials, and are one of a class of guided modes that are possible in a solid. A specific example that will be examined here are Lambwaves[2] which exist in a single layer. These can be considered as a series of modes of different order, each set being one of the two types illustrated schematically in Figure 6.2.

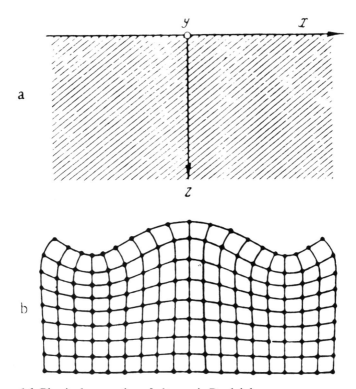

Figure 6.1 Physical properties of ultrasonic Rayleigh wave.

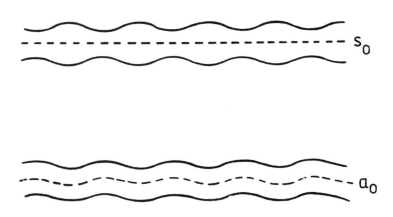

Figure 6.2 Schematic diagram of deformation of a plate caused by s_0 and a_0 waves.

The symmetric mode propagates with the centre of the layer being stationary, whereas the antisymmetric mode has displacements throughout its thickness. For layers that are thin compared to the shortest ultrasonic wavelength that is propagating, the lowest order symmetric mode (denoted s_0) travels with little distortion. Conversely, the lowest order antisymmetric mode (the a_0 mode) travels with dispersion, so that the velocity of propagation is a function of frequency, with higher frequencies travelling faster. This introduces additional complications into the study of these modes.

6.1.2 Reflection and Transmission of Ultrasonic Energy

The design of many ultrasonic sensors, and their use in many applications, depends on the transmission and reflection of ultrasonic energy at interface between two media. The amount of energy reflected or transmitted depends on the relative acoustic properties of the two media, which may be characterized by their acoustic impedance (Z). This may be defined as:

$$Z = \rho \cdot c \tag{6.1}$$

where the velocity of the longitudinal wave is c in a medium of density ρ. If the Z of the two media either side of the boundary are very different, then most of the incident ultrasonic energy will be reflected. In terms of intensity, this may be characterized in terms of coefficients of reflection (α_r) and transmission (α_t) as follows:

$$\alpha_r = [(Z_2 - Z_1) / (Z_2 + Z_1)]^2 \tag{6.2}$$

$$\alpha_t = 4Z_1 Z_2 / (Z_1 + Z_2)^2 \tag{6.3}$$

In equations (6.2) and (6.3), the wave is assumed to be travelling from medium 1 to medium 2 at normal incidence, the media being denoted by the subscripts. There are several points to note. First, if $Z_1 = Z_2$, then $\alpha_r = 0$ and $\alpha_t = 1$, and there is perfect transmission between two media of the same impedance. Conversely, if $Z_2 >> Z_1$ or vice versa, α_r tends to unity and total reflection occurs. At other relative values of Z_2 and Z_1, partial transmission and reflection results. At angles other than normal incidence, the expressions need to be modified, but the same general principles hold. Refraction can now take place on transmission, with the wave entering medium 2 at an angle θ_2, given by

$$\theta_2 = \sin^{-1} (c_2 \sin\theta / c_1) \tag{6.4}$$

which is a version of Snell's Law for acoustics.

Interesting effects occur when a layer of a given medium (2, say) of finite thickness (L) is bounded by two media (1 and 3). The transmission coefficient from medium 1 to 3 is given by

$$\alpha_t = 4Z_1 Z_3 / [(Z_3 + Z_1)^2 \cos^2 k_2 L \tag{6.5}$$
$$+ (Z_2 + Z_1 Z_3 / Z_2)^2 \sin^2 k_2 L]$$

where k_2 is the wave-number in the layer. There are several implications of this equation:

1. If L is small compared to the wavelength (λ), where $k_2 = 2\pi/\lambda$, α_t becomes

$$\alpha_t = 4Z_1 Z_3 / (Z_3 + Z_1)^2 \tag{6.6}$$

101

which is the same as that for waves travelling from medium 1 to 3. A thin layer thus transmits ultrasound well (provided Z_2 is not very small, under which conditions equation (6.6) does not hold).

2. When the layer is a multiple number of half wavelengths thick, the same conditions as in 1. apply.

3. A special case occurs when the layer is $\lambda/4$ thick or an odd multiple thereof. Under such conditions,

$$\alpha_t = 4Z_1 Z_3 / (Z_2 + Z_3 Z_1/Z_2)^2 \tag{6.7}$$

and on substitution of a particular value of $Z_2 = \sqrt{(Z_1 Z_3)}$, $\alpha_t=1$. This represents a quarter-wavelength matching layer, and can be used to provide good transmission from one medium to another of differing impedance. This concept is used extensively in transducer design, as will be shown later.

6.2 PIEZOELECTRIC TRANSDUCERS

6.2.1 Basic Design

Piezoelectric transducers are by far the most common type of sensor used for ultrasonic measurements. Typically, these devices are used over the MHz frequency range, and in this section we will examine the design of these devices, and the factors that influence their performance. Consider a transducer being used in a typical application, where the device is used as both a source and sensor of ultrasonic transients. A transducer that can be used in such a measurement is shown schematically in Figure 6.3.

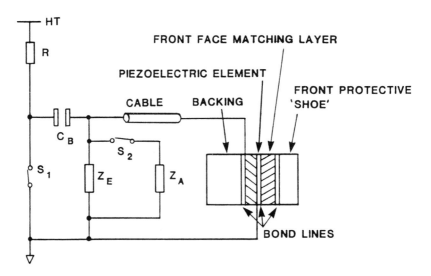

Figure 6.3 Schematic diagram of a pulse-echo transducer system with transient excitation.

The active element is a piezoelectric slab, usually fabricated from a ferroelectric material such as lead zirconate titanate (PZT). For longitudinal wave sensitivity, the material is poled at high temperature, so that when a voltage is applied between its faces, a change in dimension occurs. This is the method of ultrasonic generation used in such devices. When acting as a sensor, displacement of the two surfaces causes a charge to be generated.

If the PZT element was used in isolation, it would not act as a quantitative device because of resonances that would be established between its parallel faces. The fundamental resonance of the device occurs when the element is a half wavelength thick, as shown in Figure 6.4.

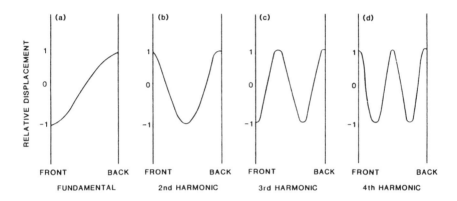

Figure 6.4 Representation of plate resonances when piezoelectric is bounded by material of lower acoustic impedance.

Under such conditions, a thickness resonance is established, with the two surfaces moving in opposite directions. Odd harmonics are also possible, but even resonances are not efficient modes of vibration.

A major design feature in piezoelectric devices is the suppression of these resonances, which distort the signal being measured or generated. This is achieved using a backing material (Figure 6.3), fabricated from a glossy material with an impedance close to that of the piezoelectric material. This is typically a polymer loaded with metal powder. Because of the similarity in impedance, a high proportion of acoustic energy is transmitted into the backing where it is absorbed, effectively suppressing any resonance. The result is a device with a frequency response centred around its fundamental thickness resonance, but with a broader response resulting from a heavier damping. This also implies a better response in the time domain, with the received signal being a more faithful representation of the signal being measured.

A final design consideration is the presence of a matching layer on the front of the transducer, to maximize transmission between the device and the medium of interest. This is designed according to equation (6.7).

6.2.2 Sensor Response

The above has described the basic design of piezoelectric transducers, but in many applications it is important to be able to predict the behaviour of the device. In particular, if the transducer is acting as a source, the variations in the emitted ultrasonic pulse shape and amplitude are of interest. These are also factors that influence the detection characteristics of the device. Hence, much work has been reported in this area.[3-5] In the following, we will look briefly at the characteristics of these devices both in liquids and at solid surfaces.

6.2.3 Response in a Liquid

The field or sensitivity of a transducer can be thought of as being within two regions. Close to the transducer, the pulse characteristics can change markedly, this region being known as the near-field. For a transducer operating at a nominal centre frequency f, resulting in a radiated wavelength λ in the liquid, an approximate distance to which this region extends is a^2/λ, where a is the radius of a circular transducer. Beyond this distance, the far-field region exists, where on moving further from the transducer face, the response drops steadily with distance.

The response of a transducer is dependent on two main factors: the transducer radius, and its electrical response. The latter is a complicated function of its construction and the associated electronics (drivers, amplifiers etc.). However, we may illustrate the general effects if a circular transducer, of radius a, is radiating into water. Assuming that the front face of the transducer moves as one cycle of a sinusoid, at a centre frequency such that $a=5\lambda$ (300 kHz in this case), the spatial variations in pressure amplitude will be those shown in Figure 6.5. The response along the axis of the beam, Figure 6.5(a), is flat until a region is reached where the response increases to a maximum, this being the near-field/far-field boundary. This area of uniform response in the near-field is also shown in the 3D plot of Figure 6.5(b). The ridge appearing to come from the edge of the transducer, towards the axis, can be thought of as due to an interference effect between a plane wave from the transducer face, and a wave from the edge (the so-called "edge-wave").

As a comparison, the field of a transducer driven with 2 cycles of the same frequency is shown in Figure 6.6. Note how now the field is more complicated, following the interaction of plane and edgewaves. In effect, the ultrasonic waves now have a narrower bandwidth in frequency, due to the elongated pulse in time, and this increases the potential for interference and complexity. Thus, as a general design rule for transducers, a wider bandwidth for the device will lead to a more uniform response within its field.

6.2.4 Transducers at a Solid Surface

Recent work[6] has shown that plane and edge wave effects can be used to describe the response of a circular transducer at a solid surface. Similar principles apply, but now the situation is complicated by the fact that both longitudinal and shear waves can be supported by the solid. Analysis has shown that the plane wave component is longitudinal only in nature, but that this interacts with both a longitudinal and shear wave which can be transmitted and/or received by the disk edge. The longitudinal mode thus behaves in a similar way to that in a liquid, as described above, whereas the shear mode is both emitted and detected primarily at the transducer edge.

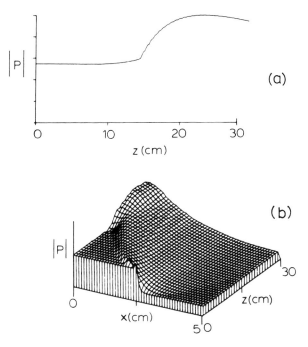

Figure 6.5 Simulated spatial variations in pressure-amplitude for a circular plate piston excited with one cycle of a sinusoid (centre frequency such that $a = 5\lambda$). (a) axial field. (b) full section.

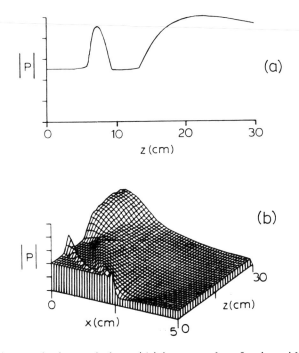

Figure 6.6 Transient excitation such that $v(t)$ is two cycles of a sinusoid.

The above analysis shows that a transducer designed to be sensitive to longitudinal waves will react primarily as expected, but will also have some sensitivity to shear waves. If it is desired to maximize the sensitivity to the latter mode, it is necessary to redesign the transducer. This is done by polarizing the piezoelectric element, such that it responds to movement parallel to the surface.

In some cases, it is required to generate the Rayleigh and Lamb modes described in an earlier section. This is difficult to achieve by direct application of a transducer to a surface. However, consider equation (6.4), presented earlier, which describes refraction at an interface between two media, where θ_2 is the refracted angle. If a wave was incident onto a solid surface from another medium, and the incident angle varied, a critical angle would be reached at which a Rayleigh wave would be generated at the surface of a thick solid. Here, c would be the Rayleigh velocity, and θ_2 would be 90° . This is the most common way of generating and detecting both this mode, and the Lamb modes that exist in layered material. Either an immersion transducer in water can be used, or a transducer mounted onto a solid wedge, that is applied to the surface. Note that in all cases, the velocity In the liquid or solid wedge must be less than that of the Rayleigh or Lamb wave of interest.

6.3 ELECTROMAGNETIC AND ELECTROSTATIC DEVICES

6.3.1 Introduction

There are many applications where it is inadvisable to use the piezoelectric transducers described in detail in the last section. Examples include situations where the instrumentation is at an elevated temperature, or where a sample is moving. In the former case, a piezoelectric material will cease to operate above its Curie temperature, which is typically 250 °C. For moving material, where the transducer is required to be stationary, the need for a couplant in the case of a piezoelectric device renders the technique inoperative. In these and other situations, it is necessary to use non-contacting sensors, which are not based on piezoelectric materials. This section will describe such devices, which can be considered under two categories: electromagnetic devices, and electrostatic (or capacitance) transducers.

6.3.2 Electromagnetic Acoustic Transducers

These devices rely on the interaction of the ultrasonic wave with a static magnetic field, imposed on the specimen usually with a rare-earth (e.g CoSm) permanent magnet. As a wave reaches the surface of the solid, which must have a sufficiently high electrical conductivity, an eddy current is induced at the surface. This can be detected by a coil which is positioned close to the surface. The form of the current induced in the coil is proportional to the velocity of the surface.

While the permanent magnet has a Curie temperature like that of a piezoelectric material, they do not rely on a couplant to the surface. They can thus be used for moving material, and in other situations where contamination of the surface is to be avoided. The bandwidth of these devices, and their sensitivity, has been the subject of much work.[7] As sensors, they are less sensitive than piezoelectric devices by several orders of magnitude. However, this is compensated for by the fact that they can have a wide bandwidth, of the order of 20 MHz. This is very difficult to achieve with piezoelectric devices of the design shown earlier, and means that they are able to detect transient motion without undue distortion of the signal. In this respect, they are a more quantitative device than piezoelectric transducers, and especially so if variability in the couplant of a contacting device is taken into account.

Recent work has shown that Electromagnetic Acoustic Transducers (EMATs) can also work at equivalent bandwidths as sources, using fast high current pulses into the coil from circuits similar to those used to drive laser diodes. In the generation mode, the coil is used to induce an eddy current at the solid surface, and the Lorentz force F is

$$F = J \times B \tag{6.8}$$

used to generate the acoustic wave by interaction of the eddy current density F with the magnetic field B.

The type of acoustic mode generated or detected by these device depends on the coil geometry and the magnetic field distribution from the magnet. For sensitivity to longitudinal waves, a magnetic field primarily parallel to the surface is required. For shearwave work, the field should be normal to the surface. Figure 6.7 shows some EMAT coil configurations for sensitivity to different modes. As examples, a spiral coil and a normal magnetic field (Figure 6.7(a)) leads to a radially polarized shear wave. A linearly polarized device would use a coil design which presented a line to the surface, such as in (d). In addition, masks can be used to screen parts of the coil from the surface, as shown by the dotted line in (e), where a circular aperture results.

6.3.3 Electrostatic Acoustic Transducers

The electrostatic type of sensor relies on a change in capacitance caused by motion of the surface relative to a fixed electrode. A high d.c. voltage is applied to the electrode, and motion of the surface causes a change in charge on the top plate, which then forms the signal. In most cases, a charge amplifier is used to convert this charge to a voltage, prior to sending the signal to subsequent instrumentation. For electrostatic reasons, these devices can only work at electrically conducting surfaces.

The sensitivity of capacitance transducers of this type depends on the electrode design (with an area A), the gap (x) between the electrode and surface, and the applied d.c. voltage (V). Assuming a parallel plate capacitor, the capacitance (C) may be written

$$C = \varepsilon_o \varepsilon_r A / x \tag{6.9}$$

which leads to a charge sensitivity of

$$dq / dx = - \varepsilon_o \varepsilon_r V A / x^2 \tag{6.10}$$

It is thus of advantage to increase V and A, and decrease the gap x. However, there are limits to this process. In the case of ultrasonic waves at a metal surface, electrode diameters of 6 mm and gaps of 20 μm are used for accurate quantitative work, but this means that the sample surface and the electrode are both polished, usually to a flatness of 1 μm. The gap also leads to a limitation on V, to prevent air breakdown. Despite this, such devices can be used to detect displacements of 10 pm or less. Figure 6.8 shows a schematic diagram of such a device, which is fitted with differential screws for careful alignment of the electrode to the surface, and a micrometer for adjustment of the gap (x). Further details of capacitance transducer design can be elsewhere.[3]

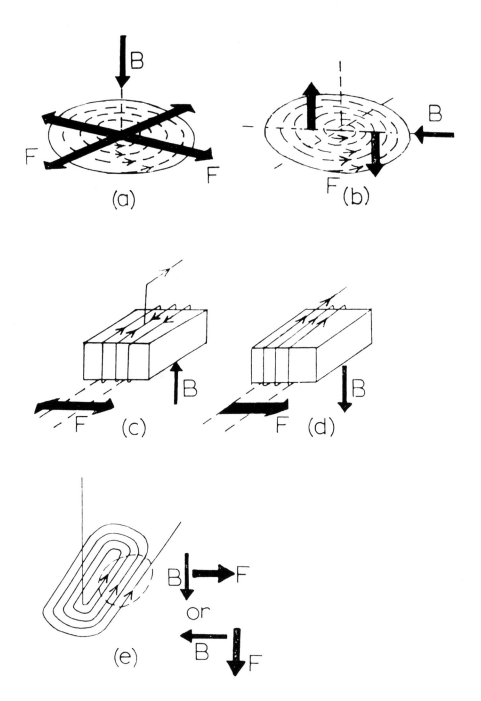

Figure 6.7 Schematic diagrams of the forces F generated by various EMAT designs as a function of eddy current density J and static magnetic flux density B.
(a) and (b) spiral pancake coils;
(c) and (d) rectangular coils;
(e) elongated pancake coils.

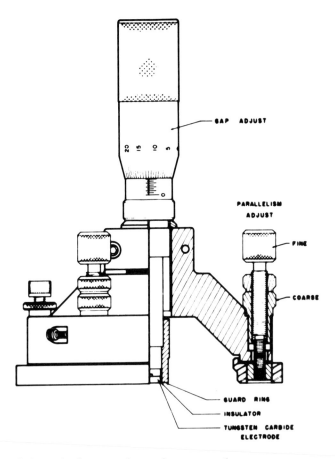

Figure 6.8 Schematic diagram of capacitance transducer.

6.4 OPTICAL DEVICES

6.4.1 Introduction

Optical devices are the most commonly used devices for the measurement of displacement and the generation of ultrasonic transients. They are non-contact in nature, and can be designed to have a wide bandwidth. Most importantly, they can often be calibrated accurately as sensors, especially those designs which rely on interferometric processes. The use of lasers for ultrasonic generation will be presented first, followed by a description of laser-based ultrasonic sensors.

6.4.2 Ultrasonic Generation by Pulsed Lasers

Pulsed lasers can be used for ultrasonic generation in a wide variety of media, including gases and vapours, liquids and solids,[8] although the present discussion will be limited to generation in solids, which has been the subject of considerable interest in recent years.[9,10] Indeed, it has only been shown in the last ten years or so that such a method can be used to generate a wide range of wave modes, to include the longitudinal, shear, Rayleigh and Lamb modes that were discussed in an earlier section. In the following, we will describe the mechanisms that can be used to generate these modes, and the properties of the source that results.

Generation mechanisms

Consider the irradiation of a solid surface by a pulsed laser beam. This laser can by one of many types, but are typically either a solid state laser or a gas laser. In the former case, the laser is Q-switched, to give a pulse which has a nominal duration in the 5-30 ns range, depending upon the optical cavity design. Examples of such lasers include the Nd:YAG laser, operating at a wavelength of 1.06 μm in the near infra-red, and the ruby laser operating in the red at 0.694 μm. Examples of the latter include the CO_2 laser, which can provide longer pulses (typically 0.1-10 μs) at high energy (up to 8 J) at far infra-red wavelengths, and excimer lasers which operate in the ultraviolet.

Assuming that the incident optical power density is insufficient to damage the surface, ultrasonic generation occurs primarily by thermal expansion mechanisms. As significant temperature rises induced by absorption of laser energy at an opaque solid occur close to the surface, ultrasonic generation will occur within this region by thermally induced stresses. These will exist primarily in directions parallel to the surface, as shown schematically in Figure 6.9(a). The forces are parallel to the surface as forces normal to it cannot exist due to boundary conditions. The transient absorption of energy leads to a step-like expansion of the solid in the directions shown, as the absorbed energy is the integral of the laser pulse.

A significant change in the source may be achieved by treatment of the solid surface with various solid and liquid coatings. For example, a thin film liquid coating can evaporate, leading to additional forces on the solid surface due to momentum transfer from evaporating material. These forces will be normal to the surface, as shown in Figure 6.9(b). Another method is the use of a constraining layer, which takes the form of a transparent solid layer bonded rigidly to the surface. Thermal expansion now results in the production of forces normal to the surface, which is no longer a stress-free boundary, as shown in Figure 6.9(c).

The sources described in Figure 6.9 are fairly simple to model, and this method is thus quantitative in this respect. As examples, Figure 6.10 shows a theoretical prediction of waveforms produced by thermal expansion at a metal surface. The longitudinal (L) and shear (S) waves are present as step-like features, whereas the Rayleigh (surface) wave is dipolar. The form of these transients has been confirmed experimentally.[8]

6.4.3 Optical Detectors

There are a variety of optical methods that can be used for the detection of ultrasonic motion of a surface. Two excellent reviews of these techniques have been written,[11,12] and only brief details are given here.

6.4.4 The Knife Edge Technique

Perhaps the simplest method is the knife-edge technique, the principle of which is shown in Figure 6.11. When an ultrasonic wave travels to a surface, or propagates along it, the laser beam is caused to deflect; a suitable optical detection system (usually a knife edge and photodiode) then senses a change in light intensity. Such a detection system is used in an ultrasonic system known as the Scanning Laser Acoustic Microscope (SLAM). The detection limit of such devices is said to be of the order of 1 pm. Such methods are under active investigation in many centres, because of the sensitivity, lack of interference from vibration and simplicity of operation.

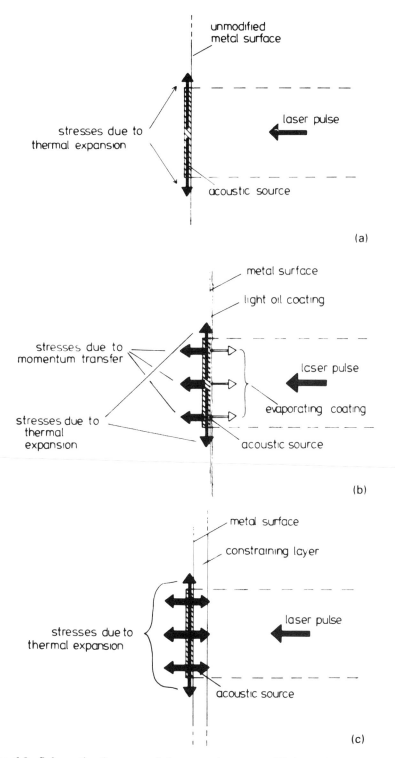

Figure 6.9 Schematic diagrams of the acoustic sources likely to be generated at a solid surface following laser irradiation. (a) Thermoelastic generation, (b) liquid on surface and (c) a constaining layer.

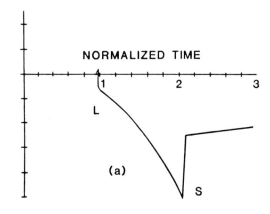

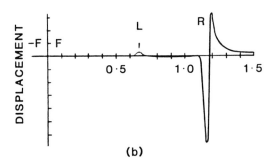

Figure 6.10 (a) Longitudinal (L) and shear (S) theoretical displacement for thermoelastic generation by a pulsed laser. (b) Equivalent Rayleigh (R) waveform.

This is the most common device used for optical detection, and such devices can be considered as being of one of two general designs: those which involve mixing a wave reflected from the surface with a reference beam, and those which measure the reflected beam directly.

The interferometers that use mixing are usually based on a Michelson interferometer design, a schematic diagram of which is shown in Figure 6.12.

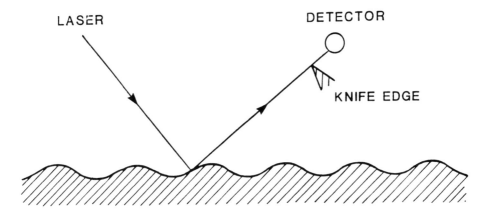

Figure 6.11 Principle of the knife-edge technique.

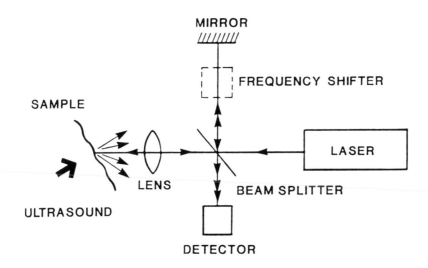

Figure 6.12 Schematic of the Michelson interferometer probe.

6.4.5 Interferometry

A beam splitter causes some of the beam to be deflected onto a reference mirror, whilst the remainder reflects from the surface. These two beams then recombine at a photodiode detector. Movement of the surface due to an ultrasonic wave causes the intensity at the detector to change, and this forms the detection signal. The resulting device is known as a homodyne interferometer. In practice, low frequency vibrations are a problem in such devices, as they have amplitudes which are much greater than those being measured at higher frequencies. This is dealt with by using a moveable reference mirror, to cancel out low frequency signals using some form of feedback. More advanced forms of processing can be undertaken, if a frequency shifter (e.g. a Bragg cell) is added as shown. This allows the low frequency noise to be extracted more readily, leading to a more stable device, the result being known as a heterodyne interferometer.

The second generic type of interferometry is based on the Doppler shift induced on the reflected laser beam by changes in velocity of the surface. An increasingly popular method for this is to use a confocal Fabry-Perot resonator to sense frequency changes. Such a resonator (or etalon) is shown schematically in Figure 6.13.

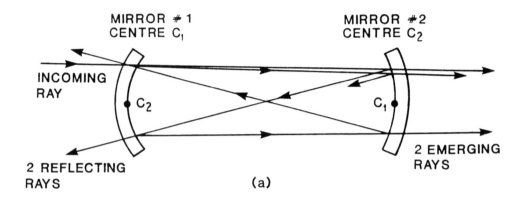

(a)

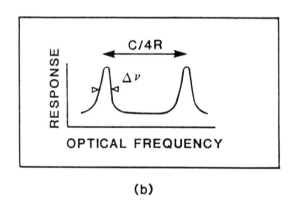

(b)

Figure 6.13   Fabry-Perot etalon. (a) Mirror configuration, (b) Transmission characteristics.

Such an etalon has maxima of transmission at well defined optical frequencies; a change in frequency in reflected light from the surface will alter the transmittance of the device. Such systems are expensive, but have the distinct advantage of being insensitive to vibration, and of being able to work surfaces with a low reflectivity.

6.5 TOTAL INSTRUMENTATION SYSTEMS

It is evident that there are a variety of methods available for ultrasonic transduction, and some of these have been described. As an illustration of these techniques, various combinations of transducers will be described. Examples will be given of systems based on pulsed laser generation, and detection by piezoelectric devices, interferometers and EMATs.

The first example that will be discussed is the ultrasonic signal generated in a plate of aluminium by a pulsed ruby laser, using the thermoelastic generation mechanisms illustrated earlier in Figure 6.9(a). The theoretical displacement waveform on the far side of the plate was shown in Figure 6.10(a). We will now demonstrate that such a signal can be detected experimentally.

The capacitance transducer shown in Figure 6.8 was used to detect the ultrasonic displacement in the aluminium plate, using a 20 mm electrode gap and a bias voltage of 200 V. The result is shown in Figure 6.14.

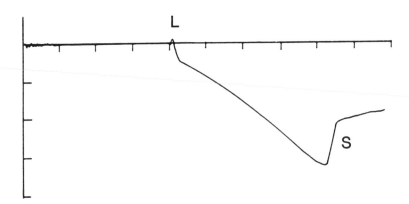

Figure 6.14   Laser generated thermoelastic signal, detected by capacitance transducer.

Note that the signal is very similar to that predicted theoretically, and demonstrates that the whole transduction system can be modelled accurately, leading to a quantitative measurement system. By way of comparison, the experiment was repeated using an EMAT detector. The result, Figure 6.15, differs from that of the displacement signal provided by the capacitance transducer. This is expected, however, as the EMAT is a velocity sensor. Inspection of Figure 6.15 shows that it is of the form expected if the displacement signal is differentiated.

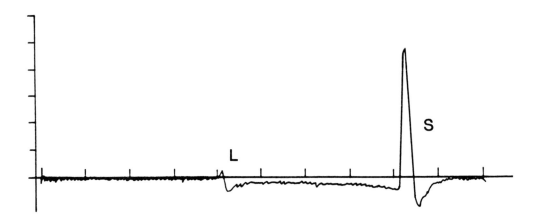

Figure 6.15 Laser generated thermoelastic signal, detected by a spiral coil EMAT.

A second example is where the sources and detectors described in section 6.3 are combined to produce an all-optical transduction system. Here, a pulsed laser is used as the ultrasonic source, and an interferometer used as a detector. The material under investigation is rolled aluminium sheet, and the aim is to propagate Lamb waves, described earlier in section 6.1, along the sheet in order to measure its properties.[13]

A schematic diagram of the instrumentation is shown in Figure 6.16.

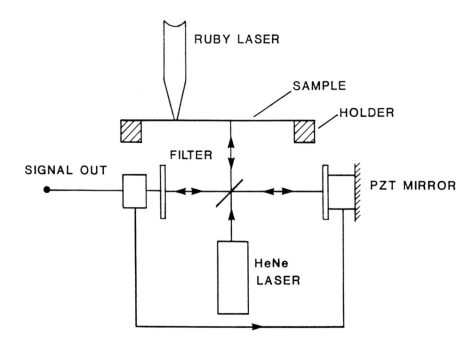

Figure 6.16 Schematic diagram of apparatus used to generate and detect transient Lamb waves.

A Q-switched ruby laser is used to generate the Lamb wave transients, which are then detected at some distance along the sheet by a Michelson interferometer. The interferometer uses a HeNe laser, with a beam diameter of 1 mm, which is focussed to 0.1 mm at the sample surface. Lamb waves generated by the pulsed laser source, using one of the mechanisms shown earlier in Figure 6.9, propagate as a lower order mode, due to the small sample thickness (25 μm). The interferometer detects the asymmetric mode preferentially, this mode exhibiting velocity dispersion. The result is the waveform shown in Figure 6.17.

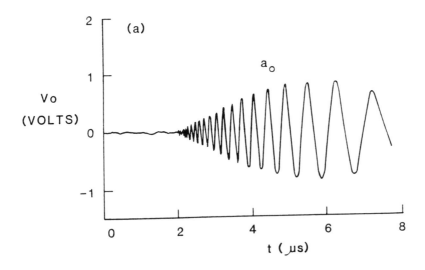

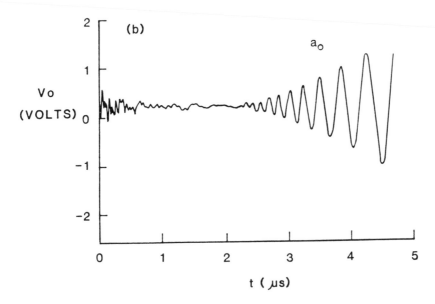

Figure 6.17 Waveforms in (a) aluminium foil and (b) metallic glass, generated by evaporation of a thin oil layer by the pulsed laser. The source and receiver were separated by 5 mm in both cases.

Of immediate interest is that the wave is not a sharp transient, but has the high frequencies arriving first (i.e. it is "chirped"). While this may appear to create complications, the dispersion responsible for this behaviour is predictable, and is a function of the elastic properties and thickness of the material. Hence, if the former is known, the latter can be calculated. This technique can measure thickness to a high degree of accuracy.

6.6 CONCLUSIONS

It has been demonstrated that many factors are involved in the design of ultrasonic sensors. In the case of piezoelectric devices, a multilayer structure is required to produce the required properties. Other designs, such as EMATs, ESATs and optical devices, can operate without contact to the surface, but all have their limitations. The transducer used will thus depend on the particular measurement being performed.

REFERENCES

1. Achenbach, J.D. (1973) *Wave propagation in elastic solids*, Amsterdam: North Holland
2. Victorov, I.A. (1967) *Rayleigh and Lamb waves* New York: Plenum
3. Sachse, W. and Hsu, N.N. (1979) *Ultrasonic transducers*. In *Physical Acoustics*, **14**, pp. 277-407. New York: Academic Press
4. Harris, G.R. (1981) Review of transient field theory for a baffled planar piston. *J. Acoust. Soc. Am.*, **70**, 186-204
5. Hutchins, D.A. and Hayward, G. (1990) *The radiated fields of ultrasonic transducers*. In *Physical Acoustics*, **19**, pp. 1-80. New York: Academic Press
6. McNab, A., Cochran, A. and Campbell, M.A. (1990) The calculation of acoustic fields in solids for transient normal surface force sources of arbitrary geometry & apodization. *J. Acoust. Soc. Am.*, **87**, 1455-1465
7. Thompson, R.B. (1990) Physical principles of measurements with EMAT transducers. In *Physical Acoustics*, **19**, pp. 157-199. New York: Academic Press
8. Hutchins, D.A. (1988) Ultrasonic generation by pulsed lasers. In *Physical Acoustics*, **18**, pp. 21-123. New York: Academic Press
9. Scruby, C.B., Dewhurst, R.J., Hutchins, D.A. and Palmer, S.B. (1982) Laser generation of ultrasound in metals. In *Research Techniques in Non-destructive Testing*, **5**, pp. 281-327. New York:Academic Press
10. Birnbaum, G. and White, G.S. (1984) Laser Techniques in NDE. In *Research Techniques in Non-destructive Testing*, **7**, pp.259-365. New York: Academic Press
11. Monchalin, J.P. (1986) Optical detection of ultrasound. *IEEE Trans Ultrason. Ferr. Freq. Contr.*, UFFC-**33**, 485-499
12. Wagner, J.W. (1990) Optical detection of ultrasound. In *Physical Acoustics*, **19**, pp. 201-266. New York: Academic Press
13. Hutchins, D.A., Lundgren, K. and Palmer, S.B. (1989) A laser study of transient Lamb waves in thin materials. *J. Acoust. Soc. Am.*, **85**, 1441-1448

Chapter 7

RECENT ADVANCES IN SOLID-STATE MICROSENSORS

JULIAN W. GARDNER

The increasing application of microelectronics in measurement and control systems is being limited by the lack of suitable sensors with a comparable specification (cost and performance) to that of microelectronic circuits. In order to obviate this problem, an enormous effort has been directed in the past ten years towards the field of sensor research. One of the fruits of this labour has been the emergence of a new class of sensors, called integrated sensors, that employ silicon-planar technology. The aim of this chapter is to present an introduction to integrated solid-state microsensors and their applications in measurement & control. Current developments of intelligent or smart sensors, and integrated microsensor arrays that employ specialized signal processing, are also discussed.

7.1 INTRODUCTION

People normally gather and process information about their local environment by using their natural senses of sight, hearing, touch and smell. In a similar manner, scientists and engineers employ instruments to gather and process information and thus perform a function, for example to identify and machine a component. This information-processing system is essentially a measurement and control system that can be represented by a functional block diagram. Figure 7.1 shows the basic elements of a generalized information-processing system.

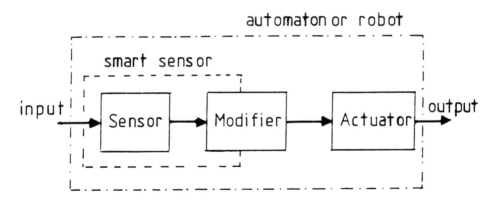

Figure 7.1 Functional block diagram of a generalized information-processing system.

Firstly, a sensor detects the input as a change in a physical or chemical quantity (e.g. position, pressure or pH) and normally converts it to an electrical signal. The type of sensor is usually classified according to the characteristic form of energy involved in the sensing process. Table 7.1 shows the main types of sensor, that is mechanical, electrical, radiant, thermal, magnetic and chemical, together with the usual objects of measurement or measurands. Secondly, the output from the sensor, usually an electrical signal, is modified by the "modifier" while the form of energy is preserved. The modifier may be a simple C-R network, operational amplifier, an analog-to-digital convertor, or even a microprocessor or transputer. Lastly the output of the modifier is fed into an actuator. Now the energy of the signal is converted by the actuator into another form which can either be detected by one of our senses, e.g. mechanical acoustic or optical energy, or fed back into the control system of an instrument. For example, Figure 7.2 shows a schematic of the control system for a motor drive that has position and velocity feedback. An optical encoder acts as the position and velocity sensor and continually feeds information into the modifiers, i.e. the adders and loop electronics. Finally, the actuator is a drive system that converts the electrical energy into rotational mechanical energy of the motor.

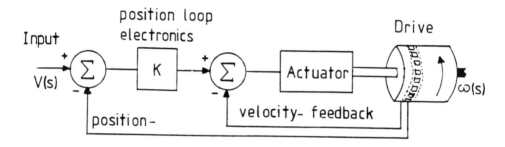

Figure 7.2 Servo-system for rotational drive with position and velocity optical sensors.

Table 7.1 Main sensor types and their measurands.

TYPE:	MEASURANDS:
Mechanical	Displacement, velocity, acceleration, force, pressure
Electrical	Voltage, current, resistance, inductance, capacitance
Radiant	Optical, IR and X-rays
Thermal	Temperature, thermopower
Magnetic	Magnetic field intensity, flux
Chemical	pH, concentration of gases, chemical process variables, ions
Biological	Antigens, lipids

Developments in the field of microelectronics over the past thirty years have led to the low cost fabrication of microcircuits using silicon-planar microtechnology. Thus electronic circuits that form the basis of modifiers can be made small to a high specification and at a low cost. The rise in the application of microelectronic devices in measurement and control created a need for comparable sensors and actuators. The partial or full integration of the modifier and sensor on a single silicon wafer produces a so called "smart sensor" (see later), while the complete integration of the system effectively produces a smart microautomaton or microrobot.

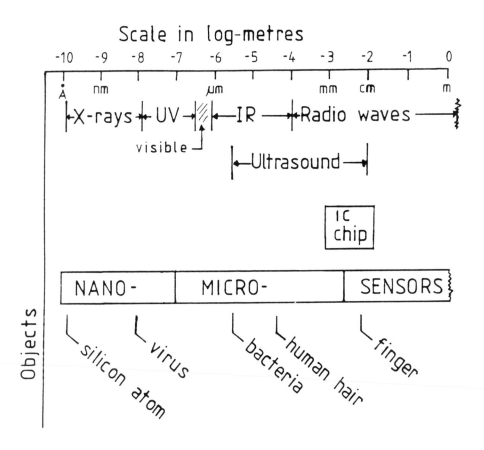

Figure 7.3 Comparative scale of sensors.

In this paper we focus upon the recent advances in solid-state microsensors, i.e sensors that employ silicon microtechnology. Figure 7.3 illustrates the comparative scale of microsensors to everyday objects. As mentioned before the field of research has moved from sensors into microsensors, of which nanosensors form the limiting case. In the case of nanosensors, their dimension approaches the nanometre level and could be used by nanorobots. It seems probable that these will employ molecular rather than silicon-based micro-electronic technology. The topic of small scale actuators is covered elsewhere and will not be discussed further except when they are integrated with the sensor and modifier on a single silicon structure.

7.2 WHY IC-BASED SENSORS?

The utilization of IC or integrated silicon-planar technology for sensor fabrication has several attractive advantages. Firstly, more is known about the processing of silicon than practically any other material due to the billions of pounds that have already been spent. Secondly, fabrication facilities already exist that could be converted to make integrated sensors. Finally, ICs have proved to have a high reliability and performance but at a low cost. Clearly the application of this technology to sensors could lead to a new type of sensor that is smaller, more reliable and much cheaper. The need in the industrial world for information-processing systems is almost insatiable. Moreover the demands vary from low-cost mass production domestic items (e.g. washing machines) through to high-cost custom made military items (e.g. heat-seeking missiles). In fact the integrated sensor market was worth about £7,000M in 1989 and could reach £25,000M by the year 2,000.

As mentioned in the introduction, sensors are usually classified by their manner of operation. However, in order to discuss integrated microsensors the emphasis here is on the technology. Consequently, the subject will be dealt with in terms of processing complexity rather than sensor type.

7.3 SILICON PROCESSING

Monolithic silicon technology has been developed during the past forty years and produces two main families of device. Firstly, Metal Oxide Semiconductor (MOS) unipolar devices such as Complimentary MOS (CMOS), Negative MOS (NMOS) and the Charge Coupled Devices (CCD) that are basically shift register filters. Secondly, bipolar devices that are either oxide-based (eg. OXIL or DI), or junction devices usually employing epitaxial layers (eg. CDI, SBC, CBIC). Conventional silicon microtechnology is well known and uses the following materials and processes:

1. Oxide, nitride and resist layers
2. Metals/alloys: Al, Au, W, NiCr
3. Polycrystalline layers
4. Epitaxial layers
5. Sublimation & sputtering
6. Photolithography
7. Dry and wet etching
8. Dicing, bonding & packaging.

Additional materials are used to fabricate integrated sensors (and actuators).
For example, new materials such as polysilicon, porous and amorphous Si, polyimide and ZnO are employed.

(a) Spacer layer deposition　　　　　(b) Base patterning etch

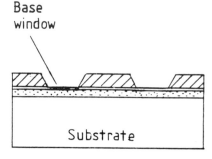

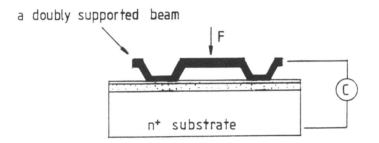

(c) Microstructure deposition

Polysilicon
layer (·3μm)

Substrate

(d) Microstructure patterning

Substrate

(e) Selective spacer etching

a doubly supported beam

F

C

n⁺ substrate

Figure 7.4 Surface micromachining of a force sensor.

The first silicon sensor was produced by Honeywell in 1962 using conventional substrate processing; it was said to be micromachined, basically using conventional techniques (wet etching, dry etching & oxidation). A critical problem to solve was that of material joining. Various glues were developed as well as the process of glass/silicon and silicon/silicon fusion bonding. The basic processes involved in surface micromachining are spacer layer deposition (sacrificial layer to support deposition), base patterning, microstructure layer deposition, pattern microstructure layer and selective etching of the spacer layer. Figure 7.4 shows a typical process sequence, in this case for a doubly-supported beam that could be used as a force sensor. The process involves five stages: first the silicon wafer is thermally processed to form oxide (250 nm) and nitride (100 nm) layers - the nitride layer acts as an etch stop. Onto this the spacer layer of phosphosilicate glass (PSG) is deposited. Next the base pattern is etched into the PCG using conventional UV photolithography. Note that if the base window is etched through to a n⁺-Si substrate then a contact window is made. Next the polysilicon microstructure layer (300 nm) is sputtered onto the base pattern. A second patterning is carried out and finally the spacer is selectively etched out to leave the free-standing microstructure - in this case a doubly supported beam from which the capacitance would be a non-linear function of the force on the beam. For a detailed discussion of surface micromachining of microsensors and microactuators see Howe.[1] A second technique used is that of bulk micromachining that involves the etching into the silicon substrate rather than using a deposited layer for the microstructure. The typical etching apparatus has been well described by Kaminsky.[2] The performance of bulk silicon micromachining is essentially determined by the choice of etchants. The selectivity of wet etchants is far superior to that attained by dry etchants and so wet etching is widely used. Anisotropic etchants are either inorganic alkaline solutions (e.g. KOH, LiOH, NH_4OH, N_2H_4:H_2O) or organic alkaline solutions (e.g. ethylene diamine, pyrocatechol & water). Maximum etch rates occur in the {100} and {110} planes with p^{++} layers used as etch stops (for a discussion of NH_4OH-based anisotropic etching see Schnakenberg[3]). Anisotropic etching of silicon can be more complicated than isotropic etching of polysilicon because calculations are needed to determine the effective etch directions, and to design masks that compensate for anomalous edge effects.

Recently there have been reports on the combination of surface and bulk micromachining in order to fabricate a submicron filter structure[4]. Despite an increase in the complexity of the machining process, this hybridization of techniques extends the overall capability.

7.4 SOLID-STATE SENSORS

7.4.1 Silicon Microsensors

Mechanical microsensors are used to measure motion related measurands such as surface roughness, displacement, and velocity, or force related measurands such as force, torque, pressure and strain. The basic principles for mechanical microsensors are the same as those for large structures when the typical dimension more or less exceeds the material grain size. Provided this condition is met then the mechanical principles are based upon the linear theory of elasticity. The piezoresistive effect is commonly used to measure strain by thin metal-foil strain gauges of a few microns in depth. However the gauge factor of silicon is much higher than that of metals (e.g. p-type is 175). Polycrystalline and amorphous silicon have also been used to make strain gauges that have a higher temperature sensitivity with compensation by a Wheatstone bridge arrangement. Table 7.2 shows the principle of mechanical microsensors that exist along with their measurands.

Table 7.2 Types of mechanical microsensors

MEASURAND:	PRINCIPLE:
Displacement, Acceleration:	Lateral microflexure of cantilever and diaphragm, capacitive plate, piezoelectric.
Pressure, Force:	Piezoelectric, piezoresistive resonant.

The use of microflexures and diaphragms to measure (or actuate) small displacements has been successful. Figure 7.5 shows three flexure systems that have been developed, simple, folded and crab-leg. These are examples of lateral flexures that offer a high lateral, but a low longitudinal spring constant. They tend to be sensitive to residual stress (folded flexure least of all) so care is needed during machining, nevertheless these devices have turned out to be surprisingly robust.

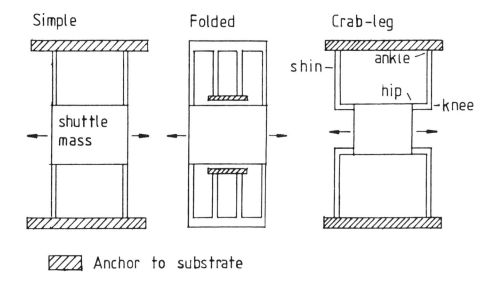

Anchor to substrate

Figure 7.5 Typical flexural systems for micromechanical sensors, the simple, folded and crab-leg flexures.

Resonant microsensors are an innovative type of mechanical sensor that involve the excitation of a mechanical structure. The principle of excitation may be electrostatic, piezoelectric, thermal, piezoresistive, or capacitive. These resonant microsensors are used to measure pressure, flow, and acceleration. For example a silicon resonant altimeter has been reported by Greenwood[5] that can detect a change in altitude of 25 cm. The basic structure is shown in Figure 7.6 where the excitation is electrostatic and the pressure modifies the resonant frequency ω_0 of the device. The sensitivity S is normally defined as the relative change in frequency ω with measurand f, $(\partial\omega/\partial f)/\omega_0$. In this case the frequency output (Ca. MHz) can be measured to a high precision. A host of other resonant microsensors have been reported recently such as the thermally driven resonant flow sensor,[6] electrostatically driven accelerometer,[7] magnetically driven double-ended tuning fork,[8] and vibratory rate gyroscope.[9]

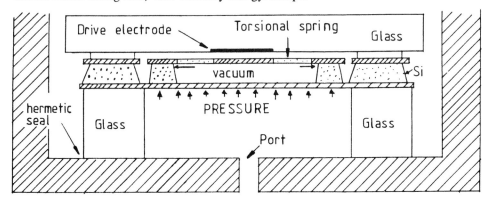

Figure 7.6 Example of a silicon resonant microsensor (From Greenwood 1984)

Flexural and resonant silicon microsensors[10] may also be designed to measure other physical parameters such as temperature (thermal) and voltage (electrical). However, there already exist several solid state Integrated Circuit (IC) devices, based upon conventional silicon technology, that can measure these and other properties, eg. diodes (temperature, strain), phototransistors (optical) and Magnetic Field Effect Transistor (MAGFETs). Figure 7.7 illustrates three microsensors that have been micromachined. Figure 7.7 shows (a)resonant pressure sensor[11] (optically driven) made from anisotropic etching (bulk machining) of silicon, (b) an SEM photograph of a silicon capacitance-type accelerometer,[12] and (c) an SEM of a twelve-stator, four-rotor-pole synchronous micromoter[13] made from surface micromachining of polysilicon.

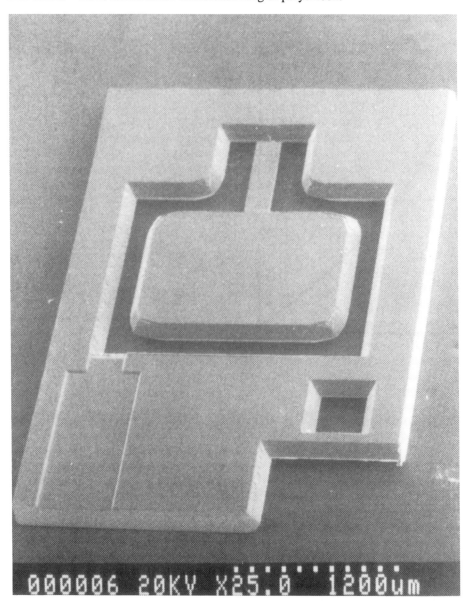

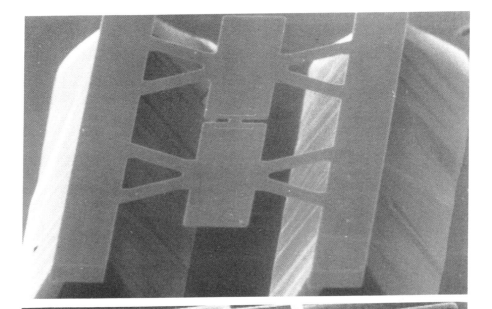

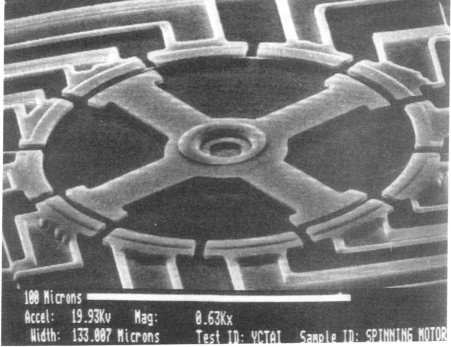

Figure 7.7 Examples of silicon micro-machined structures, (a) a resonant pressure sensor, (b) an accelerometer and (c) a motor.

More recently interest has been shown in the development of acoustical sensors for thermal (temperature), and mechanical (acceleration, pressure) applications. In this type of sensor a Surface Acoustic Wave (SAW), often called a Rayleigh wave, or a plate (Lamb) wave is excited electrically in a piezoelectric material (eg. quartz). The propagation speed or resonant frequency of the excited wave is monitored as the measurand changes. SAW devices have been reported that have a sensitivity of up to 10 kHz g^{-1} to acceleration,[14] and 400 Hz kPa^{-1} to pressure.[15] The fabrication of other microsensors often involves the use of additional processing. In this case the sensing elements use not only the conventional materials of crystalline, polycrystalline or amorphous silicon/silica, but also require additional processing such as the deposition of unconventional thin films.

7.4.2 Added Processing

The combination of unconventional thin film materials and silicon-planar technology extends considerably the scope of integrated microsensors. In the simplest case, materials can be deposited on interdigital electrodes on a silica surface to form a sensing element. Various structures have been made and applied in the field of chemical sensing. The detection of chemical signals by solid-state devices has been of great interest since the discovery in the early 1950s that certain semiconducting materials are gas sensitive.[16] A common feature of chemical (gas) sensors is that the measurand R reacts reversibly or irreversibly with the sensing materials. This reaction produces a change in the sensor state, such as temperature (exothermic reaction), mass (concentration of adsorbate) or conductivity. Figure 7.8 shows two basic types of integrated micro-electronic device, namely, a chemiresistor, and Pt-gate MOSFET. Typically, the chemiresistor consists of a thin layer (Ca. 1 μm) of SnO$_x$ reactively sputtered onto a silica substrate with an interdigital electrode pattern made by wet etching. The device shows a sensitivity to certain organic vapours such as methanol at the parts per million level in air. The Pt-gate MOSFET employs a porous gate electrode where a change in work function is detected following the introduction of a test gas. Other coatings of adsorbant or reactive materials may be used with micro-mechanical structures to make sensitive chemical sensors such as silicon resonant gravimetric microsensors,[17] piezoelectrics,[18] Schottky[17] or SAW devices.[19] The same techniques may be applied to produce biosensors, for instance an enzyme thermistor may be made by depositing a coating that contains glucose oxidase. The enzyme catalyses the reaction of β-D-glucose with oxygen and the change in temperature is readily detectable.

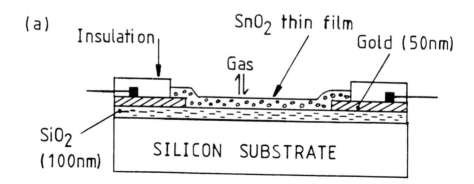

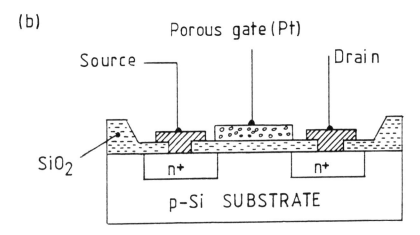

Figure 7.8 Basic designs of solid-state chemical microsensors (a) chemiresistor, (b) n-channel Pt gate MOSFET.

In conclusion, the past thirty years have seen the successful development of a range of integrated microsensors (e.g. mechanical, electrical, chemical) from the utilization of silicon-planar technology. This has resulted in low cost devices that have an active sensing area in the range of a few microns square.

7.5 DEVELOPMENTS IN INTEGRATED SENSOR ARRAYS

The use of microelectronic technology permits the fabrication of array devices at the micron scale. Clearly there are advantages to be gained from the use of microsensor arrays such as enhanced capability and reliability (through redundancy). Moreover, when we consider the use of IC modifiers the generation of a parallel structure with associated processing offers a higher level of computing speed, power and flexibility.

7.5.1 Optical

Vision is perhaps the highest form of sensing and is associated with the gathering of a large amount of data for subsequent processing. In particular, it is important in the field of automation and robotics for object recognition and other functional requirements. The development of microelectronic optical arrays has galvanised this area of research. Two types of solid-state devices are used in optical arrays for use in solid-state cameras, namely Charge-Coupled Devices (CCDs) and Charge Injected Devices (CIDs). In CCDs the incident light creates a small amount of charge at specific locations in the silicon semiconductor. Each location has a different amount of charge that is integrated over the sampling time. For example, the Fairchild CCD211 chip has a 244 by 190 element array and dissipates 100 mW when operated at 7 MHz; the element is about 25 μm square. CIDs are charge transfer devices like CCDs, but while CCDs shift the charges on each location to the end of the row CIDs may be read like ordinary RAM. The reduction in element size has led to high resolution images obtained at low power consumption with low geometrical distortion. Commercially arrays are available now with up to 4,096 by 4,096 elements on a single chip. These types of sensors are used in the optical diffractometers described in Chapter 12. More recently IBM have announced an optoelectronic device that consists of individual mirror elements (25 μm); these offer exciting possibilities for the manipulation of microbeams of light.[20]

7.5.2 Mechanical

The development of mechanical microsensor arrays lags behind that of optical arrays and is sometimes overlooked. There is clearly a need for 2-D tactile sensors in robotics in order to handle delicate objects with a high degree of precision. Recently, a 32 by 32 element (25 μm) pressure sensor array has been fabricated with silicon micromachining.[21] This tactile imager is an exciting development and should have application in precision robots in the future.

7.5.3 Chemical

Although a variety of solid-state chemical sensors are available, such as the chemiresistor and n-channel Pt gate MOSFET (see Figure 7.8), their performance is often limited by a low specificity leading to interference effects. The use of integrated sensor arrays that contain a variety of sensing materials is particularly appealing and was proposed by Zaromb and Zetter to obviate this problem.[22] Since then the principle has been used to sense a variety of gases, vapours and odours. For example, Carey and co-workers[23] have employed an array of 27 piezoelectric quartz sensors to detect and discriminate between 14 chemical vapours.

The principle also forms the basis of an instrument called an Electronic Nose that is under development at Warwick University[24]. This instrument is designed to detect odours and mimic the human olfactory system. Figure 7.9 shows the general arrangement of the artificial nose as an array of chemical microsensors followed by a modifier (current to voltage and analog-to-digital convertors) and microprocessor-based data analysis. Results obtained from a 12-element SnO_2 "nose" show that it is possible to discriminate between tobacco odours[25], and between beer odours.[26] Clearly the application potential of this instrument is enormous in the food processing industries. Provided that silicon microtechnology can be exploited, it seems likely that a commercially viable integrated odour microsensor array could be available by the turn of the 21st century. Clearly the application potential of this instrument is enormous in the food processing industries. Provided that silicon microtechnology can be exploited, it seems likely that a commercially viable integrated odour microsensor array could be available by the turn of the 21st century.

7.6 INTELLIGENT INTEGRATED MICROSENSORS

7.6.1 IC Modifiers

One important advantage of using integrated solid-state sensors is that it is possible to fabricate the modifier, i.e. the signal processing electronics, on the same substrate. The modifiers usually take the form of an interface circuit that can perform functions such as:

1. Amplifying or buffering;
2. Conditioning - linearizing, compensating;

3. Powering and sensor controlling.

The decision of how much of the modifier and display system should be fabricated on a single chip must not be taken lightly. In some cases it is essential to fabricate the modifier on a single chip, for instance in capacitive sensing with thin-film microstructures, but often the cost advantage is with a single board rather than single chip design. Moreover careful consideration of the overall measurement system requirement is necessary to decide whether the limitations are in the sensors, modifier or display system. This problem was recently addressed at an IEEE Hilton Head Workshop.[27]

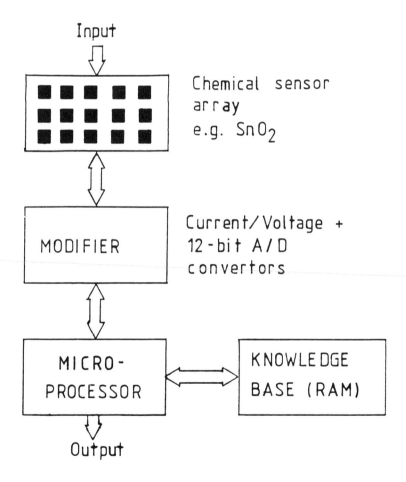

Figure 7.9 Basic design of the Warwick Electronic Nose.

Thermal considerations are often paramount and may limit the use of integrated electronics. Note that the operating temperatures of bulk and epitaxial CMOS are lower than SOI (Silicon On Insulator) technology which may reduce their application, but the cost of SOI is nearly a factor of four higher.

A considerable number of sensors containing integrated electronics have been developed. A precision micro-accelerometer has been reported that is based upon a flexural design with associated CMOS circuitry.[28] The device has a resolution of better than 1 μg at 1 Hz, and a 100 Hz bandwidth. At the current state-of-art, a magnetic force-balance micro-accelerometer[29] has been fabricated that demonstrates the current capability of on-chip integration. The device uses the principle of magnetic rebalance by the use of an external torque coil and has electronic circuitry that lies on the flexural structure. The bias stability is 10 μg per year, but at a cost of £500 its application is limited to military or aerospace needs.

7.6.2 Smart Sensors

When the electronics that process the output from the sensor and form the modifier are partially or fully integrated on a single chip the sensor is usually called an integrated smart, or intelligent, sensor. However, if the electronic circuit simply shares the same housing then it is called a hybrid smart sensor. In some ways this definition is rather misleading because a sensor that only has an integrated C-R filter or single operational amplifier may hardly be called intelligent. The main advantages of combining part or all of the modifier on a single chip are the reduction in cost base for large scale manufacture of temperature sensors, or Hall plates, reduction in circuit noise, improved sensitivity through near source amplification, standardized output, on-chip adjustment of nonlinearity and range, and reduction in size. To set against these advantages are the high costs of developing a silicon processing facility, reduced yield and reliability through increased complexity,[30] problems in thermal runaway and realization that some electronic technologies are incompatible. The interested reader is referred to a good review on sensor interfaces, smart sensors and bus systems that has recently been published.[31]

7.7 FUTURE TRENDS: VERY SMART SENSORS & NANOROBOTS

The trend towards integrated microsensor arrays in such areas as optical and tactile imaging has created a need for better processors. The improvements in speed that have been made using serial silicon technology have been dramatic over the past twenty years. Yet a comparison between current measurement systems and the biological system is rather enlightening. The basic processor in the biological system is the neuron which switches in milliseconds - compared to nanoseconds for transistors - but is capable of processing information from millions of sensors in a fraction of a second. The capabilities of machine vision, taction and olfaction systems lag far behind their biological analogs. Moreover, the loss of a single transistor in an instrument could lead to a complete functional failure whereas the biological system copes in a fault-tolerant and adaptive manner. The neuronal "clocks" tick slowly yet their power consumption is about a millionth of that of a serial silicon processor or von Neuman machine. The use of parallel processors is an obvious extension of prior art and was instigated in the mid-sixties before the similarity to the biological system was realized. Figure 7.10 shows the advances in the operating rate of computers through the use of Cellular Logic Arrays (CLAs) to process optical images.[32] The straight line shows the trend in speed of machine-coded CLAs on large general-purpose computers, and falls well short of that for parallel processors with minicomputer hosts (dotted line). In the twenty-five years between 1960 and 1985 the operating rate has improved by a factor of a million through the use of cellular parallel processing architectures. Unfortunately, these cellular arrays are a two-dimensional array of microprocessors and are extremely costly. The arrival of the transputer has gone some way towards satisfying the need for a microprocessor in parallel processing applications. Yet its applications are often limited to situations where the information can be conveniently subdivided and farmed out to each transputer. In practice this can be an inefficient way to solve a problem. In the author's opinion the most promising approach is that of "bottom-up", that is to tackle the problem by designing the appropriate architecture and software rather than using overpowered microprocessors and high level languages. This approach is basically that of neural or neuronal processing and has seen an astonishing emergence in the last ten years. The potential of parallel distributed processing was first discussed at the Massachusetts Institute of Technology (MIT)[33] in the seventies and linked the fields of computer and neuroscience. Yet it is only now that the subject is being treated seriously. Neural chips are being designed as the new generation of parallel processors to implant our instruments with features commonly associated with the biological system, ie. adaptability, fault-tolerance, distributed memory etc. I believe that these new processors or modifiers will be integrated with microsensor arrays to produce what I have called a very smart sensor. At Warwick we are investigating the use of artificial neural networks and their implementation in an electronic nose.[34]

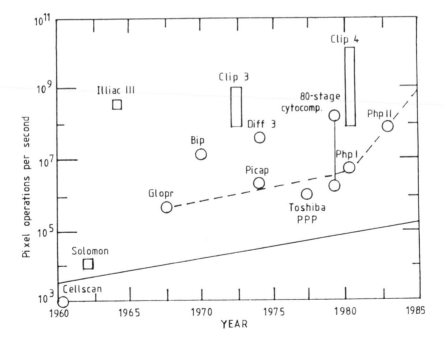

Figure 7.10 The advance in the development of parallel processors: operating rates of cellular logic machines for image analysis.

In this study the Rumelhart back propagation algorithm is used, although the latest breed of genetic algorithms that mutate to the optimal solution are being researched. This work may lead to a very smart sensor that is capable of matching the speed, sensitivity and discriminating powers of the biological system. Moreover this may lead to a move away from digital to analog VLSI computing and create the opportunity to overcome such problems as nonlinear signals, baseline drift and noise. The switch from silicon-technology to molecular electronics should enable a reduction in scale from micro to nanotechnology and produce the next generation of nanomachines.

To conclude, the ultimate goal is to integrate the sensor, modifier and actuator into a single system, that is essentially a microrobot. An advance of this kind could yield the development of micromachines capable of autonomous operations. The potential applications would then be numerous in the physical, chemical and biological sciences. For example, in medicine, micromachines could be built that travel along human arteries in order to seek and repair damaged valves in the heart. Such developments therefore offer the opportunity to revolutionize current practice, and lead to a new range of low-cost, reliable and smart micro- and nano-machines.

ACKNOWLEDGEMENTS

The author wishes to thank Professor R.S.Muller and Professor R.T.Howe of the Sensors & Actuator Centre, University of Berkeley, California for some stimulating conversations and course research material.

REFERENCES

1. Howe, R.T. (1988) Surface micromachining for microsensors and microactuators. *J. Vac. Sci. Technol.*, **B6**, 1809-1813
2. Kaminsky, G. (1985) Micromachining of silicon mechanical structures. *J. Vac. Sci. Technol.*, **B3**, 1015-1024
3. Schnakenberg, U., Benecke, W. and Lochel, B. (1990) NH_4OH-based etchants for silicon micromachining. *Sensors & Actuators*, **A23**, 1031-1035
4. Kittilsland, G., Stemme, G. and Norden, B. (1990) A submicron particle filter in silicon. *Sensors & Actuators*, **A23**, 904-907
5. Greenwood, J.C. (1984) Etched silicon vibrating sensor. *J. Phys. E: Sci. Instrum.*, **17**, 650-652
6. Lammerink, T.S.J., Elwenspoek, M., Van Ouwerkerk, R.H., Bouwstra, S. and Fluitman, J.H.J. (1990) Performance of thermally excited resonators. *Sensors & Actuators*, **A22**, 352-356.
7. Boxenhorn, B. and Grieff, P. (1990) Monolithic silicon accelerometer. *Sensors & Actuators*, **A21**, 273-277
8. Buser, R.A. and de Rooij, N.F. (1990) Very high Q-factor resonators in monocrystalline silicon. *Sensors & Actuators*, **21**, 323-327.
9. Boxenhorn, B. and Grieff, P. (1986) A vibratory micromechanical gyroscope. *Proceedings of AIAA Guidance and control conference* (Minneapolis, 1986), p1033.
10. Tang, W.C., Nguyen, T.H. and Howe, R.T. (1989) Laterally driven polysilicon resonant microstructures. *Sensors & Actuators*, **20**, 25-32

11. Pitcher, R.J., Foulds, K.W., Clements, J.A. and Naden, J.M. (1990) Optothermal drive of silicon resonators: the influence of surface coatings. *Sensors & Actuators*, **A21**, 387-390

12. Suzuki, S., Tuchitani, S., Sato, K., Ueno, S., Yokota, Y., Sato, M. and Esashi, M. (1990) Semiconductor capacitance-type accelerometer with PWM electrostatic servo technique. *Sensors & Actuators*, **A22**, 310-319

13. Tai, Y. and Muller, R.S. (1990) Frictional study of ic-processed micromotors. *Sensors & Actuators*, **A22**, 180-183

14. Hauden, D. and Loewenguth, B. (1986) Acceleration sensitivities of surface acoustic waves propagating on a cantilever quartz beam. *Appl. Phys. Lett.*, **47**, 1271-3

15. Hauden, D. (1987) Miniaturized bulk and surface acoustic wave quartz oscillators used as sensors. *IEEE Trans. Ultason., Ferroelectr. & Freq. Control*, **UFFC-34**, 253-8

16. Brattain, W.H. and Bardeen, J. (1953) Surface properties of germanium. *Bell. Syst. J.*, **32**, 1-42

17. Guilbaut, G.G. (1984) In *Methods and Phenomena*, edited by C. Lu and A. W. Czaderna. Amsterdam: Elsevier

18. Carey, W.P. and Kowalski, B.R. (1988) Monitoring a dryer operation using an array of piezoelectric crystals. *Anal. Chem.*, **60**, 541-544

19. Wohltjen, H. (1984) Chemical microsensors and microinstrumentation. *Anal. Chem.*, **56**, 87-103

20. IBM IEEE Institute November 1989

21. Sugiyama, S., Kawahata, K., Yoneda, M. and Igarashi I. (1990) Tactile image detection using a 1k-element silicon pressure sensor array. *Sensors & Actuators*, **A21**, 397-400

22. Zaromb, S. and Stetter, J.R. (1984) Theoretical basis for identification and measurement of air contaminants using an array of sensors having partially overlapping sensitivity. *Sensors & Actuators*, **6**, 225-243

23. Carey, W.P., Beebe, K.R., Kowalski, B.R., Illman, D.L. and Hirschfeld, T. (1986) Selection of adsorbates for chemical sensor arrays by pattern recognition. *Anal. Chem.*, **58**, 149-153

24. Gardner, J.W., Bartlett, P.N., Dodd, G.H. and Shurmer, H.V. (1990) The design of an artificial olfactory system. In *Chemical Information Processing* edited by D. Schild, pp.131-173. Berlin: Springer-Verlag

25. Shurmer, H.V., Gardner, J.W. and Chan, H.T. (1989) Application of discrimination techniques to alcohols and tobaccos using tin oxide sensors. *Sensors & Actuators*, **18**, 361-371

26. Shurmer, H.V., Gardner, J.W. and Corcoran, P. (1990) Intelligent vapour discrimination using a composite 12-element sensor array. *Sensors & Actuators*, **B1**, 256-260

27. Delco, *IEEE Hilton Head Workshop* (1988)

28. Rudolf, F., Jornod A., Bergqvist, J. and Leuthold, H. (1990) Precision accelerometers with μg resolution. *Sensors & Actuators*, **A22**, 297-302

29. Aske, V.H. (1987) An integrated silicon accelerometer. *Scientific Honeyweller*, **8**, 53

30. BSI *Handbook 22 Quality Assurance* (1987). London: BSI

31. Middelhoek, S. and Audet, S.A. (1989) *Silicon sensors*. Ch.8 London: Academic Press

32. Preston,K. (1983) Cellular logic computers for pattern recognition. *IEEE Computer*, **16**, 36-50
33. Rumelhart, D.E. and McClelland, J.L. (1986) *Parallel distributed processing*, Cambridge, Mass: MIT press
34. Gardner, J.W., Hines, E.L. and Wilkinson, M. (1990) Application of articicial neural networks in an electronic nose. *Meas. Sci. Technol.*, **1**, 446-451

Chapter 8

NANOTECHNOLOGY

ALBERT FRANKS

Nanotechnology is concerned with manufacturing to dimensions or tolerances in the range of 0.1 to 100 nm. It plays a key role in many areas including materials processing, mechanical engineering, optics and electronics. Technologies for manipulating materials at, or near, the molecular level are coming increasingly into use.

8.1. INTRODUCTION

What is Nanotechnology? It is a term which has entered into the general vocabulary only recently, although it was used at least as early as 1974 by Taniguchi.[1] At the National Physical Laboratory (NPL), the term nanometrology was introduced about five years ago, mainly to describe the metrology associated with the development of X-ray, optical and other very precise components. We have defined nanotechnology as the technology where dimensions and tolerances in the range of 0.1 to 100 nm (from the size of the atom to about the wavelength of light) play a critical role.

This definition is too all-embracing to be of practical value because it could include, for example, topics as diverse as X-ray crystallography, atomic physics and indeed the whole of chemistry! The field covered by nanotechnology is narrowed down, within the defined dimensional range, to include manipulation and machining, but by technological means as opposed to those used by the craftsman, and thus excludes, for example, traditional forms of glass polishing. The technology relating to fine powders also comes under the general heading of nanotechnology. Within the next few years a consensus will no doubt emerge which will roughly circumscribe the activities covered by nanotechnology, but even now there is a good feel for the hard core of the subject.

Nanotechnology is an "enabling" technology, in that it provides the basis for other technological developments, and it is also a "horizontal" or "cross-sectoral" technology in that one technique may, with slight variations, be applicable in widely differing fields. A good example of this is thin film technology, which is fundamental to electronics and optics.

In Japan, the importance attached to nanotechnology can be judged from the fact that it is one of the six topics which have been singled out for vigorous development within the ERATO project. This is a forward-looking project in which various areas have been identified as having particular technological relevance in the next decade, and in which research and development is being undertaken by a powerful combination of industry, university, foreign guest workers and even foreign institutions.

In the UK, the government has also identified nanotechnology as an important growth area, and as a topic for support under the LINK programme, which aims to bridge the gap between science and the market place by providing a significant measure of funding for collaborative ventures between academia and industry. Nanotechnology is seen to be of particular importance, and of immediate relevance, in areas such as materials science, mechanical engineering, optics, and electronics.

The field is so large that it is impracticable to select more than a few items for discussion in this chapter; the topics have been chosen to give an indication of the technologies which fall within the domain of nanotechnology and to illustrate both their cross-sectoral and enabling roles.

It is not intended to list in detail the artefacts or devices which depend on nanotechnology for their manufacture, but a short selection to illustrate the wide range would include products made of engineering (or fine) ceramics such as turbine blades, some semiconductor substrates, hydrostatic and hydrodynamic bearings, a new generation of internal combustion engines and magnetic memory discs. A wide range of materials are employed in devices such as computer and home entertainment peripherals, including magnetic disc reading heads, video cassette recorder spindles, optical disc stampers and ink jet nozzles. Optical and semiconductor components including laser gyroscope mirrors, diffraction gratings, X-ray optics, quantum well devices, GaAs Field Effect Transistors (FETs) and Semiconducting Quantum Interference Devices (SQUIDs). Some of these devices will be referred to again in context below.

8.2 MATERIALS TECHNOLOGY

This section of the paper takes a selective look aimed principally at fine powder technology, and closely follows an unpublished report prepared by PA Technology for the UK Department of Trade and Industry.

The wide scope of nanotechnology is well demonstrated in the materials field, where materials provide a means to an end, and are not an end in themselves. For example, in electronics, inhomogeneities in materials, on a very fine scale, set a limit to the nanometre-sized features which play an important part in semiconductor technology, and in a very different field, the finer the grain size of an adhesive, the thinner will be the adhesive layer, and the higher will be the bond strength.

8.2.1 Advantages of Ultrafine Powders

In general, the mechanical, thermal, electrical and magnetic properties of ceramics, sintered metals and composites are often enhanced by reducing the grain or fibre size in the starting materials. Other properties such as strength, the ductile-brittle transition, transparency, dielectric coefficient and permeability can be enhanced either by the direct influence of an ultrafine microstructure or by the advantages gained of mixing and bonding ultrafine powders.

Using the manufacturing processes described in 8.2.3, homogeneous and pure powders are produced which behave in a more controllable and predictable way than less refined ones.

Other important advantages of fine powders are that when used in the manufacture of ceramics and sintered metals, their green (i.e. unfired) density can be greatly increased. As a consequence, both the defects in the final product and the shrinkage on firing are reduced, thus minimizing the need for subsequent processing.

8.2.2 Applications

1. Thin films and coatings - the smaller the particle size, the thinner the coating can be.

2. Chromatography - the increase in specific surface area associated with small particles, allows column lengths to be reduced.

3. Electronic ceramics - reduction in grain size results in reduced dielectric thickness.

4. Strength bearing ceramics - strength increases with decreasing grain size.

5. Cutting tools - smaller grain size results in a finer cutting edge, which can enhance the surface finish.

6. Impact resistance - finer microstructure increases the toughness of high temperature steels.

7. Cements - finer grain size yields better homogeneity and density.

8. Gas sensors - finer grain size gives increased sensitivity.

9. Adhesives - finer grain size gives thinner adhesive layer and higher bond strength.

8.2.3 Processes for Producing Ultrafine Powders

Sol-gel process

This refers to the production of inorganic oxides using either metal alkoxide precursors or colloidal dispersions of hydrous oxides.[2] The starting material is processed to form a dispersible oxide and forms a sol in contact with water or dilute acid. (A sol is a dispersion of solid particles in a liquid, in which at least one dimension of the particles is between 1 nm and 1 μm). Removal of the liquid from the sol yields the gel, and the sol-gel transition controls the particle size and shape. Calcination of the gel produces the oxide. A typical particle size for the ThO_2 and ThO_2-UO_2 processes is 8 nm.

Solid state reactions

In these reactions, metal oxides are prepared from a salt by calcination. A typical reaction is the decomposition of ferrous sulphate heptahydrate ($FeSO_4.7H_2O$) to iron oxide (Fe_2O_3). The process often produces spherical particles, with the smallest ones having a diameter of about 0.25 μm.

Liquid-solid reactions

Ultrafine particles are produced by precipitation from a solution, the process being dependent on the presence of desired nuclei. For example, TiO_2 powders have been produced with particle sizes in the range of 70 nm to 0.3 μm from titanium tetraisopropoxide.[3]

Evaporation and condensation processes

Ultrafine powders may be prepared by evaporation followed by condensation of the vapour. For example, oxide powders with a particle size of less than 20 nm have been made by evaporating bulk oxide with a focused beam of electrons.[4]

Plasma processes

Powders with a particle size in the range 10-60 nm may be produced[5] by vaporizing micrometre-sized particles in an argon plasma stream for 1 to 10 ms. This technique has been applied to metals such as Al, Mo, W, Zr and Ag as well as to oxides such as Al_2O_3.

Flame hydrolysis

Flame hydrolysis is used for the manufacture of fused silica. In this process, silicon tetrachloride is heated in an oxy-hydrogen flame to give a highly dispersed silica.[6] The resulting white amorphous powder consists of spherical particles with sizes in the range of 7-40 nm.

8.3 PRECISION MACHINING AND MATERIALS PROCESSING

8.3.1 Introduction

A considerable overlap is emerging in the manufacturing methods employed in very different areas such as mechanical engineering, optics and electronics. Precision machining encompasses not only the traditional techniques such as turning, grinding, lapping and polishing refined to the nanometre level of precision, but also the application of "particle" beams, ions, electrons and X-rays. Ion beams are capable of machining virtually any material and the most frequent applications of electrons and X-rays are found in the machining or modification of resist material for lithographic purposes. The interaction of the beams with the resist material induces structural changes such as polymerization which alter the solubility of the irradiated areas.

The status of ultraprecision machining and ultrafine processing of materials has been reviewed by Taniguchi.[7] He also discusses future trends and describes a wide range of precision products which have been made by advanced technologies. (See Chapter 9 for more details on ultraprecision processing.)

8.3.2 Interpretation of Precision Machining Limits & Measurement Sensitivities

Caution should be exercised in interpreting data quoted in the literature relating to the machining and measurement of precision surfaces. It is becoming increasingly common to find it stated that the surface roughness of a machined surface has been measured to be less than, say, 0.3 nm (peak to valley or RMS), or even an order of magnitude finer than this. Since 0.3 nm is of atomic dimensions, what does this statement mean? It is a reflection on the method of measurement. A number of instruments based on optical interferometers or mechanical styli have an amplitude sensitivity of better than 0.1 nm, but their surface wavelength, or lateral resolution, may well be 1000 times worse than this. For example, an optical interferometer will have a surface wavelength resolution limited by the wavelength of light; at best the resolution may be about 500 nm. Although in the case of styli, the surface wavelength resolution improves with decreasing surface roughness (less of the stylus tip contacts the surface), a resolution of 50 nm is one of the best ever reported.[8] With one or two exceptions therefore, most of the surface roughness figures quoted in the literature are based on disproportionately large ratios of surface wavelength and amplitude resolutions, and since the value of the former is rarely given, the amplitude data is an undefined average. Examples of measuring techniques where both resolutions are subnanometre or better are the rarely used stereo-transmission electron microscopy methods[9] and scanning tunnelling microscopy, which is now being applied for metrological purposes.[10] The concept of the amplitude/surface wavelength relationship[11] is of fundamental importance in interpreting topographic measurements.

8.3.3 Techniques

Diamond turning

The Large Optics Diamond Turning Machine (LODTM) at the Lawrence Livermore National Laboratory represents a pinnacle of achievement in the field of ultraprecision machine tool engineering.[12,13] This is a vertical spindle machine with a face plate diameter of 1.6 m and a maximum tool height of 0.5 m. Despite these large dimensions, machining accuracy for form is 27.5 nm RMS and a surface roughness of 3 nm is achievable, but is dependent both on the specimen material and cutting tool.

In his review of ultra-precision diamond turning machines in Japan, Kobayashi[14] discusses the critical components of the machine tool - the spindle and the linear slide - and demonstrates that there is as yet no consensus about the optimum design: there are different views about the relative merits of cylindrical and spherical air bearing spindles and about aerostatic and hydrostatic slideways. In all the machines referred to the spindle is horizontal. In the case where a cylindrical workpiece may deform significantly under the action of gravity, it is advantageous to employ a vertical axis machine. This was the reasoning behind the design of the numerically controlled X-ray telescope diamond turning machine described by Wills-Moren et al.[15]

Grinding (fixed abrasive)

The term "fixed abrasive" denotes that a grinding wheel is employed in which the abrasive particles, such as diamond, cubic boron nitride or silicon carbide, are attached to the wheel by embedding them in a resin or a metal. The forces generated in grinding are higher than in diamond turning and usually machine tools are tailored for one or the other process. Some Japanese work is in the vanguard of precision grinding, and surface finishes of 2 nm (peak to valley) have been obtained on single crystal quartz samples using extremely stiff grinding machines;[16] the measurements having been made at NPL with an environmentally-protected Talystep measuring machine,[17] and in which the surface wavelength resolution was about 50 nm.

A versatile grinding machine for producing glass and ceramic X-ray mirrors in the form of toroidal and conicoidal sectors has been described by Lindsey et al.[18] Surface roughness amplitudes obtained by grinding on this machine are no worse than 10 nm (peak to valley).

Grinding (loose abrasive)

The most familiar "loose abrasive" grinding processes are lapping and polishing where the workpiece, which is often a hard material such as glass, is rubbed against a softer material, the lap or polisher, with an abrasive slurry between the two surfaces. The methods employed, the materials used were reviewed by Franks.[19] In many cases, the polishing process occurs as a result of the combined effects of mechanical and chemical interactions between the workpiece, slurry and polisher. Where softer materials have to be polished, and particularly where surface damage due to abrasion must be minimized (e.g. in polishing semiconductors such as GaAs or InP), chemo-mechanical methods are often employed. Mechanical polishing is required to obtain flat surfaces, while the attainment of damage-free surfaces necessitates chemical etching. This dual requirement is met by using a conventional polisher with a chemically active slurry.[20] These authors have taken one further step to reduce surface damage by hydrodynamically suspending the samples on a thin layer of etch solution above the polisher, the lift being generated by the rotation of the polishing wheel. In a related method of "float polishing", surface roughnesses of better than 0.1 nm RMS have been achieved.[21]

Loose abrasive grinding techniques can under appropriate conditions produce unrivalled accuracies both in form and surface finish when the workpiece is flat or spherical; in these configurations a random motion can be imparted to the workpiece relative to the lap. The random motion is the key to producing uniform wear after the mutual elimination of the high spots in the lap and workpiece. Surface figures or form to a few nm and surface finishes to better than 0.5 nm may be achieved.[22] Where completely random motions are not possible, for example in the development of aspherical surfaces, then the ideal conditions are compromised: laps are used which have some geometrical non-uniformity, or which have the required degree of flexibility. The problems associated with the formation of surfaces of complex shape are an important driving force in the development of super precision computer controlled machine tools. A machining method which combines the advantages of the versatility of a three-axis Computer Numerically Controlled (CNC) machine tool and loose abrasive grinding has been developed by Tsuwa *et al.*[23] The abrasive is in a slurry and is directed locally towards the workpiece by the action of a non-contacting polyurethane ball spinning at high speed, and which replaces the cutting tool in the machine. This technique has been named "elastic emission machining" and has been used to good effect in the manufacture of an X-ray mirror having a figure accuracy of 10 nm and a surface roughness of 0.5 nm RMS.[24]

Ions, electrons and X-rays

The relentless quest to reduce the size of integrated circuit elements has been a major driving force in the development of "atomic" machining techniques. But it is by no means the only one: over 20 years ago directed ion beams were used to machine aspherical optical surfaces atom layer by atom layer,[25] ion etching has been used for many years in the manufacture of X-ray gratings[26] and for thinning specimens for electron transmission microscopy.[27] In a review on machining with energetic particle beams and radiation, Taniguchi[28] refers to the use of ion beam machining for the final dimensional finishing of gauge blocks and for sharpening diamond indenters and microtome cutting blades. Points sharp to a few nanometres and sharp edges can be produced very simply by directing an ion beam at an angle of about 45° to a mechanically pre-sharpened object.[29] However, problems have been encountered in trying to produce smooth surfaces by ion beam machining because the surfaces do not necessarily erode uniformly, even in homogeneous materials like fused silica.[30]

There is a superficial analogy between ion beam and elastic emission machining, but the former is capable of much higher lateral resolution. An important emerging technology is the use of focused ion beam systems for maskless ion beam machining.

Atomic resolution can be achieved parallel to the workpiece surface by ion beam machining or "milling". Normal to the surface the resolution is limited to about 0.1 μm for acceptable removal rates.[31]

In the remainder of this section the potential of lithography will be discussed, rather than the present day industrial production reality, where only now, for example, the first semiconductor FETs with 0.35 μm geometries are being made commercially (although still for test purposes only) using synchrotron-based X-ray lithography.[32]

Both electron and X-ray lithography bear a close resemblance to optical lithography, in that interaction of the radiation with a resist material modifies the resist in such a way that a subsequent development process can distinguish between the exposed and unexposed regions, removing one by dissolution and leaving the other intact. Resists are polymer films and the energy of the incident electrons or X-rays is larger than the energy required to form or break a chemical bond. A shower of secondary electrons is produced when an X-ray photon is absorbed by the resist and these electrons are mostly responsible for the chemical changes which take place. The main difference between an X-ray and an electron exposure is that under normal conditions, the energy of the secondary electrons produced by the X-rays is much less than the energy of the electron beam (10-50 keV), so that X-ray lithography is a higher resolution technique, because of the smaller range of the scattered electrons.[33] A good review of X-ray lithography has been given by Spiller and Feder[34] and more recently by Heuberger.[35] Lateral resolutions of 0.1 μm are fairly readily achievable and resolutions of 10 nm are possible.[36]

A number of techniques are employed to enhance the resolution achievable by electron beam machining. The effects of electron scatter can be mitigated by reducing both the thickness of the resist and the substrate (to reduce electron backscatter) and to use a "contamination" resist. The latter is a carbonaceous layer formed at the point of impact of the electron beam and the specimen as a result of the interaction of the electron beam and hydrocarbon vapour in the electron beam system. The width of the carbon layer increases with exposure time but initially does not appreciably exceed the electron beam diameter. By scanning the electron beam appropriately, a resist pattern is directly written on the specimen surface. A structure is produced in the substrate by ion etching: the contaminant acting as a protective layer. These techniques can be employed to produce 10-20 nanometre-sized structures on a substrate,[37] and free-standing wires 40 nm wide by 10 nm thick by 8-120 um long.[38] Even structures as complex as gear-trains[39] and electrostatic motors[40] both no bigger than 70 μm in diameter have also been made by lithography.

An alternative electron beam machining technique dispenses with resists, and surfaces are machined directly with a resolution of 1 nm, using an intense, focused electron beam. Extremely straight holes have been drilled through 200 nm thick material with a diameter of 2.0 ± 0.2 nm, and fine lines have been cut by scanning the beam in inorganic materials such as alumina.[41]

Thin film production

The production of thin solid films, particularly for coating optical components, provides a good example of traditional nanotechnology. There is a long history of coating by chemical methods, electro-deposition, diode sputtering and vacuum evaporation, while triode and magnetron sputtering and ion beam deposition are more recent in their wide application.

Because of their importance in the production of semiconductor devices, epitaxial growth techniques are worth a special mention. Epitaxy is the growth of a thin crystalline layer on a single crystal substrate, where the atoms in the growing layer mimic the disposition of the atoms in the substrate.

The two main classes of epitaxy which have been reviewed by Stringfellow[42] are liquid-phase and vapour-phase epitaxy. The latter class includes Molecular Beam Epitaxy (MBE), which in essence is highly controlled evaporation in ultra-high vacuum. MBE may be used to grow high quality layered structures of semiconductors with monolayer precision, and it is possible to exercise independent control over both the semiconductor band-gap, by controlling the composition, and also the doping level. Pattern growth is possible through masks and on areas defined by electron beam writing.

8.4. APPLICATIONS

There is an all-pervading trend to higher precision and miniaturization, and to illustrate this a few applications will be briefly referred to in the fields of mechanical engineering, optics and electronics. It should be noted however, that the distinction between mechanical engineering and optics is becoming blurred, now that machine tools such as precision grinding machines and diamond-turning lathes are being used to produce optical components, often by personnel with a background experience in mechanical engineering rather than optics. By a similar token, mechanical engineering is also beginning to encroach on electronics particularly in the preparation of semiconductor substrates.

8.4.1. Mechanical Engineering

One of the earliest applications of diamond turning was the machining of aluminium substrates for computer memory disks, and accuracies are continuously being enhanced in order to improve storage capacity: surface finishes of 3 nm are now being achieved. In the related technologies of optical data storage and retrieval, the tolerances of the critical dimensions of the disk and reading head are about 0.25 µm. The tolerances of the component parts of the machine tools used in their manufacture, i.e. the slideways and bearings, fall well within the nanotechnology range.[43]

Some precision components falling in the manufacturing tolerance band of 5-50 nm include gauge blocks, diamond indenter tips and microtome blades, Winchester disk reading heads and ultra precision X-Y tables.[27] Examples of precision cylindrical components in two very different fields, and which are made to tolerances of about 100 nm, are bearings for mechanical gyroscopes and spindles for video cassette recorders.

The theoretical concept that brittle materials may be machined in a ductile mode has been known for some time[44] and has been demonstrated in ductile mode diamond-turning of glass and silica.[45] If this concept can be applied in practice it would be of significant practical importance because it would enable materials such as ceramics, glasses and silicon to be machined with minimal sub-surface damage, and could eliminate or substantially reduce the need for lapping and polishing. Typically, the conditions where ductile-mode machining is possible require that the depth of cut is less than 100 nm and that the normal force should fall in the range of 0.01 - 0.1 N. These machining conditions can be realized only with extremely precise and stiff machine tools such as the one described by Yoshioka,[15] and with which quartz has been ground to a surface roughness of 2 nm peak-to-valley. The significance of this experimental result is that it points the way to the direct grinding of optical components to an optical finish. The principle can be extended to other materials of significant commercial importance, such as ceramic turbine blades which at present must be subjected to tedious surface finishing procedures to remove the structure-weakening cracks produced by the conventional grinding process.

8.4.2 Optics

In some area in optics manufacture there is a clear distinction between the technological approach and the traditional craftsman's approach, particularly where precision machine tools are employed. On the other hand, in lapping and polishing, there is a large grey area where the two approaches overlap.

The large demand for infra-red optics from the 1970s onwards could not be met by the traditional suppliers, and provided a stimulus for the development and application of diamond-turning machines to optic manufacture. The technology has now progressed and the surface figure and finishes which can be obtained span a substantial proportion of the nanotechnology range.

Important applications of diamond-turned optics is in the manufacture of non-conventionally shaped optics, for example axicons and waxicons[46] and, more generally, aspherics and particularly off-axis components, such as paraboloids.

The mass production (several million per annum) of miniature aspheric lenses used in compact disk players and the associated lens moulds provides a good example of the merging of optics and precision engineering. The form accuracy must be better than 0.2 µm and the surface roughness must be below 20 nm RMS to meet the criterion for diffraction limited performance.[47]

The main demand for optics having surface roughnesses in the 0.1 nm range arises from the requirements for high energy laser mirrors, to reduce laser-induced damage,[48] and to minimize scatter in laser gyroscope optics[49] and X-ray optics.[50] This level of surface roughness can, at present, only be achieved by employing lapping and polishing techniques.

Two examples of thin film technology where nanometre or sub-nanometre accuracies are required, but for very different reasons, are the multilayer coatings deposited on laser gyroscope mirrors, where high perfection is necessary to minimize scatter, and multilayer coatings on X-ray mirrors which enable these mirrors to be employed at normal incidence, with wavelengths as short as 4 nm. Layer thicknesses are normally no more than a few nm.[51]

A novel application of electron beam lithography in optics is the manufacture of zone plates for use in X-ray microscopy.[52] These are made either by generating a master pattern in a resist which is subsequently copied into a more X-ray absorbent material or by directly drawing the pattern on a thin substrate, the pattern being formed in carbon at the point of impact of the electron beam which "cracks" the contaminant hydrocarbon film which is always present on the substrate. The latter technique is capable of drawing 20 nm wide lines.

8.4.3 Electronics

In semiconductors, nanotechnology has long been a feature in the development of layers parallel to the substrate and in the substrate surface itself, and the requirements for precision are steadily increasing with the advent of layered semiconductor structures. About one quarter of the entire semiconductor physics community is now engaged in studying aspects of these structures.[53] Normal to the layer surface, the structure is produced by lithography, and for research purposes at least, nanometre-sized features are now being developed using X-ray, electron and ion beam techniques.

Devices based on GaAs have captured a significant amount of attention because they offer the highest digital processing speeds coupled with the additional advantages over silicon of having a higher temperature tolerance and having greater radiation resistance. Quantum well devices, often made by MBE, typically consist of a multilayer stack of GaAs interleaved with an alloy such as AlGaAs, in which the layers may have thicknesses from less than 20 nm down to near atomic dimensions, to form a superlattice. Other combinations of materials, such as InP and InGaAs can also be used. Two quantum well devices are the High Electron Mobility Transistor (HEMT) and the Multiple Quantum Well Laser (MQWL) which exploit the hetero-junction between the layers. The lower resistances in the HEMT lead to higher speed operations, while in the MQWL the confinement of carriers leads to a blue shift in the lasing frequency.[54] By suitable choice of both the layer thickness and the Al/Ga ratio, it is possible to tailor the frequency of the emitted light and operate in the visible region.

The bridge-type Josephson junction offers an entirely different approach to the development of high speed digital circuits. The bridge length must be of the order of the coherence length of superconducting electrons which is about 40 nm in niobium at 4.2 K, and bridges of these dimensions have been made using electron beam lithography.[55] Electron beam lithography has also been used to fabricate GaAs MEtal Semiconductor FETs (MESFETs) with gate lengths as small as 55 nm.[56] It was found that the maximum source-drain voltage that could be pinched off in these short-gate devices was about 1 V. It was speculated that the source-drain voltage would be well in excess of 2 V if the device had been grown by MBE, rather than by vapour phase epitaxy, which would have yielded higher quality material and abrupt interfaces between the doped and undoped regions.

8.5 NANOTECHNOLOGY AT NPL & UK NATIONAL INITIATIVE ON NANOTECHNOLOGY

The stimulus for the work at NPL was provided by the requirements of X-ray optics, for use in a wide range of applications. These include the determination of the structure of polymers (by small angle X-ray scattering), the analysis of elements of low atomic number using specially developed X-ray gratings, the development of mirrors for use in synchrotrons and the development of X-ray telescopes and microscopes.[49] For many applications the tolerance requirements on surface finish are at the sub-nanometre level, and those for surface figure may in some instances (e.g. for X-ray microscopes) be in the nanometre and sub-nanometre regions.[57,58] The technology is now being more widely employed in such areas as the development of precision bearings, the finishing of semiconductor substrates and laser gyroscope mirrors.

The production of very precise components goes hand in hand with the development of the necessary metrology, and a wide range of measuring instruments has been devised to cater for the evaluation of surfaces and structures down to the 0.1 nm level.[59] Particularly noteworthy are the Nanosurf 2,[60] the polarizing interferometer,[61] the stylus profilometer,[62] the laser profilometer,[63] the Nanorond 1[64] and the X-ray interferometer.[65] This powerful array of instruments provides a measuring capability which ranges from 50 pm (picometres) to 15 mm in surface amplitude and from 50 nm to 250 mm in surface wavelength, and techniques for roundness measurements to 1 nm and displacement calibration to 10 pm, traceable to the national standards of length. Details of the performance characteristics of these and some other measuring machines in terms of their amplitude and surface wavelength measurement capabilities are presented on Stedman charts.[66,10]

The work at NPL in the field of nanotechnology has been of significant interest to a wide spectrum of representatives from industry, the universities and government, and the Laboratory has noted a marked increase in recent years for requests for technical support and advice, both from UK and foreign organizations. It became apparent that it would be timely to assess whether a national programme in support of nanotechnology should be initiated. To this end, a consultative meeting was called with industrialists and academics in November 1986 which was attended by over 150 delegates. Presentations on the applications of nanotechnology were given, and the interest of industry in a national initiative was confirmed. As a result a Forum and a Strategy Committee have been established for the purpose of overseeing the national programme, for assessing priorities and for establishing collaborative research programmes. Nanotechnology has formally been identified as a suitable area for support under the government's LINK scheme to sponsor important emerging horizontal technologies.[67]

8.6 MOLECULAR ELECTRONICS

Lithography and thin film technology are the key technologies which have made possible the continuing and relentless reduction in the size of integrated circuits, to increase both packing density and operational speed. Miniaturization has been achieved by engineering downwards from the macro to the micro scale. By simple extrapolation it will take approximately two decades for electronic switches to be reduced to the molecular dimensions. The impact of molecular biology and genetic engineering has thus provided a stimulus to attempt to engineer upwards, starting with the concept that single molecules, each acting as an electronic device in their own right might be assembled using biotechnology, to form Molecular Electronic Devices (MEDs) or even BioChip Computers (BCCs). The enzymatic reactions of DNA replication, transcription and translation in biological systems, appear to be nature's closest approach to the elements of computing and dissipate energy amounting to 10-100 kT per primitive step, which compares very favourably with the 10^{10} kT per "equivalent" programming step in a transistor-based computer, where k is Boltzmann's constant and T is absolute temperature: it is the heat dissipation problem which will limit the ultimate micro-miniaturization based on conventional technology.[68]

There has not so far been any break-through leading to the production of practical devices by upward engineering. Some possible approaches which have been suggested[69] include the production of molecular wires of polysulphurnitride (a -[NSNS]-chain being formed by alternate SCl_2 and diimide reactions initiated on a silicon substrate), the formation of three-dimensional structures using molecular epitaxial deposition, where the substrate is built up of amino acids in any specified sequence. The two techniques could be combined so that by a series of reactions conducting molecular wires would be built up in some areas, while in other areas insulation would be added. Switching and control functions could then be added and adjacent components bonded together to form a solid assembly of active and passive components. Carter[70] also discusses three possible ways of producing addressable molecular switches. One of these, is a switch based on electron tunnelling where switching takes place by changing barrier heights in cyanine using an optical input, or by the motion of a nearby soliton.

The concept of upwards engineering is of wide inter-disciplinary interest involving physics, chemistry, electrical engineering and biology, and some of the enthusiasm which has been generated in the subject, particularly in the USA, is evident in the proceedings of two workshops on Molecular Electronic Devices[71,72] and of the conference on Optical and Hybrid Computing.[73] In the UK, a report produced for the

Science and Engineering Research Council has called for a national strategy on molecular electronic research.[74] However, formidable obstacles will have to be overcome before upward engineering can become a reality. Some of these obstacles have been discussed by Haddon and Lamola,[75] and may be summarized as follows:

1. There is little structural analogy between biological systems, including the brain, and present-day computers. Although it seems clear that the brain uses conductive pathways, the active elements are ions and molecules; thus the cycle time of the brain circuitry is quite slow, the propagation velocities being of the order of 25 m/s. The fast processing capabilities of the brain are usually attributed to the exploitation of parallel processing architecture. However, since our understanding of the molecular basis of the higher brain functions such as reason is practically non-existent, we have no guide lines for developing a synthetic biocomputer.

2. The neuron is much more complex than an electronic component. Switching involves alterations or synthesis of proteins via complicated biochemical mechanisms which are not yet understood.

3. Even if it were possible, it would be unlikely that one would choose the massive and slow ions and molecules of the brain as the active carriers in a BCC. Polypeptides have large band gaps characteristic of electrical insulators, and thus there is little evidence for the general utilization of conduction band electron transport in biology.

4. As the packing density increases, there is a corresponding increase in the problems of connecting and addressing individual components, and preventing unwanted interactions (cross talk). These interactions take place by tunnelling, and if the device is electronic in nature, it may be the proximity of the components rather than their dimensions which will set the limit to miniaturization.

Advances in molecular electronics by downward engineering from the micro to the macro scale are taking place over a wide front. One fruitful approach is by way of the Langmuir-Blodgett (LB) film using a method first described by Blodgett.[76] A multilayer LB structure consists of a sequence of organic monolayers made by repeatedly dipping a substrate into a trough containing the monolayer floating on a liquid (usually water), one layer being added at a time. The classical film forming materials were the fatty acids such as stearic acid and their salts. The late 1950s saw the first widespread and commercially important application of LB films in the field of X-ray spectroscopy.[77,78] The important properties of the films which were exploited in this application were the uniform thickness of each film i.e. one molecule thick, and the range of thicknesses, say from 5 to 15 nm, which were available by changing the composition of the film material. Stacks of fifty or more films were formed on plane or curved substrates to form two-dimensional diffraction gratings for measuring the characteristic X-ray wavelengths of the elements of low atomic number for analytical purposes in instruments such as the electron probe X-ray microanalyzer.

The technology was re-invented over a decade later when there was a resurgence of activity arising from the recognition of the potential value of organic LB films in electronics. A comprehensive review of the subject is given by,[79] and is briefly summarized here.

One aspect of the more recent work has been directed towards producing films of materials more stable than the fatty acids, which melt at around 70°C, and films which incorporate molecules which can be tailored to suit specific applications. Four applications which are singled out where LB technology will be of significant importance in the medium term are non-linear physics, enhanced device processing, quantum mechanical tunnelling and sensors.

1. The highly non-linear parameters inherent in the unique molecular architecture of the films can be exploited in optoelectronics, acoustoelectrics and infra-red detectors.

2. In enhanced device processing ultra-high density integration will require thin and low defect density resists. LB films meet these demands in terms of high resolution and sensitivity: a resolution of better than 10 nm has been demonstrated using multilayers of fatty acid salts.[80] However, further work will be required to improve their etch resistance and to lower the defect densities, although in passive applications they have proved effective as thin pinhole-free insulating or passivating layers on semiconductors such as GaAs.

3. Because the thickness of LB films can be controlled down to molecular dimensions they are showing promise as insulating spacers for use in devices which exploit the phenomenon of quantum mechanical tunnelling. In photovoltaic cells, for example, the use of an LB film increased the device efficiency by 50%.[81] More generally, the incorporation of tunnelling layers should widen the scope or enhance the performance of switches, transistors and injection devices.

4. LB films have been used for sometime in sensors, such as thermal or gas sensors, and often they are based on field effect devices.[79]

Organic compounds often respond more sensitively to temperature, pressure and chemical environment and the changes induced can be made to modulate the source-drain current. The advantage of using an LB film is that its volume is small and it therefore has a fast response and recovery time.

In electronic biosensors an electrical output is produced which is related to the concentration or activity of a chemical or a biochemical; the sensor must incorporate some means of molecular recognition. Enzymes and antibodies are examples of two complex materials capable of "recognising" specific molecules and are being used in electronic biosensors.[82] Both these chemicals are proteins and both have clefts in their surfaces which permit entry and bonding of a specific foreign molecule only. This produces a reaction which may be monitored in some way. For example, if the enzyme is in solution the reaction may change the pH which is detectable by a ChemFET (i.e. a MOSFET without a gate electrode). As in the physical sensors, a thin insulator acts as a barrier between the semiconductor and its surroundings. In use, protons from the solution are absorbed on the insulator to produce a positive charge which modulates the source-drain current. LB films have not yet played any significant role in this field but because of their small volume, it is anticipated that high sensitivities and rapid response times will be achieved if biological molecules are incorporated into the layers in contact with the semiconductor substrate.

8.7 THE NEW SCANNING TECHNOLOGIES

8.7.1 The Scanning Microscopies

The early 1980s saw the first publications of a novel technique which by the middle of the decade had established itself as one of the fastest growing new technologies, namely Scanning Tunnelling Microscopy (STM)[83,84] and its derivatives. The principle underlying these techniques is that a fine probe is scanned extremely closely to a surface to yield information with high spatial resolution. The concept was first described by O'Keefe[85] who proposed that an object be scanned with a light beam that had been transmitted through an aperture much smaller than the wavelength of light. If the aperture is so close to the specimen that the light spread due to diffraction is negligibly small, then if the light transmitted through the specimen is recorded on a display system scanning in unison, the lateral resolution of the image is determined by the size of the aperture and not by the diffraction limit set by the wavelength of light. It took nearly thirty years before technology caught up with this concept and turned scanning microscopy into a practical tool.

In principle and in practice the instrumentation for scanning probe microscopy is fairly simple. The main requirements are a probe that can be brought into close proximity to the surface being studied, a scanning stage to raster the probe over the surface, a detector to provide an output relating to the interaction between probe and surface and the electronics and software to convert the output of the detector into a quantitative or pictorial form: akin to the images produced, for example in the scanning electron microscope. Very approximately, (usually within a factor of 10), the highest lateral resolution that can be achieved is equal to the specimen-probe distance, so that for sub-nanometre resolution this distance is about a nanometre. Two of the more important technological challenges that have to be met in order to do this are the elimination of relative motion, due to vibration, between the probe and specimen and the ability to scan the specimen in close proximity to the probe. The vibration problem has been tackled by making the mechanical loop between the probe and the specimen as small and as stiff as possible; aiming for resonant frequencies of the mechanical system to be at least a few kHz, and to mount the system on a well-damped platform. PieZo-electric Transducers (PZTs) are now widely used to provide the scanning motions (see Chapter 15).

The STM and the Atomic Force Microscopes (AFM) are currently the leading contenders in the field, and STMs are now being made commercially by about a dozen companies. Reviews of these and related microscopes have been published by Wickramasinghe and Pool.[86,87] In the STM, a fine tungsten tip is brought to within about a nanometre to a conducting sample and a small voltage is applied between them. The gap separation is so small that electrons from the tip can tunnel from the end of the tip to the nearest atom on the sample surface and thus generate a current. Typically, the tunnelling current of about 1 nA decreases approximately to one-tenth of its value for every 0.1 nm increase in gap separation. The tunnelling current is compared with a reference current and the error signal so generated is applied to a z-control piezo which moves the tip up or down to maintain a constant tunnelling current as the tip is rastered across the sample to record an image. The variation in voltage of the gap control piezo is proportional to the z-motion of the tip, and this signal is used to modulate the brightness of the image. Individual atoms can be clearly resolved with this instrument. In an inhomogeneous specimen, the z-motion of the tip will depend not only on the surface topography but also on the local electronic work function of the surface, since the tunnelling current is also exponentially related to the work function. This effect can be exploited to yield spectroscopic information about the surface. The STM can be used to examine thin (a few molecular layers) non-conducting specimens attached to a conducting substrate: the tunnelling current between the tip and the substrate being modulated by the intervening layer. In the AFM, a sharp tip is attached to a very delicate cantilever and the tip is brought into "contact" with the specimen surface. Contact implies that the force exerted by the cantilever is equal to the repulsive force acting on the tip due to the overlap of the electron cloud of the last atom at the very end of the tip and the specimen atoms; the cantilever is sensitive to repulsive forces as small as 10^{-8} N. The deflection of the tip is measured as it tracks over the surface e.g. by an optical method or with an STM, and a feedback loop is used to maintain a constant deflection force and thus provides an appropriate signal for imaging purposes. Atomic resolution can also be achieved with the AFM and the technique works equally well with conducting and non-conducting specimens (see Chapter 14 for details of probe microscopy).

The van der Waals microscope, also known as the attractive mode force microscope or Laser Force Microscope (LFM), because the deflection is measured with a laser probe, is a non-contacting instrument. It is based on the AFM, but in this case a piezo is used to vibrate the cantilever at about 50 kHz. The tip is in the range of 3-20 nm above the surface, and the vibration amplitude is affected by the minute variations in the attractive van der Waals forces, having magnitudes in the region of 10^{-11} N, when the probe is scanned over the surface. Non-contacting measurements have been found to be particularly useful in the microelectronics area.

The Magnetic and Electrostatic Force Microscopes (MFM and EFM) are similar to the LFM except that the probes are magnetized or carry an electric charge, and are sensitive to magnetic or electrostatic variations in the specimens under test (e.g. magnetic recording media and dopant variations in silicon). Other microscopes include the thermal and ion-conductance scanning microscopes. The former employs a 50 nm diameter thermocouple as the sensing probe and the latter has a minute electrode used, for example, for measurements of electrical activity in living cells.

The Scanning Near field Optical Microscope (SNOM) requires special mention. Two separate light waves are produced when light is transmitted through an aperture smaller than the wavelength of light. One is a spherically diverging one and the other is the evanescent wave. The two waves interfere destructively everywhere except in the immediate region of the aperture. The light beam remains collimated over a distance D, where D is the aperture diameter. If an object is placed in this region and is scanned in raster fashion, an image can be built up either using light reflected back through the aperture or by collecting the light transmitted through the specimen. The resolution achieved in such a microscope is approximately equal to D and resolutions of 20 nm can be achieved routinely.[88]

8.7.2 Scanning Tunnelling Engineering

The techniques described in the previous section are observational techniques that have produced an unprecedented step forward in our ability to observe materials on an atomic scale and even to make in situ observations, atom-by-atom, of phenomena such as corrosion and deposition.

An important broadening of STM technology is emerging through the realization that the tip can be employed for purposes other than microscopy, and that the tip may be used to modify the surface of the sample in a number of ways.

The STM tip has demonstrated its capability of drawing fine lines, which exhibit nanometre-sized structure, and hence may provide a new tool for nanometre lithography.[89] The mode of action was not properly understood, but it was suspected that under the influence of the tip a conducting carbon line had been drawn as the result of polymerizing a hydrocarbon film, the process being assisted by the catalytic activity of the tungsten tip. By extrapolating their results the authors believed that it would be possible to deposit fine conducting lines on an insulating film. The tip would operate in a gaseous environment which contained the metal atoms in such a form that they could either be pre-absorbed on the film and then be liberated from their ligands or that they would form free radicals at the location of the tip and be transferred to the film by appropriate adjustment of the tip voltage.

The modification of a surface on an even finer scale has been reported by Becker *et al.*[90] They employed an STM to examine a germanium surface, and by changing the tip from its normal STM working voltage of 1 to 4 V, produced a feature which had a diameter of 0.8 nm and which protruded about 0.1 nm from the previously atomically smooth surface (i.e. one in which the surface undulations have an amplitude of about 0.015 nm). Again the "writing" mechanism was not yet understood, but possible explanations were either that matter was transferred from the tip or that the atoms on the surface were reconfigured under the action of the high field. The former explanation was favoured because the process could be repeated only if the tip were "recharged" by bringing it into contact with a distant region of the germanium prior to writing.

Some other successful attempts at surface modification have also been reported,[91,92,93] as follows. Nanometre-scale lithography has been demonstrated based on chemical transformations initiated by the action of the tip, dimples 1 nm in size and larger have been produced by touching the surface with the tip, hillocks 20 nm in diameter and 2 nm high were made under the appropriate electrical conditions by transfer of material deposited on the tip and surfaces have been disrupted on an atomic scale by the electrostatic forces when the tip voltage exceeded a critical value. Foster *et al.* "pinned" a complex organic molecule to a graphite base by raising the tip voltage from its normal value of 30 mV to a few volts.[94] In a subsequent scan, another high voltage pulse disrupted the molecule leaving a benzene ring pinned to the surface.

Gomer has made a theoretical study of some of the mechanisms which could account for the sudden transfer of an atom from or to the tip. He concluded that thermal desorption was a possible mechanism and could be enhanced by the electric field.[95] Direct, unactivated transfer by atom tunnelling would also be possible for small gaps below 0.02 nm. In this case it was envisaged that the bulk surfaces may be separated by 0.2 to 0.4 nm, but that they were in quasi-contact via an adsorbed or otherwise protruding atom.

The significance of the experimental and theoretical work is that they demonstrate the possibility of modifying a surface at the atomic level, in a controlled manner and in any selected location. A foreseeable short-term goal is very high density information storage, which can be both read and written by the same device working in different modes.

By carefully adjusting the position and voltage of the tip Eigler *et al.* were able to pull individual atoms along a surface to form a predetermined pattern.[96] The first steps in atomic-scale writing and reading have already been taken, and if the technology can be fully implemented this would lead to the development of computers of unprecedented power and speed. Scanning tunnelling engineering would be eminently suitable for miniature wire manufacture and bonding, ultrafine component trimming and circuit repair. Enough has been accomplished even now to indicate that it is not inconceivable that experiments in quantum chemistry and molecular biology could be carried out by physical transfer of atoms, molecules and enzymes from the tip to the substrate.

REFERENCES

1. Taniguchi, N. (1974) On the basic concept of nanotechnology. *Proc. Int. Conf. Prod. Eng.* Tokyo Part 2, pp 18-23 Tokyo: JSPE
2. Woodhead, J.L. and Segal, D.L. (1984) Sol-gel processing. *Chemistry in Britain*, **20**, 310-313
3. Barringer, E.A. and Bowen, H.K. (1982) Formation, packing and sintering of monodispersed TiO_2 powders. *J. Am. Ceram. Soc.*, **65**, 199C-201C
4. Ramsey, J.D.G. and Avery, R.G. (1974) Ultrafine oxide powders prepared by electron beam evaporation. *J. Mat. Sci.*, **9**, 1681-1688
5. Fedorov, V.B., Khakimova, D.K., Petrunichev, V.A., Demidova, I.N., Salieva, O.G. and Kalita, I.V. (1981) Production and some structural characteristics of ultrafine systems. *Sov. Powder Metall. Met. Ceram.*, **20**, 601-604
6. Thompson, R. (1980) *Speciality Inorganic Chemicals*. London: Royal Soc. Chem.
7. Taniguchi, N. (1983) Current status in, and future trends of, ultraprecision machining and ultrafine materials processing. *Ann. CIRP*, **32**(2), 573-582
8. Franks, A. (1984) X-ray optics - a challenge to precision engineering. In *Proceedings of Int. Symp. for Quality Control in Prod.* (Tokyo: 1985), pp.8-17. Tokyo: Jap. Soc. Prec. Eng.
9. Butler, D.W. (1973) A stereo electron microscope technique for microtopographic measurements. *Micron*, **4**, 410-424
10. Dragoset, R.A., Young, R.D., Layer, H.P., Mielczarek, S.R., Teague, E.C. and Celotta, R.J. (1986) Scanning tunnelling microscopy applied to optical surfaces. *Opt. Lett*, **11**, 560-562
11. Stedman, M. (1987) Mapping the performance of surface-measuring instruments. *Proc. SPIE*, **803**, 138-142
12. Donaldson, R.R. and Patterson, S.R. (1983) Design and construction of a large vertical axis diamond turning machine. *Proc. SPIE*, **433**, 62-67
13. McCue, H.K. (1983) The motion control system for the large optics diamond turning machine (LODTM). *Proc. SPIE*, **433**, 68-75
14. Kobayashi, A. (1983) Recent development of ultra-precision diamond cutting machines in Japan. *Bull. Jap. Soc. Prec. Eng.*, **17**, 73-80
15. Wills-Moren, W.J., Modjarrad, H. and Read, R.F.J. (1982) Some aspects of the design and development of a large high precision CNC diamond turning machine. *Ann. CIRP*, **31**, 409-414
16. Yoshioka, J., Hashimoto, F., Miyashita, M., Kanai, A., Abo, T. and Daito, M. (1985) Ultraprecision grinding technology for brittle materials: application to surface and centerless grinding processes. In *Proceedings of Milton C Shaw Grinding Symp.* (Florida, 1985), pp. 209-227. New York: ASME
17. Lindsey, K. (1986) The assessment of ultra smooth substrates and overcoatings. *Vacuum*, **25**, 499-500
18. Lindsey, K., McCombie, A., Paul, D., Stanley, V. and Vickery, A. (1982) A machine for figuring long X-ray optical elements. In *Proceedings of RAL Symp. New Tech. in X-ray and UV Opt.*, edited by B.J. Kent and B.E. Patchett, pp. 75-83.
19. Franks, A. (1975) Materials problems in the production of high quality optical surfaces. *Mat. Sci. & Eng.*, **19**, 169-183

20. Gormley, J.V., Manfra, M.J. and Calawa, A.R. (1981) Hydroplane polishing of semiconductor crystals. *Rev. Sci. Instr.*, **52**, 1256-1259
21. Bennett, J.M., Shaffer, J.J., Shibano, Y. and Namba, Y. (1987) Float polishing of optical materials. *Appl. Opt.*, **26**, 696-703
22. Lindsey, K. (1984) The attainable limits in X-ray mirror fabrication. *Nucl. Instr. Meth. Phys. Res.*, **221**, 14-19
23. Tsuwa, H., Ikawa, N., Mori, Y. and Sugiyama, K. (1979) Numerically controlled elastic emission machining. *Ann. CIRP*, **28**, 193-197
24. Hongo, T., Mori, Y., Higashi, Y., Sugiyama, K., Yamauchi, K., Nishikawa, K. and Sakai, K. (1984) Development of high accuracy profile measuring system for focusing mirror of SOR. In *Proceedings of Symp. Quality Control in Prod.* (Tokyo, 1984), pp. 152-157. Tokyo: Jap. Soc. Prec. Eng.
25. Narodny, L.H. and Tarasevich, M. (1967) Paraboloid figured by ion bombardment. *Appl. Opt.*, **6**, 2010-?
26. Franks, A., Lindsey, K., Bennett, J.M., Speer, R.J., Turner, D. and Hunt, D.J. (1975) The theory, manufacture, structure and performance of N.P.L. X-ray gratings. *Phil. Trans. Roy. Soc. A*, **277**, 503-543
27. Franks, J. (1978) Ion beam technology applied to electron microscopy. *Adv. Electr. Electron Phys.*, **47**, 1-50
28. Taniguchi, N. (1985) Atomic bit machining by energy beam processes. *Prec. Eng.*, **7**, 145-155
29. Dietrich, H.P., Lanz, M. and Moore, D.F. (1984) Ion beam machining of very sharp points *IBM Tech. Disclosure Bull.*, **27**, 3039-3040
30. Motohiro, T. and Taga, Y. (1987) Characteristic erosion of silica by oblique argon ion bombardment. *Thin Solid Films*, **147**, 153-165
31. Watkins, R.E.J., Rockett, P., Thoms, S., Clampitt, R. and Syms, R. (1986) Focused ion beam milling. *Vacuum*, **36**, 961-967
32. Gosch, J. (1987) West Germany grabs the lead in X-ray lithography. *Electronics*, **60**(2), 78-80
33. Spears, D.L. and Smith, H.I. (1972) High resolution pattern replication using soft X-rays. *Electron. Lett.*, **8**, 102-104
34. Spiller, E. and Feder, R. (1977) X-ray lithography. *Topics in App. Phys. (X-ray optics)*, **22**, pp. 35-92 Berlin: Springer
35. Heuberger, A. (1986) X-ray lithography. *Solid State Tech.*, **29**(2), 93-101
36. Smith, H.I. (1986) A review of sub-micron lithography. *Superlattices and Microstructures*, **2**, 129-142
37. Broers, A.N. (1985) High resolution electron beam fabrication. *J. Micros.*, **139**, 139-152
38. Smith, C.G. and Ahmed, H. (1987) Fabrication and phonon transport studies in nanometer scale free-standing wires. *J.Vac. Sci. Technol.*, **B5**, 314-317
39. Mehregany, M., Gabriel, K.J. and Trimmer, W.S.N. (1988) Integrated fabrication of polysilicon mechanisms. *IEEE Trans. Electron Devices*, **35**, 719-723
40. Fan, L.S., Tai, Y.C. and Muller, R.S. (1988) IC-processed electrostatic micro-motors. In *Proceedings of IEEE Int. Conf. Electron Devices*, pp. 666-669
41. Humphreys, C.J., Salisbury, I.G., Berger, S.D., Timsit, R.J. and Mochel, M.E. (1985) Nanometre scale electron beam lithography. In Electron microscopy and analysis, *Inst. Phys. Conf. Ser.*, **78**, , 1-6
42. Stringfellow, G.B. (1982) Epitaxy. *Rep. Prog. Phys.*, **45**, 469-525

43. McKeown, P.A. (1986) High precision manufacturing and the British economy. *Proc. Inst. Mech. Eng.*, **200**, 1-19

44. Puttick, K.E. (1980) The correlation of fracture transitions. *J. Phys. D: Appl. Phys.*, **13**, 2249-2262

45. Puttick, K.E. and Franks, A. (1990) The physics of ductile-brittle machining transitions: single-point theory and experiment. *Jap. Soc. Prec. Eng.*, **56**, 788-792

46. Arnold, J.B., Sladkey, R.E., Steger, P.J. and Woodall, N.D. (1977) Machining nonconventional-shaped optics. *Opt. Eng.*, **16**, 347-354

47. Andrea, J. (1986) Mass-production of diffraction limited replicated objective lenses for compact disk players. *Proc. SPIE*, **645**, 45-48

48. Bennett, H.E., Guenther, A.H., Milam, D. and Newman, B.E. (1984) Laser induced damage in optics materials: fourteenth ASTM symposium. *Appl. Opt.*, **23**, 3782-3795

49. Thomas, N.L. (1978) Low-scatter, low-loss mirrors for laser gyros. *Proc. SPIE*, **157**, 41-48

50. Franks, A. (1977) X-ray optics. *Sci. Prog.*, **64**, 371-422

51. Barbee, T. (1985) Multilayers for X-ray optics. *Proc. SPIE*, **563**, 2-28

52. Michette, A.G. (1986) *Optical systems for soft X-rays*. New York: Plenum Press

53. Kelly, M.J. (1986) Nanometre physics and microelectronics. *Phys. Bull.*, **37**, 67-69

54. Kelly, M.J., Davies, R.A., Long, A.P., Couch, N.R. and Kerr, T.M. (1986) Quantum semiconductor devices with microwave applications. *GEC J. of Res.*, **4**, 157-162

55. Sugano, T., Okabe, Y., Tamura, H., Miyake, H. and Takatsu, M. (1984) Bridge type Josephson junctions as high speed digital devices. In *Proceedings of Int. Symp. on Nanometre Structure Electronics Toyonaka, Japan Microelectr. Eng.*, **2**, 175-182

56. Patrick, W., Mackie, W.S., Beaumont, S.R., Wilkinson, C.D.W. and Oxley, C.H. (1985) Very short gate-length GaAs MESFETs. *IEEE Electr. Dev. Lett.*, **6**, 471-472

57. Franks, A. and Gale, B. (1984) Grazing incidence optics for X-ray microscopy. In *X-ray microscopy*, edited by G. Schmahl and D. Rudolph, *Springer Ser. Opt. Sci*, **43**, pp. 129-138

58. Franks, A. and Gale, B. (1985) The development of single and multilayered Wolter X-ray microscopes. *Proc. SPIE*, **563**, 81-89

59. Franks, A. (1986) The metrology of grazing incidence optics at the National Physical Laboratory. *Proc. SPIE*, **640**, 170-174

60. Lindsey, K., Smith, S.T. and Robbie, C.J. (1988) Subnanometre surface texture and profile measurement with nanosurf 2. *Ann. CIRP*, **37**, 519-522

61. Downs, M.J., McGivern, W.H. and Ferguson, H.J. (1985) Optical system for measuring the profiles of super-smooth surfaces. *Prec. Eng.*, **7**, 211-215

62. Stedman, M. and Stanley, V.W. (1979) Machine for the rapid and accurate measurement of profile. *Proc. SPIE*, **163**, 99-102

63. Ennos, A.E. and Virdee, M.S. (1983) Precision measurement of surface form by laser autocollimation *Proc. SPIE*, **398**, 252-257

64. Stedman, M. (1986) The metrology of X-ray optical components: mapping the limits of measuring instruments. In *Proceedings of SERC/RAL Workshop on Advanced Technology Reflectors for Space Instrumentation*, pp. 254-258

65. Bowen, D.K., Chetwynd, D.G. and Davies, S.T. (1985) Calibration of surface roughness transducers at Angstrom levels, using X-ray interferometry. *Proc. SPIE*, **563**, 412-419

66. Stedman, M. (1986) The precision measurement of roundness. *Wear*, **109**, 367-373

67. Franks, A. (1987) Nanotechnology opportunities. *J. Phys. E: Sci. Instrum.*, **20**, 237
68. Bennett, C.H. (1982) Thermodynamics of computation - a review. *Int. J. Theor. Phys.*, **21**, 905-940
69. Feynman, R.P. (1985) Quantum mechanical computing. *Opt. News*, **11**(2), 11-20
70. Carter, F.L. (1983) Molecular level fabrication techniques and molecular electronic devices. *J. Vac. Sci. Technol.*, **B1**, 959-968
71. Carter, F.L. (Editor) (1982) *Molecular Electronic Devices*. New York: Dekker
72. Carter, F.L. (Editor) (1986) *Molecular Electronic Devices II*. New York: Dekker
73. Stu, H.H. (Editor) (1986) Optical and hybrid computing. *Proc. SPIE*, ?, 634-?
74. Roberts, G.G. (1985) Biochip research needs more money. *New Sci.*, **108**, 24-?
75. Haddon, R.C. and Lamola, A.A. (1985) The molecular electronic device and the biochip computer: present status. *Proc. Nat. Acad. Sci.*, **82**, 1874-1878
76. Blodgett, K.B. (1935) Films built by depositing monomolecular layers on a solid surface. *J. Am. Chem. Soc.*, **57**, 1007-1022
77. Henke, B.L. (1964) X-ray fluorescence analysis for sodium, fluorine, oxygen, nitrogen, carbon and boron. *Adv. X-ray Anal.*, **7**, 460-488. New York: Plenum Press
78. Henke, B.L. (1965) Some notes on ultrasoft X-ray fluorescence analysis - 10 to 100 Å Region. *Adv. X-ray Anal*, **8**, 269-284. New York: Plenum Press
79. Roberts, G.G. (1985) An applied science perspective of Langmuir-Blodgett films. *Adv. in Phys.*, **34**, 475-512
80. Broers, A.N. and Pomerantz, M. (1983) Rapid writing of fine lines in Langmuir-Blodgett films using electron beams. *Thin Solid Films*, **99**, 323-329
81. Roberts, G.G., Petty, M.C. and Dharmadasa, I.M. (1981) Photovoltaic properties of cadmium-telluride/Langmuir-film solar cells. *Proc. IEE Part I*, **128**, 197-201
82. Marshman, C.E. (1986) Electronic biosensors. *Phys. Bull.*, **37**, 296-299
83. Binning, G. and Rohrer, H. (1985) The scanning tunnelling microscope. *Sci. Amer.*, **253**, 40-46
84. Binnig, G. and Rohrer, H. (1986) Scanning tunnelling microscopy. *IBM J. Res. Develop.*, **30**, 355-369
85. O'Keefe, J.A. (1956) Resolving power of visible light, *JOSA*, **46**, 359
86. Wickramasinghe, H.K. (1990) Scanning probe microscopy: current status and future trends. *J. Vac. Sci. Technol.*, **A8**, 363-368
87. Pool, R. (1990) The children of the STM. *Science*, **247**, 634-636
88. Pohl, D.W., Fischer, U.C. and Durig, U.T. (1988) Scanning near-field optical microscopy (SNOM) *J. Micros.*, **152**, 853-861
89. Ringger, M., Hidber, H.R., Schlogl, R., Oelhafen, P. and Guntherodt, H.J. (1985) Nanometer lithography with the scanning tunnelling microscope. *Appl. Phys. Lett.*, **46**, 832-834
90. Becker, R.S., Golovchenko, J.A. and Swartzentruber, B.S. (1987) Atomic-scale surface modifications using a tunnelling microscope. *Nature*, **325**, 419-421
91. McCord, M.A. and Pease, R.F.W. (1986) Lithography with the scanning tunnelling microscope. *J. Vac. Sci. Technol.*, **B4**, 86-88
92. Abraham, D.W., Mamin, H.J., Ganz, E. and Clarke, J. (1986) Surface modification with the scanning tunnelling microscope. *IBM J. Res. Develop.*, **30**, 492-499
93. Ringger, M., Corb, B.W., Hidber, H.R., Schlogl, R., Wiesendanger, R., Stemmer, A., Rosenthaler, L., Brunner, A.J., Oelhafen, P.C. and Guntherodt, H.J. (1986) STM activity at the University of Basel. *IBM J. Res. Develop.*, **30**, 500-508

94. Foster, J.S., Frommer, J.E. and Arnett, P.C. (1988) Molecular manipulation using a tunnelling microscope. *Nature*, **331**, 324-326

95. Gomer, R. (1986) Possible mechanisms of atom transfer in scanning tunnelling microscopy. *IBM J. Res. Develop.*, **30**, 428-430

96. Eigler, D.M. and Schweizer, E.K. (1990) Positioning single atoms with a scanning tunnelling microscope. *Nature*, **344**, 524-526

Chapter 9

USE OF ENERGY BEAMS FOR ULTRA-HIGH PRECISION PROCESSING OF MATERIALS

SAMUEL T. DAVIES

Developments in ultra-high precision (UHP) processing of materials has meant that, frequently, one or more of the critical dimensions involved is of nanometre order. Traditional tools for materials processing are mostly inadequate for these purposes and, increasingly, the tools used for such nanotechnology applications are based on energy beams of various types. This chapter gives an outline of energy beam processing, illustrated with examples of typical and emerging areas of application, and discusses the potential and implications of energy beam techniques for nanotechnology.

9.1 INTRODUCTION

The last decade has seen a rapid growth in the application of energy beams to the processing of engineering materials, mainly as a means of progressively increasing achievable levels of accuracy and precision, or occasionally speed or ease of processing.[1,2] In this context, the term "energy beam"[3] is taken to mean a controllable, directional, flux of photons or atomic or molecular particles having the ability to modify the physical state of the surface on which they impinge. Energetic photon beams, electron beams, inert ion beams, chemically reactive ion beams, cluster ion beams, plasma beams, neutral atom beams and molecular beams are all included in this category. In contrast, processing methods that rely on the mechanical properties of fine suspensions of microscopic particles, for example, in order to bring about material modification or removal are excluded. Such processing methods are metallurgical polishing, progressive mechanical and chemical polishing, and so-called elastic emission machining. Also material processing by purely wet chemical etching methods is excluded.

It is, of course, outside the scope of this chapter to give a comprehensive review of all the energy beam applications alluded to above. Rather, selected UHP processes utilizing photon, electron and ion beams are emphasized; these being indicative of current research interests and future possibilities.

9.2 PHOTON BEAM PROCESSING

Energy beam processing with photon beams relies mainly on photons with wavelengths in the 1 to 1000 nm range, or energies from approximately 10 to 10^4 eV, and utilizes both laser (coherent) and non-laser (incoherent) radiation. Motivation for on-going research has come largely from the drive to further reduce the geometrical size of structures for active microelectronic devices, but also from the need for miniaturization of features on passive devices such as diffraction gratings and zone plates for X-UV optics.

9.2.1 Incoherent Radiation

Photolithography

Photolithography remains the dominant method of ultra high resolution pattern transfer in present-day manufacture of Integrated Circuits (ICs). The technique of illuminating a photosensitive resist by means of a flood beam through a mask, followed by etching in order to fabricate circuit features with sub-micrometre resolution, is likely to remain important in the immediate future[4] in spite of competing or complementary technologies which offer superior resolution.

The move to projection printing and away from contact and proximity printing (which although offering higher resolution for a given wavelength are susceptible to mechanical damage of the photomask) has resulted in mercury arc UV "g-line" (436 nm) or "i-line" (365 nm) tools now being photolithographic workhorses. Resolutions or line widths in production using these methods are approaching 500 nm, and seem likely to be reduced to 250 nm or lower. However, the resulting process complexity and high cost mean that deep-UV tools and/or X-ray lithography will almost certainly begin to dominate in this regime.

The resolution of modern diffraction limited projection step-and-repeat systems is given approximately by :

$$r ~ 0.5 ~ \lambda / (NA)$$ (9.1)

where λ is the wavelength of the radiation and NA is the numerical aperture of the lens.

The depth of focus (d) however is also related to the numerical aperture as :

$$d ~ \pm ~ 1/2 (NA)^2$$ (9.2)

thus indicating that higher resolution cannot be obtained by increasing the numerical aperture, as the depth of focus rapidly becomes too small to be realistic for a mechanically aligned system. The only possibility, therefore, is to move to shorter wavelengths.

X-ray Lithography

Limitations on photolithography are ultimately imposed by the wavelength used, so even deep UV radiation will be limited to feature sizes of some 250 nm. X-rays wavelengths, however, can be fractions of a nanometre and are thus capable of extremely high resolution patterning for lithography. X-rays with wavelengths in the range 0.2 to 5 nm are expected[5] to be capable of achieving a resolution of ~ 100 nm in a production as opposed to a laboratory environment.

Much effort has already gone into the development of X-ray sources (plasma, laser plasma, synchrotron, compact synchrotron etc.), while stepper and resist technology have advanced to the level where they are now viable; the critical problems remaining are in mask fabrication technology and in tools for mask defect elimination. The latter is a serious problem due to the impossibility of producing totally defect free masks and due to defects introduced in-process. However recent developments in mask inspection show that 400 - 500 nm size defects can be detected with high probability and in addition focused ion beam mask repair systems (see section 9.4.2) have ample precision (in principle) to allow correction of such defects.

The masks used in X-ray lithography consist of an X-ray transparent membrane supporting a patterned X-ray absorbing film of thin metal. The thickness of the film is determined by the X-ray wavelength being used, the absorption coefficient of the metal and the contrast necessary in the resist to form an image. Typically, gold or tungsten of thickness 500 nm is used as an absorber. The masks are usually generated by electron beam lithography and plasma etching techniques. The supporting membrane must be dimensionally extremely stable, rugged and ideally transparent to visible light for optical alignment. Polyimide coated boron nitride of total thickness 12 μm has been successfully used for this purpose. The mask is placed in close proximity to a substrate coated with a sensitive resist and illuminated with a distant point source of X-rays which projects the shadow of the X-ray absorber onto the resist. A proximity gap of typically 50 μm between mask and wafer protects the mask from mechanical damage. The method is therefore similar to optical proximity projection, as shown in Figure 9.1.

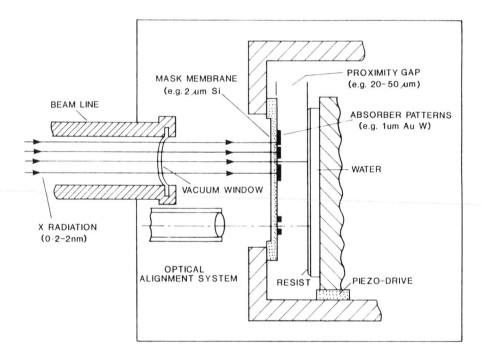

Figure 9.1 X-ray lithography using synchrotron radiation (after Heuberger 1988).

In practice, in addition to diffraction effects, resolution will be further limited by the geometric effect of finite source size resulting in penumbral blurring (δ):

$$\delta = (\phi g) / L \qquad (9.3)$$

with ϕ the source size, g the proximity gap and L the distance from source to mask. For synchrotron sources, typical values of ϕ (5 mm), g (50 μm) and L (50 m) mean that this is negligible, but for laboratory sources with typically values of 3 mm, 50 μm and 50 cm respectively, the penumbral blur can be ~ 0.2 μm. The effect of beam divergence also produces a similar geometrical contribution in the form of a lateral magnification or run-out error.

9.2.2 Coherent Radiation

Lithography

Lithography using excimer lasers has been the subject of much recent development work. The excimer (**excited dimer**) laser is a pulsed gas laser capable of emitting radiation in the deep UV. A suitable excited dimer is that of a noble and halogen gas, existing only in the excited state, which on dissociation emits UV radiation. Examples of excimer lasers together with wavelengths of emission are given in Table 9.1.

Table 9.1 Types of excimer lasers.

Excimer	Laser Wavelength
Argon Fluoride (ArF)	193 nm
Krypton Fluoride (KrF)	248 nm
Xenon Chloride (XeCl)	308 nm
Xenon Fluoride (XeF)	351 nm

Typical UV sources provide only a few mW of power, whereas excimer lasers can provide peak power in excess of 10 mW, with pulse widths of some 10 ns. Areas of 1 cm by 1 cm can be illuminated with excellent uniformity (~ 5%) thus making the excimer laser ideal for step-and-repeat systems.

Micromachining

Q-switched Nd:YAG lasers are widely used for micromachining operations in semiconductor device processing. The ability to deliver high power densities to localized areas at high speeds is an important feature. Trimming of thin film resistors on mixed signal ICs is a typical application; for example, trimming of thin film resistors to linearize ADCs and DACs. More recently laser programming of Application Specific Integrated Circuits (ASICs) and laser activation of redundancy circuitry in high capacity memory devices has been developed. In the latter, the laser beam vaporizes a conductive link to disconnect a faulty row or column of memory cells.

Also the XeCl excimer laser operating at 308 nm has proved to be suitable for a variety of micromachining operations. Advantages include long gas lifetime and relatively long lifetime for the beam delivery optics compared with shorter wavelengths of operation. Most materials of interest exhibit strong absorption (50 to 90%) at around 300 nm, leading to high efficiency. Material removal takes place by a process of photoablation whereby the absorbed radiation excites target molecules and breaks chemical bonds. The ablated material is ejected at high velocity, rapidly cools, recombines or undergoes reaction with ambient chemical species. A disadvantage is that debris may fall back and be redeposited close to the point of beam incidence, but this can sometimes be removed by application of a suitable solvent. High power densities (> 10 mW cm^{-2}), short pulse durations (~ tens of nanoseconds) and controllable pulse frequency help give adequate control over material removal. Automated excimer laser micromachining systems have been developed to cut thin (~ 100 nm) metal interconnects on SiO_2, Si, GaAs and polymers with minimum damage to the interface regions.

Microdeposition

Deposition of material at localized sites with submicron spatial resolution can be initiated by irradiation from a focused laser beam. Material from the gas, liquid or solid phase can be deposited by "direct writing" onto a substrate, opening up a host of possibilities for microstructure and nanostructure fabrication. The process can also be initiated by focused electron or ion beams as discussed in sections 9.3.3 and 9.4.2.

For gas-phase reactants, two techniques are used for deposition: thermochemical or pyrolytic deposition and photochemical or photolytic deposition. In pyrolytic deposition, the substrate is heated by the incident beam and the gas above is decomposed to subsequently deposit on the substrate. In photolytic deposition, the gas is directly dissociated by the incident beam and deposition on the substrate takes place.

Deposition using focused laser beams has been applied to the repair of clear (i.e. absence of chrome) photomask defects by pyrolytic deposition. Clear defect repairs present a formidable challenge as an opaque material must be placed accurately over the region without contaminating adjacent areas and furthermore it must adhere to the substrate sufficiently well to prevent removal in subsequent processing. Clear photomask defect repair by laser photolysis was demonstrated by Ehrlich et al.,[6] using the second harmonic of an argon ion laser at 257 nm to decompose the organometallic molecule $Cd(CH_3)_2$ and to deposit Cd atoms onto the clear defect site. This method for various reasons however was not suitable for routine application. Clear defect repair using laser pyrolysis has also been demonstrated[7] using a visible cw argon ion laser delivering a peak 20 mW of power into a 2 μm^2 area to decompose an organometallic compound and deposit a metallic film. However, the gases used to date have invariably been metal alkyls and carbonyls which are toxic and dangerous to handle, thus resulting in a search for more suitable reactants.

9.3 ELECTRON BEAM PROCESSING

9.3.1 Lithography

Electrons can readily be focused to spots of diameter < 10 nm and can be deflected and modulated at high speeds with electrostatic or magnetic fields. This has resulted in exploitation for direct-write lithography, as the energy or dose delivered to a specific area of resist on a wafer can be precisely controlled[8]. Resist exposure can be achieved either by scanning the focused electron beam under computer control or by illuminating an area of resist through a mask with a flood beam.

In the former case the accuracy of positioning required is about 1 part in 10^6 thus placing great demands on the electron deflection system. DAC nonlinearities, errors in the positioning of the substrate, wafer bow and electromagnetic interference mean that in practice the range of deflection is limited to about 1 mm and as this is less than the field of view required to address a whole wafer, a mechanical table must move the substrate in order to expose the whole area. Alignment marks are used to ensure the relative position of beam and table are within the required accuracy either by aligning on separate registration marks for each table position or by using laser interferometers to measure the table position following electron beam registration. The beam can be either raster scanned or vector scanned to generate the required pattern. In raster scanning, the whole field of view is scanned with the beam blanked as necessary to expose the desired areas. In vector scanning, the beam is deflected as necessary to generate the required pattern. The main uses of electron beam lithography are to directly expose resist (direct-write lithography) or to create a high resolution mask which can then be used for photolithography. The major drawback in direct-write lithography is low throughput: in general it is too slow for most production manufacturing. Development and production of specialized devices is viable, however, and the latest generation e-beam direct write machines incorporate features such as variable shaped beam and high current electron optics together with sophisticated deflection control and high speed mechanical and vacuum subsystems in order to increase the speed of processing.

9.3.2 Micromachining

A remarkable use of electron beams for micromachining has been demonstrated by Humphreys.[9] Using an extremely fine focused (~ 0.5 nm) beam, holes with nanometre diameters have been drilled through thin films of materials such as alumina, lithium fluoride, various ceramics and semiconductors. In some instances round holes can be produced; in others square or triangular holes result, related to the crystallography of the material. In an impressive demonstration of the information storage possibilities, text consisting of dot matrix characters ~ 10 nm high has been drilled into an alumina substrate by this method. Such a storage density could result in the whole of the Encyclopedia Brittanica being recorded on a pinhead! The mechanism of hole formation is in itself intriguing: the details are still uncertain but simple melting, as in laser machining, is not thought to be significant. The rate of drilling may even increase with decreasing temperature. In drilling crystalline alumina for example, a cavity forms at both entry and exit surfaces and progresses inwards, whilst in amorphous alumina a cavity forms in the centre and grows outwards.

9.3.3 Microdeposition

Very high resolution structures can be fabricated on suitable substrates by the direct write technique of using a finely focused electron beam to decompose an organometallic or hydrocarbon gas.[10] Such structures have been investigated for information recording, for etch masks in reactive ion etching, for masks for light or X-rays, for nanofabrication of devices and to generate three-dimensional nanostructures. Area deposits, lines and dots have been observed, depositing tungsten or gold on solid substrates, membranes and across physical holes. Deposition rates of a few nm s^{-1} for dots and of order 100 nm/(coulomb/cm^2) for areas are observed. The technique is applicable to repair of clear defects on photolithography or X-ray lithography masks and also to the repair of clear defects in ion beam stencil masks.

9.4 ION BEAM PROCESSING

Ion beam processing of materials can be divided into non-spatially selective (using a broad or showered ion beam) or spatially selective (using a fine focused ion beam or showered beam in conjunction with a delimiting mask) methods. In Figure 9.2. areas of application are shown as regions on an ion energy versus ion current plot.

9.4.1 Showered Ion Beams

Theory

Low energy (50 to 1500 eV) ions are widely used for the purpose of Showered Ion Beam Machining or Milling (SIBM). This is due primarily to the fact that they are a means of ultra-precision material removal in the form of an universal etchant with the capability of milling essentially any material regardless of the mechanical properties. SIBM is of particular interest in processing hard, brittle materials where removal by conventional mechanical machining is invariably problematical.

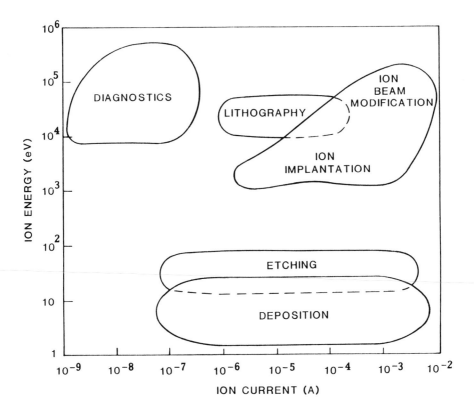

Figure 9.2 Ion beam applications in UHP processing (after Muray and Muray 1985).[11]

The most important parameter when considering material removal rates by SIBM is the sputter yield $S(\theta)$. The sputter yield has a strong dependence on incident ion direction, with most materials showing a monotonic increase in $S(\theta)$ with increasing angle of incidence (measured from the surface normal), a broad maximum in the range 40° to 60° and a steady decrease as the angle of incidence approaches 90°. The shape of the curve can be understood simply from the collisional nature of ion milling : at angles in the vicinity of 45° maximum momentum transfer occurs between incident ions and target atoms which increases the probability of ejecting a target atom.

The sputter yield is also a strong function of energy: in the range 20 to 100 eV it is small, increasing exponentially from 10^{-4} to 10^{-1} atoms/ion. From 100 eV to 500 eV it increases linearly with energy but beyond 500 eV less rapidly, reaching a maximum between 10 and 100 keV. Beyond this maximum, ion implantation rather than sputtering occurs. Hence for this reason SIBM is usually carried out in the energy interval between 100 and 1000 eV. $S(\theta)$ also depends on the type of incident ion, tending to increase with atomic weight, as would be expected from momentum transfer, but also increasing as electron shells are filled for elements in each row of the periodic table due to the rate of energy transfer to the target increasing with electronic scattering of the incident ions. As a result, inert gas ions exhibit the most favourable sputter yields which also fortuitously means that no chemical effects compete with the sputter etch process. Consequently, argon is most widely used as a working gas for SIBM; the increase in sputter yield for Kr and Xe is only marginal in the energy range of 100 to 1000 eV.

Finally, the target material itself influences the sputter yield with a dependence both on position in the periodic table and on the inverse heat of sublimation. Data on sputter yields and etch rates, at normal ion incidence and 1 mA cm^{-2} beam current density, for various materials have been compiled by Commonwealth Scientific Corporation.[12] Values for selected elements and compounds are given in Table 9.2. The range of values of etch rates quoted reflects the experimental variations observed using different equipment. For this reason, calibration for a given application is essential.

Table 9.2 Sputtering and etching rates for various materials.

Materials	Sputtering Rate (nm/min)	Etching Rate (nm/min)
Metals:		
Aluminium	64	73
Gold	155	100-170
Tungsten	34	19-38
Titanium	34	15-38
Molybdenum	47	22-54
Copper	88	45-110
Dielectrics:		
SiO_2	40	26-40
Al_2O_3	13	80-130
$LiNbO_3$	40	39-42
Semiconductors:		
Si	34	20-38
Ge	92	90-100
GaAs (110)	150	63-160
Resists:		
PMMA	•	55-58
PSG	•	16-30
Riston	•	24-26

The etching rate or material removal rate $V(\theta)$ can be calculated from the sputter yield rate as:

$$V(\theta) = 9.6 \times 10^{24} (S(\theta)/n)\cos\theta \quad (nm/min)/(mA/cm^{-2}) \tag{9.4}$$

where n is the atomic density of the target in atoms cm^{-3} and θ is the angle between the surface normal and the direction of the incident beam. Figure 9.3 shows $V(\theta)$ plotted as a function of θ for various materials.

The depth of the damaged layer produced by ion bombardment is also frequently of interest. A common assumption is that the depth of damage corresponds to the depth of ion penetration, but total dose can also be significant. A good fit to such data as is available is given by :

$$I = 0.1 \ W_t \ E^{2/3}/\rho_t \ (\ Z_i^{1/4} + Z_t^{1/4}) \tag{9.5}$$

where I is the depth of damage in nm, W_t and ρ_t are the atomic weight and density of the target element respectively, E is the ion energy in eV, and Z_i and Z_t are the atomic numbers of ion and target materials.

Instrumentation for SIBM

An instrument for SIBM, using a Kaufman type ion source[14] has been described by Davies and Bowen.[15] Since its introduction, the permanent magnet Kaufman source has been widely used for dry etching applications on account of the control over ion energy and current density and its capability to produce an uncontaminated and relatively collimated ion beam.[16] A schematic of the instrument is shown in Figure 9.4.

In normal operation the workchamber is evacuated to 10^{-5} torr and backfilled with argon at a pressure of approximately 5×10^{-4} torr and gas flow rate of 5 cm^3 min^{-1}. Electrons emitted from a hot filament initiate a plasma discharge when a voltage just exceeding the first ionisation potential of argon is applied between cathode and anode. An axial magnetic field is normally applied to increase the ionisation probability. The ion beam is then extracted from the plasma through a set of dual graphite grids. Graphite is chosen for its low sputter yield and low thermal expansion. The positively charged inner (screen) grid contains the plasma in the discharge chamber and the negatively charged outer (accelerator) grid extracts and accelerates the ion beam. The source provides a uniform beam of 3 cm diameter with energy and current density variable up to 1500 eV and 25 mA cm^{-2}, respectively.

Applications

Applications of SIBM are diverse and too numerous to discuss in detail within the scope of this chapter. They have included pattern transfer for semiconductor device fabrication (see, for example, Lee 1984 and references therein),[17] fabrication of grating structures for surface acoustic wave devices and integrated optical devices,[18] fabrication of magnetic bubble memories,[19] fabrication of holographic gratings,[20] surface texturing for optical and solar cell applications,[21] micromachining of diamond;[22-24] patterning, thinning and polishing of high-T_c superconducting thin films,[25] surface smoothing for X-ray multilayer substrates[26] and figuring of large optics.[27]

SIBM has been used for precision 3-D profiling in the processing of diamonds used for contact surface metrology styli. Algorithms used to extract surface parameters from instruments such as the Rank Taylor Hobson Form Talysurf assume the tip of the stylus to be geometrically perfect. However in practice this is rarely the case. Ideally, the industry standard 2 μm diamond stylus has a 2 μm radius hemispherical tip with radius tolerance not exceeding 500 nm and form error not exceeding 100 nm. These specifications cannot be met with conventional lapping and polishing methods. Figure 9.5a shows a typical stylus prepared in this way, while Figure 9.5b shows the results of controlled SIBM to produce the required profile.

By exploiting the characteristics of the sputter yield curve for diamond, it is possible to control the tip geometry in a variety of ways. For example, Figure 9.6a shows the production of an asymmetrical stylus, while Figure 9.6b shows the production of a finely pointed tip (estimated radius 10 to 15 nm).

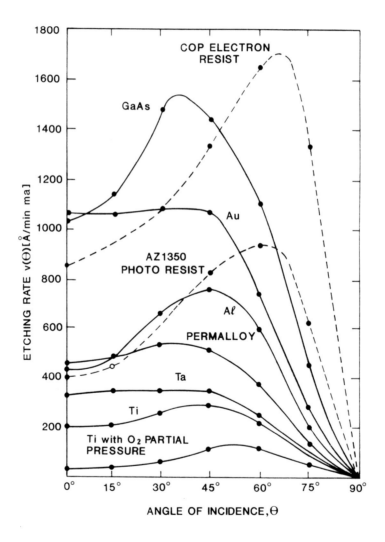

Figure 9.3 Ion beam etching rate as a function of angle (after Somekh 1976)[13].

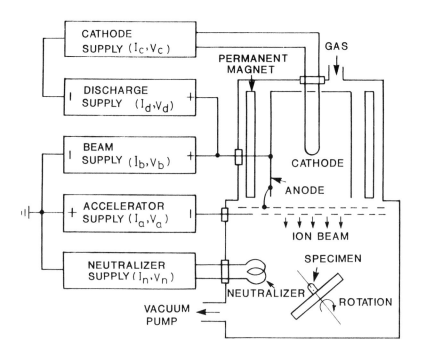

Figure 9.4 Schematic of instrumentation for SIBM.

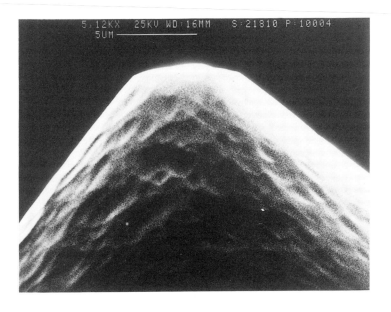

Figure 9.5a Diamond stylus finished by mechanical lapping and polishing.

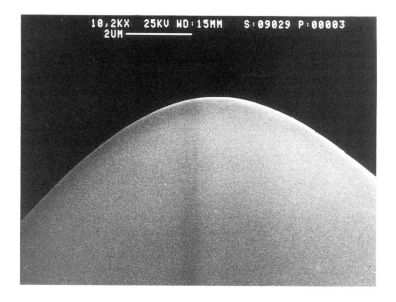

Figure 9.5b Diamond stylus finished by SIBM. Two axis motion was used during ion bombardment.

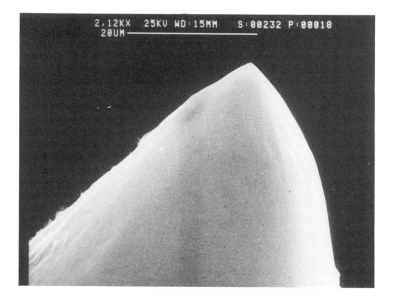

Figure 9.6a Diamond stylus with asymmetric tip produced by control of tilt angle.

Figure 9.6b Diamond stylus with extremely sharp tip.

9.4.2 Focused Ion Beams

Ion beam columns based on liquid metal ion sources have lead directly to the development of Focused Ion Beam (FIB) technology capable of producing ion probe diameters of < 10 nm. Such technology has opened up a host of exciting new possibilities in ultra-precision materials processing. Applications currently include direct-write lithography and ion implantation, microfabrication, microscopy and microanalysis.[28,29]

Liquid Metal Ion Source (LMIS)

The introduction of the LMIS[30,31] brought about a surge of interest in focused ion beams. The implications for the applications mentioned above were soon recognized and prototype FIB systems were developed by several groups (for example, Seliger *et al.*1979;[32] Ishitani, Tamura and Todokoro 1982).[33]

A typical implementation of the LMIS consists of a fine tungsten capillary of diameter 20 - 200 μm, or of a capillary with a tungsten needle protruding through, with an extraction electrode having a circular aperture placed in front of and axial with the arrangement.[34,35] The capillary is filled with a suitable metal (usually gallium or indium, but operation with a wide range of other elemental or compound metals has been reported) and a voltage generally between 5 - 10 kV applied between capillary and extractor electrode. On heating the metal above its melting point the liquid metal at the capillary tip is pulled by balance of electrostatic and surface tension forces into a cone having apex of extremely small radius. The theoretical solution for the static case of a conducting fluid situated near a charged plane[36] predicts the formation of a cone (Taylor cone) with half-angle 49.3°, provided that the applied voltage exceeds a critical voltage V_c, where V_c is given[37] by:

$$V_c = 1.432 \times 10^3 \ \gamma^{1/2} \ R_0^{1/2} \tag{9.6}$$

γ is the surface tension of the liquid (dynes cm^{-1}) and R_o is the distance from capillary tip to extractor (cm). V_c is then the voltage for onset of ion emission in volts.

The apex of the Taylor cone is some 10 to 50 nm resulting in electric fields of order 10 V nm^{-1} which are high enough to cause field evaporation of the liquid metal and produce ion currents in the range 1 to 100 μA. Due to the extremely small source size the brightness β and angular ion current density $dI/d\Omega$ are both large and of the order of 10^6 A cm^{-2} sr^{-1} and 10 - 50 μA sr^{-1}, respectively. Coulomb interaction of emitted ions causes an energy spread of 5 to 10 eV.

As the source emits a beam of high divergence, suitable ion optics following the extraction aperture must be provided to produce a fine focused beam. The small source size means that ultimate probe diameter is limited by lens aberrations and for spot sizes < 10 μm chromatic aberration is the dominant factor. An aperture defines the lens acceptance angle which for submicrometre diameter beams must be around 10^{-3} rad. For performance limited by chromatic aberration the focused spot diameter (d) is given[38] by:

$$d ~ \sim ~ C_c ~ \alpha ~ \Delta E / E \qquad (9.7)$$

where,
C_c is the chromatic aberration coefficient of the lens;
α is the lens acceptance angle;
ΔE is the ion energy spread
and E is the ion energy.

The total current (I) in the focused spot is given by:

$$I ~ = ~ \alpha^2 . dI / d\Omega \qquad (9.8)$$

and the current density (J) given by:

$$J ~ = ~ (4M^2 / E) . ~ dI / d\Omega ~ (E / C_c)^2 \qquad (A~m^{-2}) \qquad (9.9)$$

where M is the lens magnification.

Instrumentation for FIBM

Most instrumentation for FIBM comprises as a bare minimum a suitable FIB column, a means for rastering the ion beam, a means of detecting sample current or secondary ion-induced particles and a method of forming an image of the sample. Figure 9.7. shows a schematic of the FIBM instrumentation developed at the University of Warwick, based around a 10 keV ion microprobe column.

The liquid metal anode is held at a positive voltage in the range 8.5 to 10.0 keV relative to an earthed extraction electrode E. Regulation of emission current is possible by varying the potential of grid electrode G. The extracted beam traverses a drift tube DT and is collimated by an aperture mounted upstream of a beam blanking electrode. Focusing is achieved by a pentode lens stack L, consisting of two identical three-element, voltage-symmetric lenses placed back-to-back. The ion optics focuses the beam to a micrometre sized spot at a working distance of between 30 and 40 mm from the final lens element. Beam deflector plates beneath the lens stack allow a maximum field of view of 4 mm by 4 mm to be scanned.

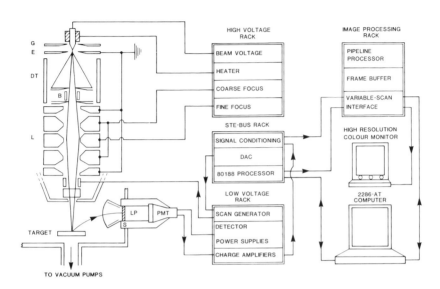

Figure 9.7 Schematic of instrumentation for FIBM.

Beam operating voltage and current and also the lens focusing voltages are controlled manually. Beam scanning is controlled by software from a PC-AT computer via a STE-bus sub-system having a dedicated 80188 processor, 4-channel DAC and analogue signal conditioning electronics. Parameters downloaded from the host machine allow raster or vector pattern generation. Programmable raster scanning enables the required area of the sample to be imaged and vector writing allows any required micromachining operation to be carried out.

Imaging of the workpiece *in situ* may be done by collecting either secondary electrons or secondary ions by means of a scintillation detector S. The YAG scintillator is aluminized on the front face and normally biased to a voltage of $^+5$ kV. When detecting ions, a cone surrounding the scintillator is biased at a voltage between $^-5$ and $^-$10 kV. Ions striking the cone then produce electrons which are detected by the scintillator. A screen placed in front of the cone can be biased to shield the ion beam from the electric fields of the cone and scintillator and thereby avoid unwanted deflections. A photomultiplier tube PMT is coupled by a light pipe LP to the scintillator and the PMT signal fed to a charge amplifier whose output is buffered and clamped to standard video levels before being digitized by the image processor system.

The image processing system comprises a variable-scan interface, frame buffer and pipeline processor. The variable scan interface allows capture of images from an analogue or digital video source and is provided with line and frame synchronization and pixel clock signals from the STE-bus controller. The variable-scan interface contains two on-board memories which buffer the digitized images in order to match the timing

173

of the asynchronous video source to the synchronous operation of the frame buffer and the pipeline processor. The frame buffer consists of a single 512 by 512 by 16 store and two 512 by 512 by 8 frame stores. Image data transferred into the frame buffer can then be processed recursively by the pipeline processor, with operations such as frame summation and convolution performed in real time to 16-bit accuracy.

Micromachining and Microdeposition

Focused Ion Beam Micromachining (FIBM) and Focused Ion Beam Deposition (FIBD) are currently attracting much attention as extremely high resolution nanofabrication tools. FIBM and FIBD are capable of producing artefacts, that often could not be produced in any other way, by virtue of the highly localized and controllable nature of the sputter machining or deposition process.[39]

Commercially, the most advanced application at the present time is probably photolithographic chrome-on-glass mask repair, with several systems currently available being capable of tackling both opaque (unwanted chrome areas) and clear defects. These machines are typically equipped with sophisticated CAE and pattern recognition tools allowing automated inspection, determination of optimum repair strategy, repair and verification with the minimum of operator intervention. Opaque defects are corrected by FIBM; unwanted chrome is removed by sputter erosion. A potential problem is "gallium staining" of the underlying glass. In order to avoid this, an antistaining gas can be used to combine with and remove the Ga^+ ions or alternatively very low energy ions can be used to avoid implantation. Ion beam induced decomposition of a suitable hydrocarbon gas such as styrene has been used to correct clear defects by FIBD. Prewett[40] has observed deposited thicknesses of ~100 nm for ion doses of ~4 × 10^{17} ions cm^{-2}. Proprietary gases used on commercial systems can improve the deposition rate by a factor of two. Attention has also been directed at the use of FIBD for depositing both tungsten and gold for X-ray lithography mask repair.

Many other applications of FIBM/FIBD have been explored, but as yet most are at an early stage of development. Device repair and modification has been reported, in particular reconfiguration of ASICs. The motivation is to have the capability of correcting device design errors at the first prototype testing stage, hence avoiding the need to iterate around the design-manufacture-test loop, and potentially leading to enormous savings in cost and time.

Other uses of FIBM for device diagnostics have included microsectioning of an area of interest, followed by SEM examination. Figure 9.8. shows some typical examples of FIBM operations. In contrast with laser milling they are characterized in general by an absence of local heating and an absence of redeposition and debris in the vicinity of the milled feature.

Advanced applications of FIBM/FIBD in integrated optics also look certain to be developed. Harriott et al.[41] pioneered the use of FIBM for on-chip fabrication of laser mirrors. More recently diode laser output mirrors, coupled cavity oscillating mirrors and parabolic mirrors in GaAs lasers have been described.[42]

Lithography

In lithography, the use of a scanning focused ion beam is similar to the use of a focused electron beam but owing to the much larger mass of the ion, avoids resolution limiting proximity effects due to lateral scattering. Resolution is higher compared with electron lithography as the resist is exposed only within ~5 nm of the ion path. Increased resist sensitivity is also possible due to the efficient absorption of the ion energy by the resist, giving higher sensitivity compared with X-rays or electrons. Conventional e-beam polymer resists such as PMMA, PBS or COP can be used.

Figure 9.8a FIB groove milled across IC bonding pad.

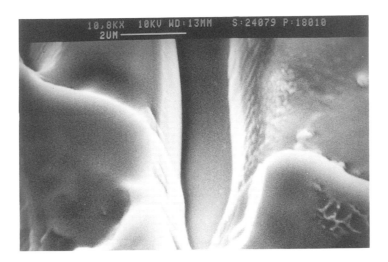

Figure 9.8b Trench milled part through IC interconnect.

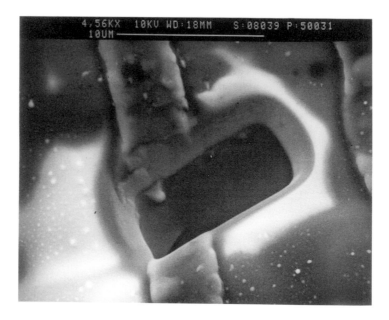

Figure 9.8c Rectangular trench showing vertical walls and absence of redeposited material.

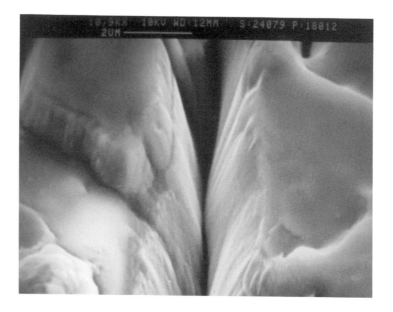

Figure 9.8d Narrow groove milled through IC interconnect. The interface between the metal and semiconductor is clearly visible.

As scanning focused ion beam lithography is a serial exposure process, and hence suffers problems of low throughput, parallel exposure schemes such as image projection and stencil masking are also under investigation as well as advanced multiple focused beam machines.

9.5 SUMMARY AND CONCLUSIONS

Energy beam methods are being widely exploited to address a variety of tasks where critical processing dimensions fall within the nanotechnology domain. The major application areas of microlithography, micromachining and microdeposition have been briefly reviewed and some well developed, in addition to emerging, techniques using energetic photon, electron and ion beams described. The inherent suitability of energy beam methods for fabrication applications in nanotechnology has been emphasized. The use of energy beam methods, in particular those utilizing focused beam technology, may be expected to increase as the dimensions of structures both active and passive continues to be reduced.

9.6. REFERENCES

1. Taniguchi, N. (1983) Current status in, and future trends of, ultraprecision machining and ultrafine materials processing. *Annals of the CIRP*, **32**(2), 1-10
2. Snoeys, R., Staelens, F. and Dekeyser, W. (1986) Current trends in non-conventional material removal processes. *Annals of the CIRP*, **35**(2), 467-480
3. Taniguchi, N., Ikeda, M., Miyamoto, I. and Miyazaki, T. (1989) In *Energy-Beam Processing of Materials*. Oxford: Oxford University Press.
4. Wilczynski, J.S. (1987) Optical lithographic tools: current status and future potential. *Journal of Vacuum Science and Technology B*, **5** (1), 288-292
5. Heuberger, A. (1988). X-ray lithography. *Journal of Vacuum Science and Technology B*, **6** (1), 107-121
6. Ehrlich, D.J., Osgood, R.M., Silversmith, D.J. and Deutch, T.F. (1980) One-step repair of transparent defects in hard-surface photolithographic masks via laser photodeposition. *IEEE Electron Devices Letters*, **1**, 101-103
7. Oprysko, M.M., Beranek, M.W. and Young, P.L. (1985) Visible-laser repair of clear defects in photomasks. *IEEE Electron Devices Letters*, **6**, 344-346
8. Hohn, F.J. (1988) Electron beam lithography : Its applications. *Journal of Vacuum Science and Technology B*, **7**(6), 1405-1411
9. Humphreys, C.J. (1990) Fabricating holes and lines on a nanometre scale and nanomachining using a 0.5 nm diameter electron beam. In *Proceedings of Joint Warwick/Tokyo Symposium on Nanotechnology*, University of Warwick, UK, August 1990
10. Koops, H.W.P., Weiel, R., Kern, D.P. and Baum, T.H. (1988) High-resolution electron-beam induced deposition. *Journal of Vacuum Science and Technology B*, **6**, 477-481
11. Muray, A.J. and Muray, J.J. (1985) Microfabrication with ion beams. *Vacuum*, **35**, 467-477
12. *Commonwealth Scientific Corporation* 500 Pendleton Street, Alexandria, VA 22314, USA. Ion beam etch and sputtering rates.
13. Somekh, S. (1976) Introduction to ion and plasma etching. *Journal of Vacuum Science and Technology*, **13**(5), 1003-1007
14. Kaufman, H.R. (1978) Technology of ion beam sources used in sputtering. *Journal of Vacuum Science and Technology*, **15**(2), 272-276
15. Davies, S.T. and Bowen, D.K. (1987) An apparatus for batch fabrication of micromechanical elements by ion beam machining. *Journal of Vacuum Science and Technology B*, **5**(1), 337-341

16. Kaufman, H.R. and Robinson, R.S. (1989) Broad-beam ion source technology and applications. *Vacuum*, **39**, 1175-1180
17. Lee, R.E. (1984) Ion-beam etching(milling). In *VLSI Electronics: Microstructure Science,* Vol.8, pp. 341-364. Academic Press
18. Johnson, L.F. and Ingersoll, K.A. (1981) Generation of surface gratings with periods < 1000 A. *Applied Physics Letters*, **38**(7), 532-534
19. Melliar-Smith, C.M. (1976) Ion etching for pattern delineation. *Journal of Vacuum Science and Technology*, **13**(5), 1008-1022
20. Darbyshire, D.A., Overbury, A.P. and Pitt,C.W. (1986) Ion and plasma assisted etching of holographic gratings. *Vacuum,* **36**, 55-60
21. Rossnagel, S.M. and Robinson, R.S. (1982) Optical properties of ion beam microtextured surfaces. *Journal of Vacuum Science and Technology*, **20**(3), 336-337
22. Miyamoto, I. and Davies, S.T. (1988) Computer simulation of profile changes of hemi-spherical diamond styli during ion beam machining. *Annals of the CIRP,* **37**(1), 171-174
23. Davies, S.T. (1989) Ion beam machining of diamond. *Industrial Diamond Review,* **5**, 201-203
24. Miyamoto, I., Ezawa, T. and Nishimura, K. (1990) Ion beam machining of single-point diamond tools for nano-precision turning. *Nanotechnology,* **1**(1), 44-49
25. Hebard, A.F., Fleming, R.M., Short, K.T., White, A.E., Rice, C.E., Levi, A.F.J. and Eick,R.H.(1989). Ion beam thinning and polishing of YBa2Cu3O7 films. *Applied Physics Letters*, **55**(18), 1915-1917
26. Honda, M. (1990) Surface smoothing of tungsten thin film by low energy ion beam processing. In *Proceedings of ERATO Symposium*, University of Warwick, UK, August 1990.
27. Allen, N.J., Klein, R. and Romig, H. (1990) Demonstration of an ion figuring process for large optic fabrication. In *Proceedings of 1990 Annual Meeting of American Society for Precision Engineering*, Rochester, New York, September 1990.
28. Melngailis, J. (1987) Focused ion beam technology and applications. *Journal of Vacuum Science and Technology B,* **5**(2), 469-495
29. Namba, S. (1987) Current work on focused ion beams in Japan. *Microelectronic Engineering,* **6**, 315-326
30. Clampitt, R., Aitken, K.L. and Jefferies, D.K. (1975) Intense field-emission ion source of liquid metals. *Journal of Vacuum Science and Technology*, **12**(6), 1208
31. Krohn, V.E. and Ringo, G.R. (1975) Ion source of high brightness using liquid metal. *Applied Physics Letters,* **27**(9), 479-481
32. Seliger, R.L., Ward, J.W., Wang, V. and Kubena, R.L. (1979) A high-intensity scanning ion probe with submicrometre spot size. *Applied Physics Letters*, **34**(5), 310-312
33. Ishitani, T., Tamura, H. and Todokoro, H. (1982) Scanning microbeam using a liquid metal ion source. *Journal of Vacuum Science and Technology*, **20**(1), 80-83
34. Prewett, P.D. (1984) Focused ion beam systems for materials analysis and modification. *Vacuum*, **34**(10), 931-939
35. Prewett, P.D. (1985) Liquid metal ion sources for FIB microfabrication systems - recent advances. *Nuclear Instruments and Methods in Physics Research*, **B6**, 135-142

36. Taylor, G.I. (1964) Disintegration of water drops in an electric field. *Proceedings of the Royal Society of London*, A **280**, 383
37. Gomer, R. (1979) On the mechanism of liquid metal electron and ion sources. *Applied Physics*, **19**, 365-375
38. Wagner, A. (1983) Applications of focused ion beams to microlithography. *Solid State Technology*, May 1983, 97-103
39. Rubloff, G.W. (1988) Maskless selected area processing. *Journal of Vacuum Science and Technology B*, **7**(6), 1454-1461
40. Prewett, P.D. (1989) Focused ion beams in microfabrication. *Colloque de Physique, Colloque C8*, **50**(11), 179-190
41. Harriott, L.R., Scotti, K.D., Cummings, K. and Ambrose, A. (1986) Micromachining of optical structures with focused ion beams. *Applied Physics Letters*, **48**, 1704-1706
42. Ximen, H., Defreez, R.K., Orloff, J., Elliot, R.A., Evans, G.A., Carlson, N.W., *et al.* (1990). Focused ion beam micromachined three-dimensional features by means of a digital scan. *Journal of Vacuum Science and Technology* (in press)

Table 10.1 Criteria for the grinding of brittle materials in ductile mode of material removal and motion copying mode of surface generation.

Type I: Fine abrasive wheel, grain size $<d_c$	
Wheel runout	$< d_c$
Feed resolution	$< d_c$
Work and wheel support stiffness	Sufficiently high for setting depth of cut less than d_c under grinding load
Type II: Large abrasive wheel, grain size $>d_c$	
Wheel runout	$< d_c$
Feed resolution	$< d_c$
Height distribution of cutting points	$< d_c$
Work and wheel support stiffness	Sufficiently high for setting depth of cut less than d_c under grinding load

Table 10.1 gives the design criteria for grinding machines and wheels that need to be satisfied to realize the two functions of motion copying in surface generation and ductile mode in material removal. Type I specifies the case of applying a fine abrasive grinding wheel, of which the grain size is not more than the d_c value to avoid the risk of chipped or dropped grain from the wheel resulting in cracks in the machined surface. Type II specifies the case of applying a coarse abrasive wheel where the grain size is more than the d_c value. This does not assume any chipping or dropping of abrasive grain to result in cracks on the workpiece surface.

Table 10.2 Design specifications of machine tools and abrasive wheels.

Feed resolution	$d_c/10$
Straightness of carriage	d_c/travel range
Work and wheel support stiffness	No more deformation than the feed resolution under grinding load
Vibration level of work and wheel support systems	$< d_c$
Truing accuracies	$< d_c$
Height distribution of cutting points	$< d_c$

For example, a value of d_c of 100 nm and a grinding load of 10 N produces motion errors under the grinding load of less than 10 nm, and a stiffness of the work and wheel support system greater than the grinding-to-feed resolution ratio, i.e. 1 N/nm (10N/10nm).

Table 10.2 gives the design specifications of a grinding machine and abrasive wheel in terms of the d_c value for motion accuracies, feed resolution, truing accuracies and height distribution of cutting points and stiffness. In the worked example for stiffness, in the case of $d_c = 100$ nm and a grinding load of 10 N, a stiffness of 1 N/nm is required for depressing the resultant deformation of the wheel head feed system less than 10 nm.

183

Technologies for supporting these ultraprecision and high stiffness machine tools lie within the field of nanotechnology. The necessary control technique for precision instruments and machines described is presented below for nanometre target accuracies.

10.2 INTRODUCTION OF CONCEPT OF FORCE OPERATED ACTUATOR

10.2.1 Function of Centres and Workrest - Proposal of Workrest with Force Operated Actuator[3]

Conventional work support centres have the following functions:
1. To command the position of workpiece;
2. To supply the support force to keep the workpiece against the applied grinding load.

When the workpiece is slender, deformation by bending caused by the grinding load cannot be considered to be negligible; the workrest gives additional support to the workpiece to compensate the deflection.

Functional errors occurring with the workrest are:
1. Estimation of the error of deflection;
2. Adjustment error;
3. Elastic deformation of the workrest caused by the grinding load.

The structure of a possible workrest fitted with a force actuator to replace the conventional workrest is shown in Figure 10.3a. The grinding load is detected by the force sensors attached to the centres and the compensation unit is balanced with the load to result in zero deflection of the workpiece.

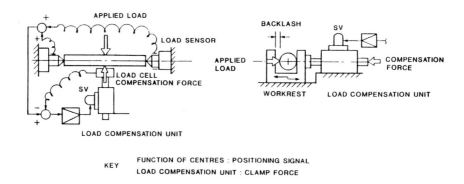

Figure 10.3a Proposed workrest fitted with force actuator.

Figure 10.3b shows the function of the workrest with load compensation unit as a support device for the workpiece. Here a small clearance or backlash between the workrest and the workpiece is assumed.

184

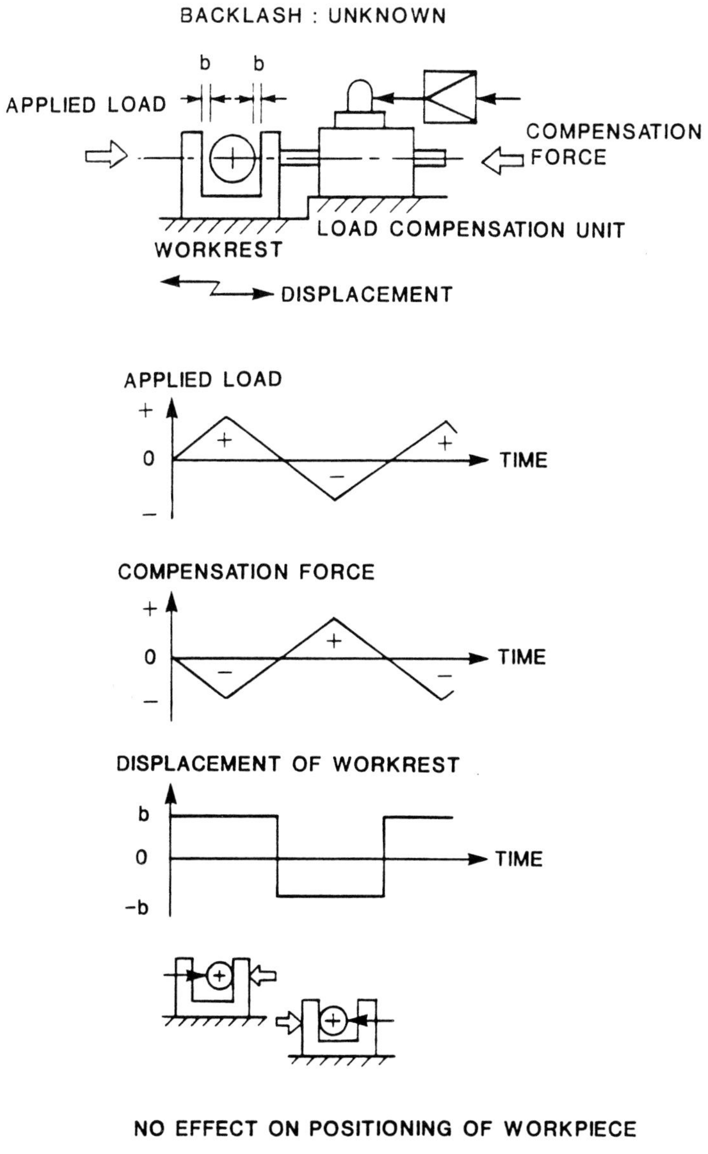
BACKLASH : UNKNOWN
b
b
APPLIED LOAD
COMPENSATION FORCE
LOAD COMPENSATION UNIT
WORKREST
DISPLACEMENT
APPLIED LOAD
+
0
−
TIME
COMPENSATION FORCE
+
0
−
TIME
DISPLACEMENT OF WORKREST
b
0
-b
TIME
NO EFFECT ON POSITIONING OF WORKPIECE

Figure 10.3b Clamp capability of workrest with backlash.

Conventional workrest adjustment is made to compensate for the residual deflection of the workpiece or the functional errors of the workrest and is unable to cope with the dynamic condition, Figure 10.4.

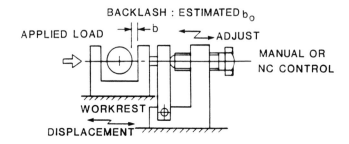

APPLIED LCAD

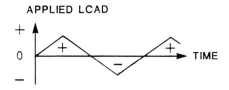

DISPLACEMENT OF WORKREST/ADJUSTMENT

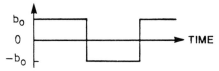

RESIDUAL LOAD ON CENTRES/DEFLECTION OF WORKPIECE

Figure 10.4 Clamp capability of the conventional workrest with backlash.

10.2.2 Function of Feedscrew and Nut for Feeding Carriage - Proposal of Load Compensation Unit[4]

The feedscrew and nut functions in positioning the carriage are:
1. To command the amount of feed;
2. To supply the feed force to drive.

Causes of a deterioration in the feed accuracies of the carriage are:
1. Backlash between feedscrew and nut;
2. Axial deformation of the feedscrew;
3. Friction between the feedscrew and the nut;
4. Friction between the carriage and the guideways.
The conventional strategies for overcoming these problems are:

186

1. Preliminary axial load applied to eliminate the backlash between the feedscrew and the nut;
2. Preliminary tensile force applied to the feedscrew;
3. Application of an anti-friction ball-nut and screw;
4. Application of an anti-friction bearing or hydro/aero-static bearing guideways.

The concept common to these strategies is to eliminate any backlash and to reduce any friction in the feed system. Elimination of the effect of backlash can be achieved by the application of a force operated actuator or load compensation unit to a feed system; the former technique has been demonstrated in section 10.2.1.

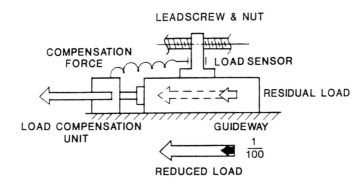

Figure 10.5 Feed with leadscrew and load compensation unit.

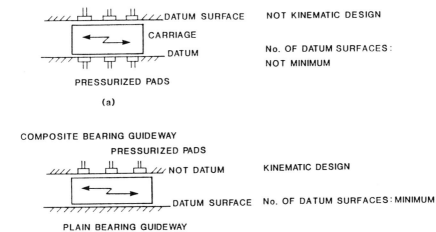

Figure 10.6 Guideway system of carriage.

The application of anti-friction bearing or pressurized bearing guideways for reducing friction between carriage and guideways, results in the need to consider the requirements for sufficiently high stiffness, statically and dynamically, of the feed system against vibrational disturbances These requirements are difficult to meet. Friction is very useful as a means of improving effective dynamic stiffness over a wide frequency range. For this reason, a feed system with load compensation unit as shown in Figure 10.5 is proposed. In this way, assuming a loop gain of K in load compensation system, an effective stiffness of the feed system of feedscrew and nut increases by a factor of K.

From the above discussions it can be seen that the feed system of feedscrew and nut with load compensation unit has the following merits:

1. To eliminate the decrease of feed accuracy caused by the backlash;
2. To reduce the feed load on the feedscrew and nut;
3. To eliminate the deterioration of feed accuracy caused by the friction between the carriage and the guideways.

10.2.3 Kinematic Design of Guideway - Proposal of Composite Bearing Guideway[5]

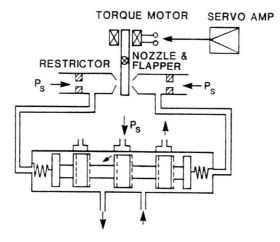

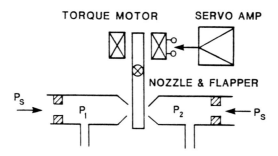

Figure 10.7 Flow rate control servo system.

Figure 10.6a shows the guideway system of carriage with pressurized bearings generally applied to ultraprecision instruments and machine tools. There are two datum surfaces, whereas considering the number of degrees of freedom for motion in a plane there should only be one datum surface to satisfy the kinematic principle.

In Figure 10.6b the structure of a proposed composite bearing guideway is shown, in which a plain bearing guideway is a datum to support the carriage. The pressurized pads are placed to give the contact pressure distribution between the carriage and the plain bearing. According to Pollard[6], surfaces in contact kinematically have to satisfy the requirement of "functional surfaces". That is, "functional surfaces" are made to fit over their whole length during the total range of the relative motion between the coupled members. In terms of this definition, the function of the pressurized pads is to give the defined contact pressure distribution over surfaces in contact to favourably clamp or feed the carriage.

10.2.4 Components of Force Operated Actuator

A conventional servo system comprises a flow rate control servo valve and a flow rate operated linear motor. An example of which is shown in Figure 10.7. Figure 10.8 gives an example of a pressure control servo valve. For linear hydraulic motors, there are two kinds of motor, one is flow rate operated and the other is force operated[7]. With the linear hydraulic motor conventionally used, for correct functioning no leakage may occur, therefore oil seals are fitted. However, the friction between the cylinder and the spool caused by the seals reduces the positioning accuracy of servo systems using this type of motor.

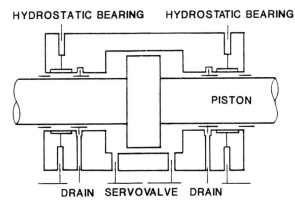

Figure 10.8 Pressure control servo valve.

With a force operated linear motor, leakage may occur without affecting the function. A servo system comprising the pressure control servo valve and the force operated linear motor achieves higher resolution of feed and higher feed stiffness compared with a conventional servo system comprising a flow rate control servo valve and flow rate operated linear motor.

10.3 APPLICATIONS OF FORCE OPERATED ACTUATORS

10.3.1 Workrest with Force Operated Actuator

Figure 10.9 shows the construction of a workrest incorporating a force operated actuator. Sensors attached on the centres detect the grinding load and the load cell on the workrest is the feedback transducer for the compensation force. The load compensation unit comprises a pressure control servo valve and a force operated actuator.

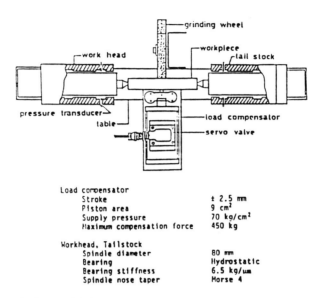

Figure 10.9 Workrest with force operated actuator.

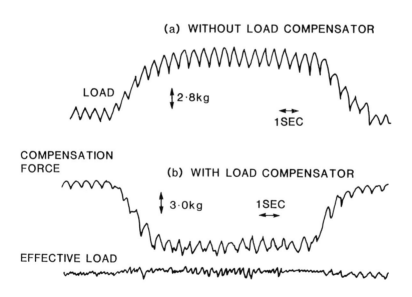

Figure 10.10 Grinding load - plunge grinding.

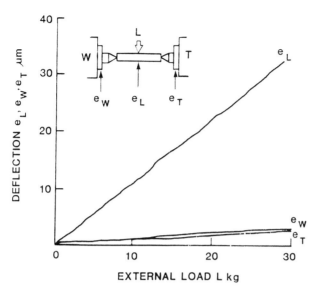

Figure 10.11 Deflections at spindle, workpiece and tailstock.

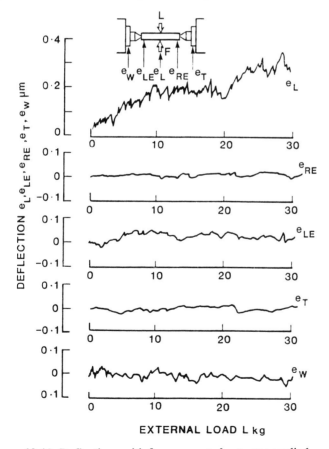

Figure 10.12 Deflections with force operated actuator applied.

Figure 10.10 shows an example of the normal component of grinding load, and the residual grinding load on the workpiece in the case of plunge grinding at the middle of the workpiece supported on both the centres. The periodic component of the load synchronized with the rotation of the workpiece is caused by an imbalance of the driving torque applied to the workpiece.

The deflection at the main spindle, the workpiece and the spindle of the tailstock - e_w, W_1 and e_T are shown in Figure 10.11 when a load L is applied to the workpiece supported on the centres. The lowest stiffness is about 1 kgf/μm at the middle of the workpiece.

When the workrest with force operated actuator is applied to this case, the deflections can be reduced to about 1/100 as shown in Figure 10.12. In this case, the loop gain of load compensation system is set at a value of 100. Test results where the workpiece is supported by a chuck on the main spindle are shown in Figure 10.13 without load compensation and in Figure 10.14 with load compensation applied. The same effect on the stiffness of using the proposed workrest is observed.

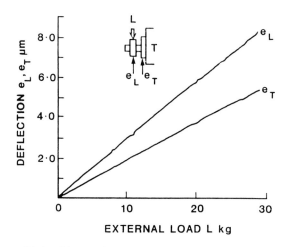

Figure 10.13 Chuck without load compensation.

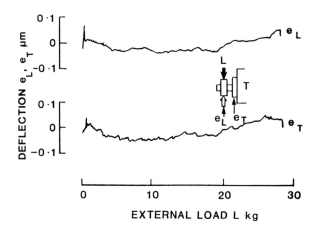

Figure 10.14 Chuck with load compensation applied.

192

Test results of the grinding accuracy as affected by using the workrest without or with compensation are shown in Figures 10.15a and b. Figure 10.15a shows the roundness error of ground workpiece without workrest compensation; some chatter vibration is observed on the record. Figure 10.15b shows the roundness record of the ground workpiece with the compensated workrest. The waviness caused by the chatter vibration is eliminated. This is a reflection of the increased stiffness of the work support system. Figure 10.16 is the result of the workpiece with non uniform compliance in deflection.

The workpiece has partially parallel flats as shown in the figure, which results from a variation of compliance in deflection for angular position. The variation of compliance of workpiece gives an oval shape of roundness as shown in the figure in the case of no compensation force. The workrest with compensation force can eliminate the effect of variation of compliance on the roundness error as shown in the figure.

```
Grinding conditions
    Infeed/rev. work: 5 μm
    grinding width: 20 mm
    work rev. times for sparkout: 5
    grinding wheel speed: 60 m/sec
    speed ratio: 100
    work: SCM-3
    grinding wheel: SA60M
```

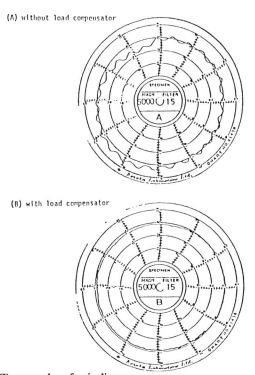

Figure 10.15 Test results of grinding accuracy.

10.3.2 Feed of Carriage on Plain Bearing Guideways with Linear Actuator and Load Compensation Unit[4]

A typical example of the conventional feed characteristics of a 345 kg carriage with feedscrew and nut for the feed rate of 30 μm/min, is given in Figure 10.17a, and shows initial stick and stick-slip phenomena. Figure 10.17b is an example for 152 μm/min, which shows the same initial stick of about 100 μm but with no stick-slip.

The feed characteristics of a 345 kg carriage on the same test rig with load compensation unit for the feed rate of 28 μm/min are shown in Figure 10.18. The initial stick and stick/slip phenomena have almost disappeared and the residual load on the feedscrew is reduced to a few Newtons. The feed characteristics of 845 kg carriage on the same test rig with load compensation unit for the feed rate of 4 μm/s and a travel range of ± 40 μm are shown in Figure 10.19. The residual load on the feedscrew is about 10 N and the compensation force is about 1 kN. In Figure 10.20, similar feed characteristics for a step feed of 0.125 μm/step are shown.

A surface grinder has been designed incorporating a vertical feed to the grinding head, composite bearing guideways and load compensation unit. Test results for a vertical step feed of 0.1 μm/step have shown that almost zero lost motion is observed at the turning point compared to the step size. An improved design of grinding machine has been constructed with a similar feed mechanism except that the feedscrew has been replaced by a force operated actuator. Two force operated actuators are provided at the top and at the bottom of the grinding head carriage respectively and the ratio of the driving forces at the two feed points is adjustable. The test results of step feed of the vertical spindle of 0.1 μm/step and 0.005 μm/step are shown in Figure 10.21.

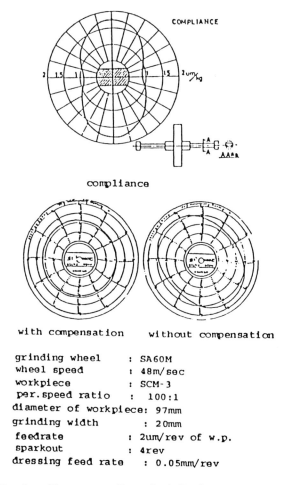

COMPLIANCE

compliance

with compensation   without compensation

```
grinding wheel      : SA60M
wheel speed         : 48m/sec
workpiece           : SCM-3
per.speed ratio     :  100:1
diameter of workpiece: 97mm
grinding width        : 20mm
feedrate            : 2um/rev of w.p.
sparkout            : 4rev
dressing feed rate  : 0.05mm/rev
```

Figure 10.16 Results with non-compliance in deflection.

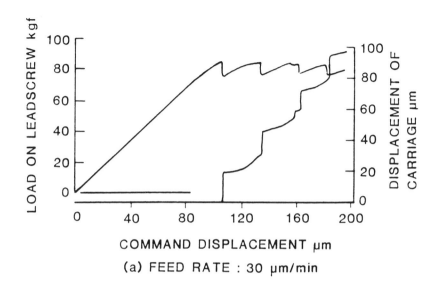

(a) FEED RATE : 30 μm/min

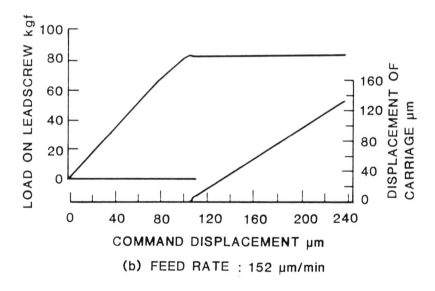

(b) FEED RATE : 152 μm/min

WEIGHT OF CARRIAGE : 345kgf
GUIDEWAY : V–FLAT PLAIN BEARING
CONTACT PRESSURE : 1·2kgf/cm²
LEADSCREW : TRAPZOIDAL, 6 mm PITCH,
40mm DIA.

Figure 10.17 Typical example of conventional feed characteristics of a 345 kg carriage.

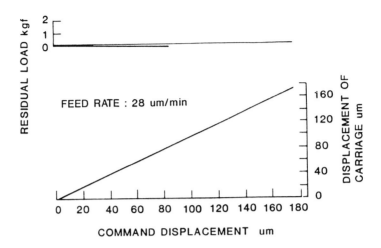

Figure 10.18 The feed characteristics of a 345 kg carriage with load compensation.

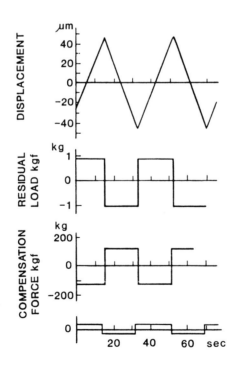

Figure 10.19 Feed characteristics of a 845 kg carriage with load compensation.

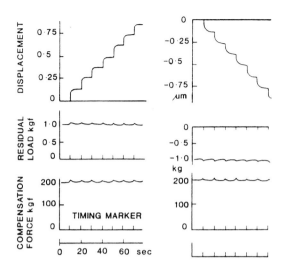

Figure 10.20 Feed characteristics for a step feed of 0.125 µm/step.

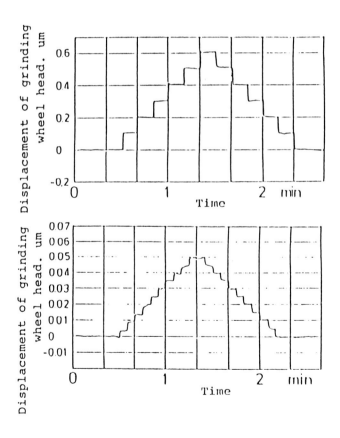

Figure 10.21 Vertical feed characteristics of wheelhead: 0.1 µm/step and 0.005 µm/step

10.3.3 Feed Characteristics of Force Operated Linear Actuator without External Load[7]

As described in section 10.2.4, there are two types of positioning servo control systems. One system has a flow rate servo valve with flow rate operated linear motor and the alternative system has a pressure control servo valve with a force operated linear motor. For a specified command signal, in the case of a flow rate servo system the flow rate into the two cavities is zero and for a pressure control system the pressure difference in the two cavities is zero. The limitations of these systems have already been seen and as an alternative the force operated hydraulic linear actuator design proposed. The spool rods are hydrostatically supported and it is not necessary to fit seals.

To assess the effectiveness of the system, a test rig has been built to measure the feed characteristics. On the rig an eddy current type reference sensor has been provided in addition to a capacitive feedback sensor. The displacements recorded are shown in Figures 10.22a to 10.22e for step feeds of (a) 25 nm/step, (b) 5 nm/step, (c) 1 nm/step, (d) 0.25 nm/step and (e) 0.05 nm/step. It was found that the reference sensor was not sufficiently sensitive to record displacements of less than 1 nm/step.

(a)

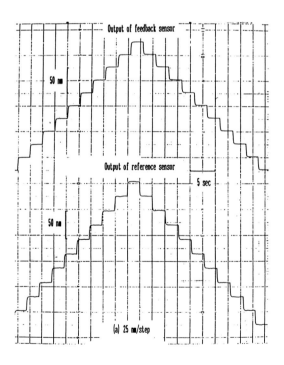

Output of feedback sensor

50 nm

Output of reference sensor

5 sec

50 nm

(a) 25 nm/step

(b)

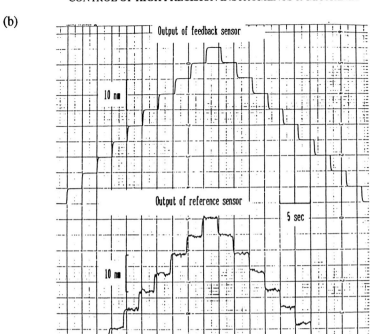

(c)

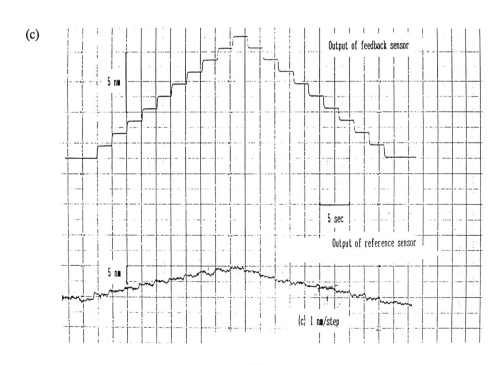

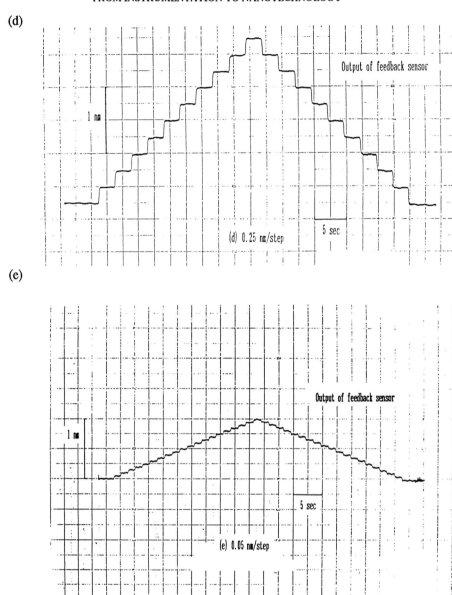

Figures 10.22a-e Feed characteristics of force operated actuator (a) 25 nm/step, (b) 5 nm/step, (c) 1 nm/step, (d) 0.25 nm/step and (e) 0.0 5nm/step.

The resultant displacement of the servo system from the application of static loads up to 200 N and 500 N is seen in Figure 10.23. An arrangement of a test rig for measuring the static stiffness of the servo system is shown in Figure 10.24 At a loop gain close to the critical stability of the system the static stiffness is about 60 N/nm which almost coincides with the analytical value of the model. Figures 10.25a and b show the analytical result and the experimental result of frequency response characteristics of the force operated actuator in the system. Analysis of the dynamic stiffness of the actuator, which is a function of parameters such as clearance, travel range and mass of the actuator is shown in Figure 10.26.

Table 10.3 shows the main specification of vertical spindle surface grinding machine.

Grinding Wheel	175mm diameter cup wheel
Wheel speed	30 ~ 3000 rpm
Working size of rotary table	420mm diameter

Composite Bearing Guideway

Front	Plain bearing area 310 cm^2
Rear	Hydrostatic bearing area 18m^2 × 12

Wheelhead Positioning

Actuators for positioning and for load compensation	Hydraulic cylinder bore 66 mm diameter
Support of cylinder	Hydrostatic bearing

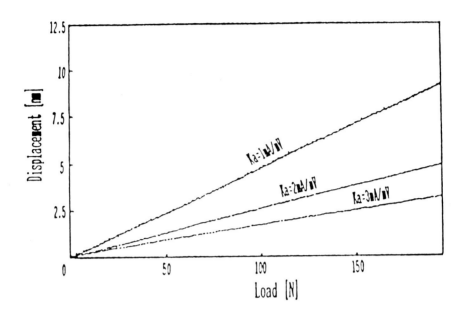

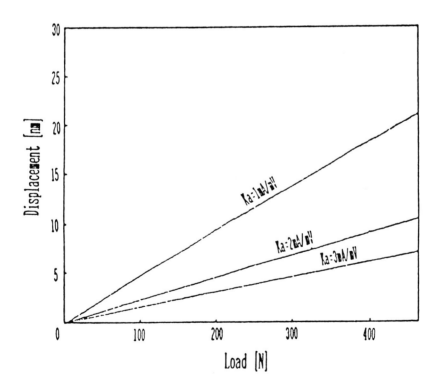

Figure 10.23 Stiffness of force operated actuator.

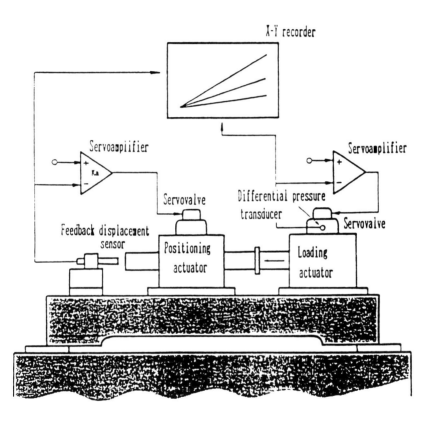

Figure 10.24 Arrangement of test rig for measuring static stiffness of actuator.

(a) Analysis

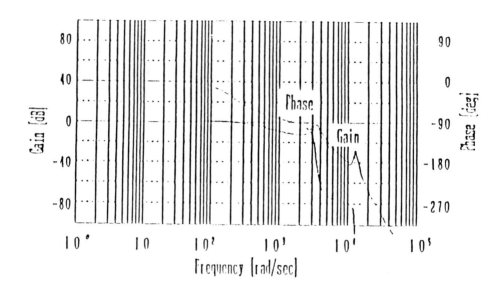

(b) Experiment

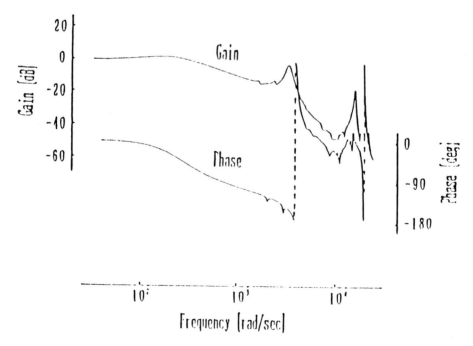

Figure 10.25 Frequency response characteristics of force operated actuator.

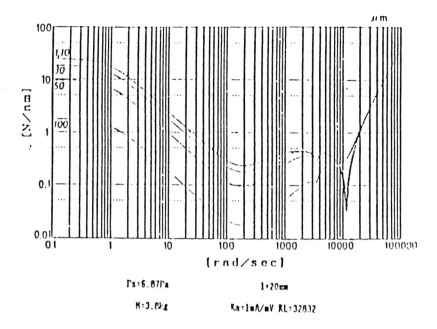

Figure 10.26 Dynamic stiffness of force operated actuator - analysis.

10.3.4 Feed of a 200 kg Carriage on Plain Bearing Guideways with Force Operated Linear Actuator[8]

A plain bearing is generally not applied to an ultraprecision diamond turning machine due to its frictional load, which is an obstacle to an improved resolution of feeding a carriage. On the other hand, the frictional load is very useful to achieve an equivalent high damping effect at higher frequency.

The high stiffness and high feed resolution of the force operated actuator described in section 10.3.3 can be expected to realize a feed mechanism of a heavy carriage on plain bearing guideways with high effective damping of its frictional load and with the high feed resolution or better than state of the art ultraprecision diamond turning machines.

10.3.5 Active Control of Ultraprecision and High Stiffness Rotating Spindle System Applying Iterative Measurement and Compensation Algorithm[9]

Runout of a rotating spindle supported by pressurized bearing pads is caused by roundness error of the coupled part of spindle and also a variation of applied load on the spindle. For realizing a spindle system with runout accuracy to nanometer under a variation of load, an active control of ultraprecision and high stiffness rotating spindle system applying iterative measurement and compensation algorithm is proposed. This is composed of two parts:

1. Actuators for actively compensating the spindle runout, and
2. measuring system of the roundness error of spindle.

In the first part, two coupled force operated actuators can be applied instead of four pressurized bearing pads for supporting the spindle as compensation actuators of the spindle runout.

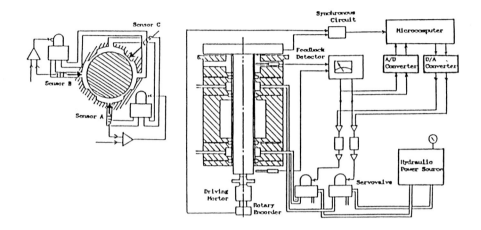

Figure 10.27 Compensation mechanism of spindle runout.

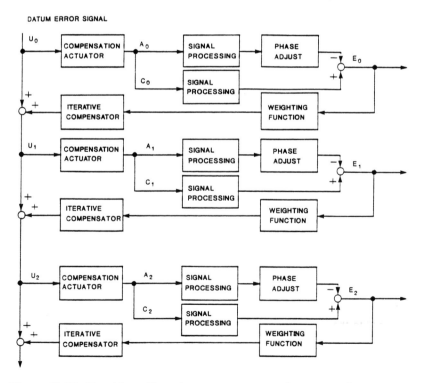

Figure 10.28 Algorithm of iterative measurement and compensation.

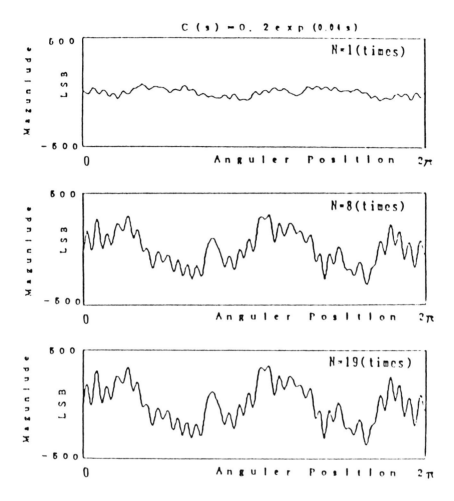

Figure 10.29 Output signal A after N times learning.

For detecting the roundness error of the datum part of spindle, on which peripheral displacement sensors are placed, a multipoint roundness error measuring system is applied.

The construction of compensation mechanism is shown in Figure 10.27 for spindle runout composed of the above two parts. The vertical spindle is supported on two hydrostatic bearings and displacement sensors of A, B and C, are the feedback sensors for the compensation actuators, with a displacement sensor D as a reference outside of the closed system, which is placed at the opposite side to sensor B. The angular position of the sensor C to the sensor A is set at 115°, which enables coverage of the detectable frequency range of up to 20 lobes of the roundness error. The roundness error of the datum part of spindle is sequentially derived from the outputs of the A and C by the learning control and the compensation of the spindle runout as shown in Figure 10.28.

Figure 10.29 shows the test results of the sequentially derived roundness errors for a number of learning and compensation iterations. In this case the output signal involves the spindle runout and also the datum error of the spindle where it is in contact with the sensor. The converged output at N = 19 means that the spindle is fully compensated. This gives an output that is the datum error or roundness error of the periphery of the spindle at the point where it is in contact with the sensor.

Figure 10.30 is another description of the test results to settle down to the final runout reading after sequential learning.

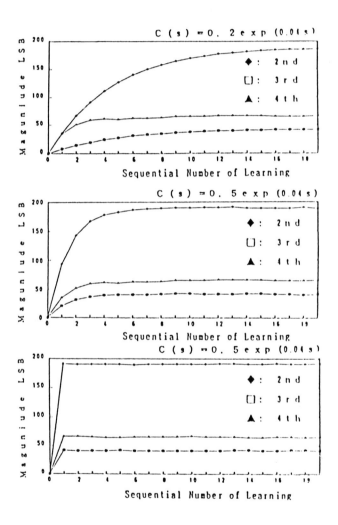

Figure 10.30 Magnitude of harmonic components of output A.

The top figure is the case of a feedback gain K of 0.2, the middle figure is 0.5 and the bottom figure has a feedback gain of unity at N=1 followed by a feedback gain of 0.5 at N≥2. One Least Significant Bit (LSB) is 0.2 nm.

Table 10.4 Result of runout error compensation.

Waviness component	Output Signal (LSB)			Difference (LSB)	
	A	C	D	E_{AC}	E_{AD}
1	0	25	6	25	6
2	186	190	191	4	1
3	41	41	40	1	1
4	64	65	65	2	3
5	41	43	43	1	1
6	41	42	51	1	10
7	15	15	15	1	0
8	82	85	90	2	8
9	10	11	10	1	1
10	9	10	10	1	0
11	12	13	14	0	1
12	15	16	16	1	1
13	18	20	19	1	1
14	34	35	34	0	7
15	17	18	17	2	0
16	17	18	24	1	9
17	7	7	8	1	1
18	9	10	9	1	2
19	14	14	15	0	0
20	1	2	2	1	1

Table 10.4 is a comparison of the output signals of the sensors of A, C and D at the final state. E_{AC} is the difference between the sensors A and C, and E_{AD} the difference between the sensors A and D. The discrepancy between the output of the sensors means the residual runout of the spindle system from the active control for compensation. E_{AC} shows the maximum difference of 4 LSB except the amplitude of one lobe of 25 LSB and E_{AD} the maximum difference of 10 LSB.

The output signals of A, C and D should progressively converge to the datum error at the final stage of compensation plus the residual runout of the spindle. Table 10.4 expresses the outputs of A, C and D as frequency components determined by Fourier analysis and are composed of errors of roundness and the minute residual runout errors.

10.4 DESIGN OF MACHINE TOOLS EQUIPPED WITH FORCE OPERATED LINEAR ACTUATOR

The concept of the force operated actuator is applied to the design of centreless grinding machine for ductile grinding of brittle materials.[10] The characteristics of this machine are based on the following design concepts:

1. Alignment of grinding point, guideways of grinding wheel head slide and driving point of feed in a common plane that is, the Abbe's offset being minimized.

2. Realization of functional surfaces in the kinematic principle by the use of composite bearing guideway.
3. Utilization of force operated actuator for feeding the grinding wheel head slide.
4. Grinding wheel hydrostatically supported on stationary spindle, which is effective to improve the stiffness of spindle at the grinding load point and also to reduce the rotating mass to result in an increase of predominant resonant frequency of the wheel support system.

Figure 10.31 shows the test result of feeding the grinding wheel head slide of this machine a step feed of 5 nm/step, no lost motion is observed at turning point in feed in spite of the existence of frictional load of about 300 N between the slide and the guideways.

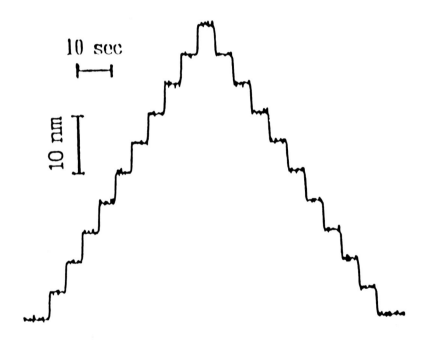

Figure 10.31 Stepwise feed of grinding wheel head slide.

REFERENCES

1. Miyashita, M. (1990) Nanogrinding technology. *SPIE*, **1333** (in press).
2. Bifano, T., Blake, P., Dow, T. and Scattergood, R.O. (1987) Precision machining of ceramic materials. *Intersociety symposium on machining ceramic materials*, ACerS., ASME and AES in Pittsburgh. p99.
3. Miyashita, M., and Kanai, A. (1976) Development of the load compensator for the cylindrical grinding machines. *CIRP*, **25**(1)
4. Miyashita, M. and Yoshioka, J. (1981) Development of ultraprecision machine tools for microcutting of brittle materials. *Bull JSPE*, **16**(1)

5. Yoshioka, J., Hashimoto, F., Miyashita, M., Kanai, A., Abo, T. and Daito, M. (1985) Ultraprecision grinding technology for brittle materials. *ASME, PED* **16**, 209

6. Pollard, A.F.C. (1951) *The kinematic design of couplings in instrument mechanisms.* Hilger & Watts Ltd. p16

7. Kanai, A., Yoshioka, J., Sano, H. and Miyashita, M (1987) Nanometer positioning actuator based on the principles of force control. *ASPE Annual Meeting* in Columbus.

8. Kanai, A., Sano, H., Yoshioka, J. and Miyashita, M (1991) Positioning of a 200 kg carriage on plain bearing guideways to nanometer accuracy with a force operated linear actuator. *Nanotechnology* (in press).

9. Takahashi, M., Kanai, A. and Miyashita, M. (1990) Active control of ultraprecision and high stiffness rotating spindle system applying iterative measurement and compensation algorithm. *Proc. JSPE* (in Japanese).

10. Yoshioka, J., Miyashita, M., Kanai, A., Daito, M. and Hashimoto F. (1990) Design of centerless grinding machine for ductile mode grinding of brittle materials. *Proc. ASPE.*

Chapter 11

OPTICAL METROLOGY: THE PRECISION MEASUREMENT OF DISPLACEMENT USING OPTICAL INTERFEROMETRY

MICHAEL J. DOWNS

Laser interferometers are widely used in the free atmosphere for the precision measurement of length. In order to achieve the high potential accuracy of these instruments, not only must the frequency of the radiation source be calibrated and the system carefully aligned, but also suitable corrections have to be made due to variations in the refractive index of air. The various problems of using interferometers in this particular application will be discussed, and techniques will be described for realizing the optimum performances from these systems .

11.1 INTRODUCTION

At the Conference Generale des Poids et Mesures in 1983 the metre was redefined as the distance travelled by light in free space during $1/c$ of a second, where c is the defined speed of light ($299,792,458$ ms^{-1}).

This standard can be realized by optical interferometers[1] and these instruments are widely employed for the precision measurement of length, making use of lasers as precision reference standards of frequency. The coherence of laser sources permits fringe counting systems with ranges up to 50 metres in the free atmosphere. These devices are readily available commercially and experience has shown that their frequencies do not usually change by more than a few parts in 10^8 over the lifetime of the laser tube (20,000 h).

The most commonly used interferometric length measuring systems currently employed are based on the Michelson interferometer and use bidirectional fringe counting techniques to correct automatically for vibration and retraced motion, ensuring that the fringe count truly represents the displacement of the moving reflector. These instruments measure length in terms of wavelength of the radiation from the light source, the most widely employed source being a red helium-neon laser ($\lambda \approx 633$ nm). The frequency of a typical frequency stabilized helium-neon laser can be calibrated and maintained to a few parts in 10^8 over the lifetime of the laser. This would readily satisfy the accuracy required for precision length measurement; however invariably these interferometers are used for measurement in the free atmosphere and to realize the full potential accuracy of these systems, not only must they be carefully aligned but the wavelength of the radiation must be corrected for variations in the refractive index of air.

By eliminating the problems due to the stray reflections from the interferometer beamsplitter the National Physical Laboratory (NPL) length measuring interferometer allows the optimum performance to be obtained from the instrument, whilst in addition enabling the required alignment of the optical and mechanical axes to be achieved.

The limitations to the accuracy and resolution achievable by interferometry in the precision measurement of length in the free atmosphere (shown in Table 11.1) will be described and it will be shown that providing the correct optical and mechanical calibration and alignment procedures are adhered to, then an absolute measurement accuracy of better than 1 part in 10^7 can be achieved from these systems in a controlled environment.

11.2 RADIATION SOURCES AND FREQUENCY CALIBRATION

The radiation sources most widely used for interferometry are frequency stabilized helium-neon lasers which are now readily available commercially. Their intense collimated beams are ideally suited to laser interferometry and their narrow bandwidths provide, in theory, an interferometric measurement range of hundreds of metres. In practice however, the inhomogeneity of the media in the optical paths transversed by the two beams restricts this measurement range. When these systems are used over a range of 30 metres or more, atmospheric variations in the optical path can cause up to 1 fringe of tilt across the wavefront of the light beam causing a complete loss of contrast in an interferometer and total failure of the system.

Helium-neon lasers have three main limitations when used in interferometric applications: the possibility of multiple modes of oscillation, frequency instability and sensitivity to optical back coupling into the cavity.[2]

By careful optical design, helium-neon lasers can be restricted to axial modes, i.e. the off-axis modes are prevented. The axial modes have a frequency spacing of $c/2L$ (where c is the speed of light and L is the cavity length) and the number of modes can be directly controlled by the separation of the laser mirrors. By making the mirror spacing sufficiently short, for example, the axial mode spacing is made larger than the Doppler width of the transition, shown in Figure 11.1, then only a single axial mode is emitted.

The three main types of frequency stabilized laser commercially available are shown schematically in Figure 11.1. The Lamb dip[3] and Zeeman split types[4,5] shown in Figure 11.1(a) and (b) respectively, use short tubes supporting only a single frequency. The two mode type shown in Figure 11.1(c) employs a tube length capable of supporting two axial modes.

The frequency stabilization of all these lasers is achieved by maintaining a constant distance between the mirrors employed for the cavity, automatically correcting for any thermal expansion changes that may occur. The mirrors may be separated from the discharge tube or, as is now more usual, form an integral part of the tube providing a durable permanently aligned structure. In the case of the mirrors separated from the discharge tube, their positions may be readily controlled using piezo-electric devices. When the mirrors are an integral part of the tube, a number of techniques are employed for cavity length control and these include the mechanical stretching of the tube with piezo-electric devices and the thermal techniques listed below:

1. Using a fan for controlled cooling of the tube.
2. Heaters to regulate the tube length by thermal expansion.
3. Varying the electrical power in the discharge to control the expansion of the tube.

All these techniques normally involve small acceptable variations in the output power of the laser.

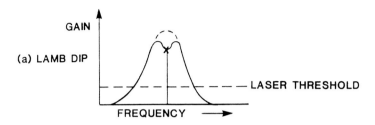

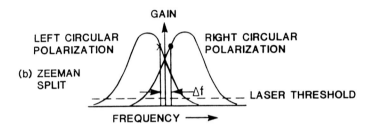

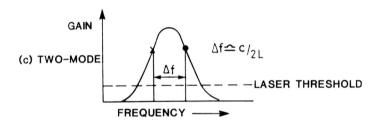

Figure 11.1 Laser doppler profiles.

11.3 CAVITY LENGTH SERVO-SYSTEMS

11.3.1 Lamb Dip

At the centre of the Doppler profile, less power is obtainable within approximately one natural line width of the Doppler line centre than is possible towards the edges of the line as shown in Figure 11.1a. This effect was first described by Lamb and more recently by Javan and Szoke. By examining the power output of the cavity and applying a small length modulation the laser can be frequency locked to the "Lamb dip". In practice the Lamb dip in the visible Helium-Neon laser is about 200 MHz wide and 5-10% deep.

Measurements at the International Bureau of Weights and Measures (BIPM)[4] show the stabilized vacuum wavelength of a new Lamb dip laser device, using isotope ^{20}Ne, to be 632.991410 nm, drifting to 632.991430 nm 3 years later near the end of the plasma tube life.

11.3.2 Zeeman Split

Zeeman split lasers use either axial or transverse magnetic fields to split the emission into two polarized components. In the case of the axial field (shown in Figure 11.1b) these components are orthogonal, circularly polarized components, the Doppler profile of each being frequency shifted. The components have equal intensities only when they are symmetrically disposed about the central frequency of the laser and by using a polarizing beam splitter and photodetectors to measure their relative amplitudes, a servo signal can be obtained to control the cavity length.

The two components from such a Zeeman split laser, due to refractive index differences in the laser gas, have slightly different optical frequencies, shown in Figure 11.1b, and are ideally suited as sources for the dual frequency heterodyne interferometer systems previously described. A commercially available example of this type of laser is produced by Hewlett Packard in the USA.

11.3.3 Two-Mode

The "two-mode" method of stabilization[6,7,8] is shown schematically in Figure 11.1c. The relative intensity of adjacent, linear, orthogonally polarized modes is measured. The cavity length is then controlled to maintain a fixed intensity ratio between the two modes. The control system is continuous and there is no frequency modulation of the laser output. One or both of the modes can then be selected using a polarizing beam splitter. It is worth noting that in practice the two frequencies sit approximately 250 MHz (5 parts in 10^7) on either side of the laser frequency at the centre of the Doppler profile. All three types of frequency stabilized laser sources described have long term frequency stabilities of better than a few parts in 10^8 and, when calibrated, can be used in precision length measurement applications requiring an absolute accuracy of 1 part in 10^7.

It is vitally important with all stabilized lasers that no light is reflected back into the laser cavity as reflections of the order of 0.01% of the original laser intensity can severely disturb the frequency stabilization of the laser and can even prevent the stabilization mechanism from functioning. In order to minimize the problems due to stray reflections from the outer surface of the output mirror some manufacturers use mirrors with anti-reflection coating and also wedge the mirror approximately 1.5 degrees, accepting the slight divergence of the output beam that this causes.

In order to realize an absolute accuracy of 1 part in 10^8, it is essential to calibrate the frequency of all stabilized radiation sources, as variations in the constituents of the gas and the pressure can cause significant variations in the laser frequency. However, in practice, once the frequency has been calibrated for a typical laser, it will not vary by more than 1 part in 10^8 over a period of a year.

The frequency calibration curve shown in Figure 11.2 was produced by the Wavelength Standards group at the NPL for a two-mode 117 Spectra-Physics laser, used as a radiation source for a length interferometer. This curve was obtained by beating the laser with an absorption stabilized Helium-Neon laser specifically developed as a national wavelength standard. The saturated absorption iodine stabilized Helium-Neon laser used for this calibration employed an inter-cavity iodine cell, the frequency stability and reproducibility of this type of laser[9] being better than 1 part in 10^{10}.

11.4 MECHANICAL ALIGNMENT OF LENGTH MEASURING INTERFEROMETER

The six degrees of freedom affecting the position of an object in a single axis measuring system are shown in Figure 11.3. (It is worth noting that in a 3 axis system there are 18 degrees of freedom together with the orthogonalities of the axes, making a total of 21 sources of error).

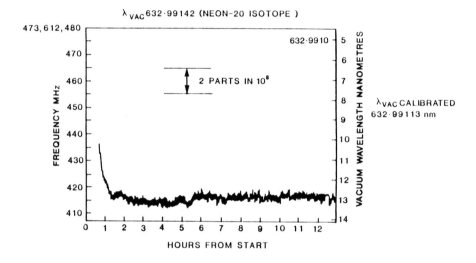

Figure 11.2 Laser frequency calibration curve. Spectra physics 117 (two mode stabilization)

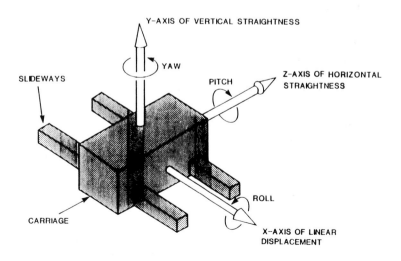

Figure 11.3 The six degrees of freedom of a single axis positioning system.

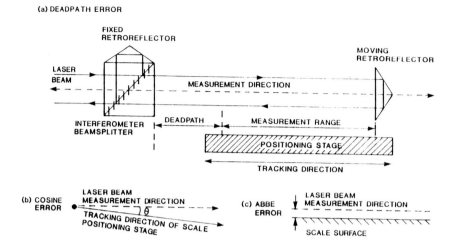

Figure 11.4 Direction and position alignment criteria.

The two fundamental errors generated by a single axis machine are Abbe and cosine errors. In addition there is the less significant "deadpath" error. These are shown diagrammatically in Figure 11.4. The effects of these can be reduced to an acceptable level by careful alignment and tracking of the measurement system.

11.4.1 Abbe Error

Abbe errors are generated by angular motions of the measurements stage when the displacement measurement is taken at a location which is offset from the actual displacement to be measured (Figure 11.4c). They can be almost completely eliminated by aligning the measurement point of the system to be coincident with the measurement direction of an interferometer which is a line through the apex of the corner cube retroreflector and parallel to the laser beam direction. The tolerance to this alignment is shown in Table 11.1. Alignment of the measurement point and direction of the system can be achieved using either the actual apex of the retroreflector or, if it is enclosed, a reference datum on its housing. It should be noted that yaw affects horizontal Abbe errors and pitch affects vertical Abbe errors. There is no Abbe sensitivity to roll.

The pitch and yaw of a commercial 150 mm range scale positioning system was measured with a Rank Taylor Hobson photoelectric autocollimator by mounting a mirror on the substrate stage. The results of measurements from 20 runs over the 150 mm range of the stage indicated a pitch and yaw of 2.1 seconds of arc, both approximately in the centre of the movement with a repeatability of $\pm$ 0.25 seconds arc. These angular movements would result in a maximum Abbe error in the system with a 0.1 mm Abbe offset of 0.001 μm (1 part in 10^8). Reducing this error to an acceptable level is of course only achievable by the use of a high quality tracking stage.

11.4.2 Cosine Error

If the axis of motion of the measurement stage is not aligned with the axis of the laser beam, an error is generated between the measured distance and the actual distance travelled. This is normally referred to as the "cosine error" as it is directly proportional to the cosine of the angle between these two axes. The cosine error always causes the

218

interferometer to read shorter than the actual distance travelled. In order to achieve the accuracy of alignment required, it is necessary to employ a position sensitive quadrant photodetector system as shown in Figure 11.5.

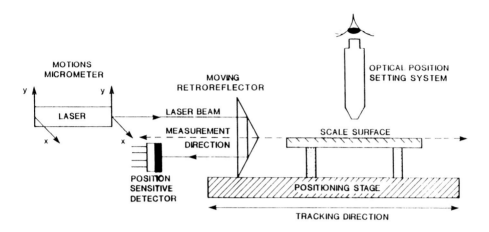

Figure 11.5 Length measuring system.

Using the detector in this configuration doubles the measurement sensitivity to displacements resulting from tracking direction and straightness. As shown in Figure 11.5, the laser beam direction can be controlled by vertical and horizontal micrometer motions attached to both the front and rear of the laser tube; these not only provide angular adjustment but also enable the position of the laser beam to be controlled and maintained in the optimum position through the centre of the interferometer block. In an ideal system, the laser positioning and directional controls would be independent of each other.

The position sensitive cell was calibrated using an xy micrometer motion and the tracking alignment achieved using the micro-control system was 20 μm in 150 mm. This is equivalent to an angle between the laser beam direction and the tracking direction of the stage of 27.5 seconds of arc resulting in a cosine error of 0.0015 μm in 150 mm (1 part in 10^8). It will be appreciated that the non-linearity of the cosine function would give an error of 1 part in 10^7 for an angle of 92 arc seconds.

11.4.3 Deadpath Error

"Deadpath" errors result from a difference in the optical paths of the two arms of the interferometer at the zero position as shown in Figure 11.4(a). They are generated by atmospheric changes during the measurement cycle causing anomalous path differences in the uncompensated light path. The deadpath error in the system can be minimized by appropriately positioning the interferometer block. It will be appreciated however that where a deadpath is inevitable in a system, it is not required to be known to a high precision as the error is typically very small. In single axis systems the technique of taking a "zero" reading before and after a scale measurement gives an indication not only of the repeatability and setting accuracy of the system, but also of whether any deadpath or thermal expansion errors have been generated during the measurement cycle.

11.5 POLARIZING INTERFEROMETERS

Many interferometric systems employ polarization techniques to derive the electrical signals required for reversible fringe counting from their optical outputs. These signals should be sinusoidally related to path differences and, ideally, they should be in phase quadrature, equal in amplitude and their mean D.C. levels zero. In practice the signals are not ideal and, when resolving to sub-nanometric precision, these imperfections impose a limit on the accuracy achievable by the interferometer system. Thin film polarizing beamsplitter designs providing sufficient isolation between the two orthogonally polarized beams are unlikely ever to be available and it is sometimes difficult to maintain the required alignment of the polarization azimuths of the optical components. However it is possible to correct systems for non-ideal optical signals electronically. This is achieved by scanning the optical path in the interferometer through at least one fringe and examining the phases, amplitudes and D.C. levels of the signals both to compute any necessary changes and also to confirm the sinusoidal quality of the interferometer signals. Birch[13,17] has used this technique and applied software corrections, based on mathematical solutions proposed by Heydemann[14] to the NPL interference refractometer system.[11] The instrument achieved a measurement linearity in optical path length of 0.1 nm.

11.6 UNWANTED REFLECTIONS

Stray reflections are another severe systematic limitation to achieving both accuracy and resolution in interferometers. With a laser source these unwanted beams are coherent, so that even one tenth of a percent of the beam energy can cause an anomalous variation in the interferometer signal and a non-linearity error of 1.6 nm in the optical path length measured.

The NPL length measuring interferometer utilizes a plate beamsplitter and it is standard practice with this type of interferometer to minimize the effects of reflections from the non-beamsplitting surface by both employing a standard anti-reflection coating on the surface and by slightly wedging the beamsplitter plate. The latter practice is the most efficient way of solving the problem. From the equations of Rowley[15] it may be shown that the beam divergence caused by a wedge only 1.1/2 minutes of arc introduces sufficient fringes across the aperture of 1 mm diameter Gaussian distribution beam for any stray reflections falling on the photodetectors to have less than nanometric influence on the phase of the interferogram.

Although wedging removes the problems due to stray reflections, it presents another problem in that it effectively turns the beamsplitter into a weak prism. The resulting beam divergence, together with the displacement caused by the 45° angle of incidence on to the plate impose the condition that the beamsplitter must be in position when the optical beam is aligned to the mechanical axis of movement. In practice this makes the alignment procedure extremely difficult and, in addition, prevents the interchange of beamsplitters other than those fabricated with a specified thickness and wedge angle. In order to realize accurate sub-nanometric resolution and to facilitate the operation of the NPL interferometer, a beamsplitter and compensator plate system has been designed. This enables the optimum alignment of the optical and mechanical axes of the system to be achieved before the interferometer block is introduced into the system as it leaves the alignment totally unaffected. The technique involves fabricating a beamsplitter plate of twice the required size. This is then cut into two equal parts which are used in the optical configuration shown in Figure 11.6. One plate acts as a compensator, cancelling out the beam displacement and deviation introduced by the beamsplitter plate. It will be appreciated that it is important to introduce some means of orientation identification onto the plate, for example by slightly chamfering two corners at one end, before it is cut.

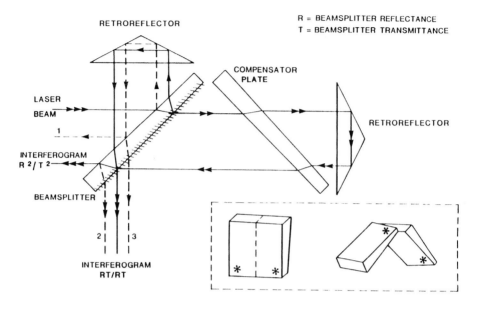

Figure 11.6 Michelson interferometer with compensator plate.

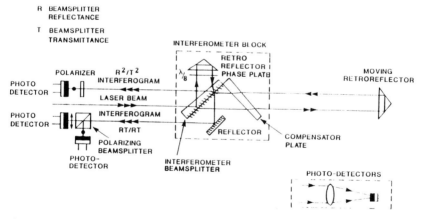

Figure 11.7 Length measuring interferometer.

This arrangement is completely insensitive to the thickness and wedge angle of the original plate. If the direction of the wedge in the beamsplitter plate is confined to the direction in which the beams are displaced (as shown in Figure 11.6), the interferometer system is also chromatically corrected when the outgoing and reflected light beams are symmetrically disposed about the centre of the beamsplitter plates. This would be an advantage if a multi-wavelength source were to be employed. The main stray reflections from the non-beamsplitting interference labelled 1, 2 and 3, are shown in Figure 11.6 by the dashed ray paths. Beams 2 and 3 are deviated, but it is important to note that the beamsplitter wedge does not introduce any angular deviation between reflection 1 and the main interferogram. Although this is a potential source of error the beam undergoes two relatively low intensity reflections, of the order of 1%, and it is significantly displaced so that it will not fall on the photodetector, provided that a beamsplitter plate several mm thick is employed.

The NPL interferometer has also been modified by the addition of a reflector, as shown in Figure 11.7 to allow both interferograms to be examined remotely from the interferometer block.

11.7 AIR REFRACTIVITY CORRECTION

When interferometer systems are used in the free atmosphere, they measure displacements in terms of the wavelength in air rather than the vacuum wavelength of the radiation concerned. It is common practice to make the appropriate corrections for changes in the refractive index of air using Edlen's[10] equation which involves measuring parameters such as pressure, temperature and humidity. This technique, however, is expensive and has been shown to be in error due to the instability of the sensors and variations in the composition of the atmosphere. In addition, it is difficult to maintain the calibration of the sensors to the required accuracy. The sensitivity of the refractive index or the air to these parameters is shown in Table 11.1.

Table 11.1 Interferometer Data

Parameter	Change equivalent to 0.01 µm error in 1 metre length (1 part in 10^8)
Wavelength	
Vacuum Wavelength	1 part in 10^8 (equivalent to a frequency change of 5 MHz)
Interferometer Alignment	
Cosine Error	Single axis 29 seconds/arc (simultaneously in two axes 21 seconds/arc per axis)
Abbe Error	2 seconds/arc (0.1 mm offset)
Atmospheric Wavelength Correction	
a) Calculated	
Refractive Index Equation	1 part in 10^8
Absolute Air Temperature	0.01 °C
Atmospheric Pressure	3.73 Pa (0.03 mm Hg)
Relative Humidity	1%
Carbon Dioxide	67 ppm (normal level 340 ppm)
b) Directly Measured by Refractometer	
Cell Length	10 µm (32 µm long cell)
Differential Temperature	0.01 °C

Two other techniques are currently used to minimize the problems caused by the media in which these interferometers are measuring; measurement in an atmosphere of helium and in a vacuum. However, they are difficult to implement in practice and the most common techniques are either the calculation or measurement of the refractive index of the atmosphere.

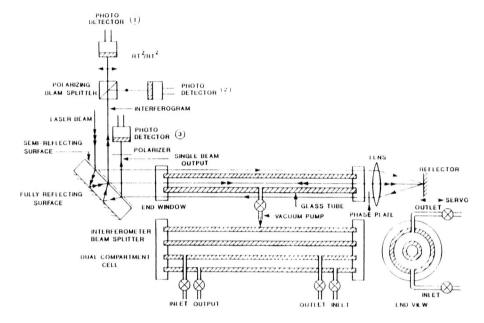

Figure 11.8 NPL gas refractometer.

The approach taken at NPL is to use an interference refractometer[11] the optical configuration of which is shown in Figure 11.8, to provide directly an accurate measurement of the refractive index of the atmosphere and to use this to maintain the overall accuracy of the length measuring interferometer.

Using this instrument and the appropriate measurement technique, the only calibration required to achieve and maintain an accurate value for the refractive index of the atmosphere is the relatively crude one of the length of the gas cell employed in the refractometer. The technique of using an interference refractometer, with an accuracy of ± 1 part in 10^8, in conjunction with a length measuring interferometer enables the absolute accuracy of the complete measuring system to be maintained to better than ± 1 part in 10^7.

The performance of any length measuring system can of course be monitored by the periodic measurement of a calibrated standard and optical scales are being developed at NPL that would be suitable for this purpose.

An extensive study of calculated and measured refractive index values of the atmosphere carried out at NPL[12] have indicated that, where an uncertainty of ± 1 part in 10^7 is required in air refractive index, accurate parametric measurements and a suitable equation may be used provided the only significant air contaminant is carbon dioxide for which corrections should be applied. Where an uncertainty of less than ± 1 part in 10^7 is required the use of a refractometer is recommended.

11.8 OPTICAL SCALES

In order to achieve precision length measurements on optical scales at NPL, photo-electric position location techniques capable of setting on the edge of a feature on a scale to 0.01 μm are employed. A typical optical scale consisting of lines produced in 0.1 μm thick chromium using conventional photo-lithographic techniques on either low expansion glass or silica substrate. The achievement of accurate feature position location when measuring scales necessitates the use of reasonably high magnification in the optical system, in order to increase the sensitivity to setting on the intensity profile of the diffraction broadened image of the feature. Provided that the features are within the depth of focus of the setting system the major limitations to this setting accuracy are their 'raggedness' and variations in the shapes of their edges.

11.9 MECHANICAL MOUNTING

Scales are normally supported using one of the more traditional methods, the most common of these being the Airy points. The Airy support points are two points on a standard scale of length l, each distant from its centre point by $l/2\sqrt{3}$. When it is supported at these two points, varying flexure of bar, arising from changes in the intensity of gravity or small errors in the exact point of support has least effect on the horizontally projected length of the bar.

It can be shown that on its neutral axis the reduction in length of a scale of length l due to a deflection d, resulting from a lack of flatness or bending, is approximately $2d^2/l$. This formula is only applicable on the neutral axis and where thick scales are involved their thickness and cross-section must be taken into consideration. It can be shown that for a scale of thickness t the change in length is approximately $4dt/l$.

Another limitation to the resolution in the precision length measurement of thin optical scales is that the mounting restraints used to prevent scales from moving during measurement can cause significant mechanical deformation of the scale producing errors in the measured length. The mechanical forces generated between the restraints at the ends of the scale either stretch or compress the scale. The problem can be solved using a mounting technique in which one of the restraints used at the ends of the scale is designed to be incapable of sustaining a significant sideways force. For example, a clamp can be used at one end of the scale to stop movement, while a pad on the end of a PTFE pad on the end of a compression spring will hold the other end of the scale down flat whilst preventing any significant forces along the length of the scale.

The technique of having one of the mounts with lateral flexibility could form the basis of a design for all mechanical substrate restraints. For example, the technique would also be applicable to vacuum restraints if one of the constraints were a bellows.

11.10 SUMMARY AND CONCLUSIONS

The limitations to the accuracy and resolution achievable by interferometry in the precision measurement of length in the free atmosphere have been described and a number of aspects of the physics of these systems, when used in this particular application, have been discussed.

The calibrated frequencies of the Helium-Neon laser radiation sources used in these instruments typically change by only 1 or 2 parts in 10^8 in the life-time of the laser tube, and providing the correct optical and mechanical alignment procedures are employed then an absolute measurement accuracy of better than $\pm$ 1 part in 10^7 can be achieved from these systems in a thermally controlled enclosure. The enclosure minimizes the problems caused by air turbulence or inhomogeneity of the air in the optical measuring path. These variations in the atmosphere not only result in noise in the system due to changes in optical path, but also from fluctuations in the wavefront causing the contrast in the interferometer to vary.

When these optical systems are used on length measurement applications in the free atmosphere, corrections for the atmosphere wavelength variations can now be made in any one of the two ways described. First by measuring the parameters of pressure, temperature and humidity and calculating the refractive index of the air using the Edlen's equation with the modification to the water term developed at NPL[12], and secondly by measuring the refractive index directly using an interference refractometer. The first technique, involving the corrected Edlen's formula, has a theoretical accuracy of approaching $\pm$ 3 parts in 10^8 but is dependent on the constituents of the atmosphere, whereas the refractometer technique is independent of the constituents and is capable of an accuracy approaching $\pm$ 1 part in 10^8. However, care must be taken to ensure that the temperature of the gas specimen in the refractometer is the same as that in the path of the length interferometer or, alternatively, that a suitable correction is applied by accurately measuring any temperature differences that may exist. The measurement of differential temperature avoids the problems encountered in absolute temperature measurement of self-heating in the sensors and non-linearities in the sensor electronics. In addition, it minimizes any errors generated by the thermal response delays of the sensors in environments where the air temperature is varying.[17]

An absolute accuracy of better than $\pm$ 1 part in 10^7 in the measurement of length can be achieved by these systems in the free atmosphere, with nanometric resolution in certain specialized applications through the use of fringe fractioning electronics. In the more common interferometric applications such as the measurement of optical scales, a setting resolution of $\pm$ 0.02 μm is typical but to merit such precision, care must be taken in mechanical stability of the scale with respect to both the material and its physical dimensions.

REFERENCES

1. Rowley, W.R.C. (1972) Interferometric Measurement of Length and Distance. *Alta Frequenza*, **IXLI**, 887-896
2. Duanrdo, A.J., Wang, S.C. and Hug, W. (1976) Polarization Properties of Internal Mirror He-Ne Lasers. *SPIE*, **88**, Polarized Light, 34-49
3. Javan, A. and Stokes, A. (1963) *Isotope Shift and Saturation Behaviour of the 1.15 m Transition in Neon*. Report No.NSG-330, MIT, Cambridge
4. Baer, T., Kowalski, F.V. and Hall, J.L. (1980) Frequency Calibration of a 0.633 μm He-Ne Longitudinal Zeeman Laser. *Applied Optics*, **19**, 173-177
5. Takasaki, H., Umeda, N. and Tsukiji, M. (1980) Stabilised Transverse Zeeman Laser as a New Light Source for Optical Measurement. *Applied Optics*, **19**, 435-441
6. Balhorn, R., Kunzmann, H. and Lebowsky, D.C. (1972) Frequency Stabilization of Internal Mirror Helium-Neon Lasers. *Applied Optics*, **11**, 742-744
7. Bennett, S.J., Ward, R.E. and Wilson, D.C. (1973) Comments on: Frequency Stabilisation of Internal Mirror He-Ne Lasers. *Applied Optics*, , 1406
8. Ciddor, P.E. and Duffy, R.M. (1983) Two-Mode Frequency Stabilised He-Ne (633 nm) Lasers: Studies of Short and Long Term Stabilities. *J. Phys. E: Sci. Instrum*, **16**, 1223-1227

9. Wallard, A.J. (1973) The Frequency Stabilisation of Gas Lasers. *J. Physc. E: Sci. Instrum*, **6**, 793-807

10. Edlen, B. (1966) The Refraction Index of Air. *Metrologia*, **2**, 71-80

11. Downs, M.J. and Birch, K.P. (1983) Bi-Directional Fringe Counting Interference Refractometer. *Precision Engineering*, **5**, 105-110

12. Birch, K.P. and Downs, M.J. (1988) The Results of a Comparison between Calculated and Measured Values of the Refractive Index of Air. *J. Physics E: Sci. Instrum*, **?**, 694-695

13. Birch, K.P. (1988) The Precise Determination of Refractometric Parameters of Atmospheric Gases, *Ph.D. Thesis*, Southampton University, UK

14. Heydermann, P.L.M. (1981) Determination and Correction of Quadrature Fringe Measurement Errors in Interferometers. *Applied Optics*, **20**, 3382-3384

15. Rowey, W.R.C. (1969) Signal Strength in Two-beam Interferometers with Laser Illumination. *Optica Acta*, **16**, 159-168

16. Downs, M.J., Ferriss, D.H. and Ward, R.E. (1990) Improving the Accuracy of the Temperature Measurement of Gases by Correction for the Response Delays in the Thermal Sensors. *Meas. Sci. Technol*, **1**, 717-719

17. Birch, K.P. (1990) Optical fringe subdivision with nanometric accuracy. *Precision Engineering*, **12**, 195-198

Chapter 12

OPTICAL DIFFRACTION FOR SURFACE ROUGHNESS MEASUREMENT

JAN H. RAKELS

The use of light scattering or diffraction for the determination of the surface roughness of machined metal components has been investigated by many researchers. As a result, a number of theoretical formulations relating the RMS height R_q to the scattered light intensity distribution have been derived for periodic and random surface lays. This has led to the manufacture of commercial surface finish sensors, based upon the principle of light scattering. Using scalar scattering theory, formal relationships are derived, linking: the specular reflectance to the RMS height parameter R_q, and the standard deviation of the scattered light to the RMS slope parameter Δ_q.

12.1 INTRODUCTION

The use of optical diffraction to determine the surface roughness of machined metal components has been widely investigated. As a result, a number of theoretical formulations of the rough surface scattering problem have been derived for the cases of surfaces with random and periodic surface height distributions. Most of these formulations relate the RMS surface roughness (height) parameter R_q to the specular reflectance as in the Total Integrated Scatter (TIS)[1,2] and the Angularly Resolved Scatter (ARS)[3,4] techniques. TIS is normally applied to random surface height distributions, and ARS to periodic surface profiles. These quantitative relationships are valid for surface roughness values R_q up to approximately 0.1λ under normal incidence illumination with a wavelength of λ.

One should not be under the impression that the terms scattering and diffraction relate to two different phenomena. Both are the same "animal", however, normally we understand diffraction as scattering caused by periodic structures like diffraction gratings without realizing that it is just a special case of light scattering. The same misconception exists in optics about physical and geometrical optics, here again, geometrical optics is just a subset of physical optics.

This chapter first introduces the Fresnel-Kirchhoff scalar diffraction integral and their two approximations, namely the Fraunhofer and Fresnel formulations. The Fraunhofer approximation is equivalent to the Fourier transform (see chapter 3). The Fresnel approximation can be changed into a Fourier transform by employing a converging lens, as explained in section 12.4. In section 12.7 relationships are derived connecting surface roughness parameters to the light intensity distribution, produced by reflecting a plane wavefront from a rough surface. Also three practical implementations of this technique are discussed with their relative merits.

12.2 DEFINITION OF THE SURFACE FINISH PARAMETERS

The RMS surface height and slope parameters, respectively R_q and Δ_q, are defined as:

$$R_q = \sqrt{\frac{1}{L} \int_0^L h^2(x)\,dx} \tag{12.1}$$

and

$$\Delta_q = \sqrt{\frac{1}{L} \int_0^L \left\{ \frac{d}{dx} h(x) \right\}^2 dx} \tag{12.2}$$

where $h(x)$ is the surface profile and L is the sampling length in the x-direction. Please note that these parameters are defined for surface profiles and not areas.

12.3 DIFFRACTION

Before discussing the use of diffraction in surface metrology, a brief account will be given of the history of diffraction.

What is now called diffraction was reported by Grimaldi (1618-1663). He noticed the deviation of light from rectilinear propagation, something he called "diffractio". This effect is a general characteristic of wave phenomena, occurring whenever a portion of a wavefront, be it sound or light, is obstructed in some way. Grimaldi had observed bands of light within the shadow of a rod illuminated by a small source. Hooke (1635-1703) later also observed diffraction effects. He was the first to study coloured interference patterns, which brought him to propose a wave theory of light. The wave theory of light was not favoured very much by Newton (1642-1727). Perhaps his main reason for rejecting a wave theory as it stood then was the blatant problem of explaining rectilinear propagation in terms of waves which spread out in all directions, and he became more committed to the emission (corpuscular) theory. At the same time that Newton was emphasizing the emission theory in England, Huygens (1629-1695) was greatly extending the wave theory on the continent. However, the great weight of Newton's opinion stopped the further evolution of the wave theory. Young (1773-1829) revived the wave theory, but did not meet much appreciation because of Newton's opposition to it.

Unaware of Young's efforts, Fresnel (1788-1827) began the revival of wave theory in France. Fresnel synthesized the concepts of Huygens' wave description and the "interference principle". The corresponding Huygens-Fresnel principle states that every unobstructed point of a wavefront, at a given instant in time, serves as a source of spherical secondary wavelets. The amplitude of the optical field at any point beyond is the superposition of all these wavelets (considering their amplitudes and relative phases). Under the criticism of Laplace (1749-1827) and Biot (1774-1862), who advocated the emission theory, Fresnel's theory took on a more mathematical emphasis. He was able to calculate the diffraction patterns from various obstacles and accounted for rectilinear propagation in homogeneous isotropic media, thus dispelling Newton's main objection to the wave theory.

Kirchhoff (1824-1887) developed a more rigorous theory based directly on the solution of the differential wave equation and Green's theorem (1793-1841). His refined analysis lent credence to the assumptions of Fresnel and led to an even more precise formulation of Huygen's principle as a consequence of the wave equation. Whilst the integral theorem of Kirchhoff embodies the basic idea of the Huygens-Fresnel principle, the laws governing the contributions from different elements of the obstacle are more complicated than Fresnel assumed. Kirchhoff showed, however, that in many cases the theory may be reduced to an approximate but much simpler form, which is essentially equivalent to Fresnel's formulation, but which in addition gives an explicit formula for the obliquity factor that remained undetermined in Fresnel's theory. (Note this obliquity factor is (1+cosα), being the angle between the incident and transmission directions). Even so, the Kirchhoff theory itself is an approximation which is valid for sufficient small wavelength, i.e. when the diffracting objects have dimensions which are large in comparison with the illuminating wavelength. Kirchhoff's theory, however, works very well even though it deals with scalar waves and is insensitive to the fact that light is a transverse vector field.

12.3.1 Introduction to Simple Diffraction Theory

Fundamental to the understanding of physical optics and diffraction is the concept of light as a wave disturbance. Consider, for example, the passage of a light beam through a small opening in an opaque screen. The ray optics description of this situation leads to the conclusion that the size of the spot of light observed on a second screen some distance from the first will be simply proportional to the size of the hole. Such a proportionality law holds quite well for fairly large holes but does not apply at all for small holes. In fact, if the transition from illuminated to unilluminated areas are examined carefully, the geometric predictions do not hold even for large holes. Furthermore, as the hole is made smaller, the observed spot of light will actually increase as the diameter of the hole decreases. Quite clearly, simple geometric predictions are inadequate, and a physical optics approach is necessary.

From the point of view of basic physics, the wave nature of light is fundamental, stemming from the consideration that light is an electromagnetic disturbance and hence is propagated by the wave vector equations, which are readily derived from Maxwell's equations. This approach involves quite a level of mathematically complexity, which will be omitted here. Those interested in this aspect should find Stratton[5] a very readable treatment. In this chapter a much more pragmatic point of view is assumed, namely that a large class of optical phenomena can be accurately described by the hypothesis that light is a scalar, monochromatic wave.

12.3.2 Wave Propagation

The basic problem of diffraction is the determination of the manner in which a wave propagates from one plane to another. Let the light wave be expressed by the electric component $E(x, y, z, t)$ of an electromagnetic wave. For justification see Wiener's experiment[6] on standing waves produced by reflected light. In a region where there is no electromagnetic source, E satisfies the 3-D wave equation.

$$\nabla^2 E(x, y, z, t) = \frac{1}{c^2} \frac{d^2 E(x, y, z, t)}{dt^2} \tag{12.3}$$

where c is the velocity of light in a vacuum.

For monochromatic waves, $E(x,y,z,t)$ separates to a form

$$E(x,y,z,t) = V(x,y,z)\exp[-j2\pi\nu t] \tag{12.4}$$

where ν is the frequency of the wave and $V(x,y,z)$ describes the spatial variation of the amplitude and phase of the disturbance. By substituting this monochromatic form into the general wave equation, the time dependence is eliminated and the spatial part of the disturbance is seen to satisfy the Helmholtz equation

$$\nabla^2 V(x,y,z) + \left(\frac{2\pi\nu}{c}\right)^2 V(x,y,z) = 0 \tag{12.5}$$

with c being the speed of light as before. A simplified theory will now be adopted which focuses on only one of the vector components of $V(x,y,z)$. This component will be represented by the scalar quantity $v(x,y,z)$, which will satisfy the scalar wave equation

$$\nabla^2 v + k^2 v = 0 \tag{12.6}$$

where $k = 2\pi/\lambda$. Equation (12.6) may be rigorously solved using Green's theorem[6] but here a solution will be derived based upon Huygens' principle. I.e., the solution is constructed from the following principle: "A geometric point source of light will give rise to a spherical wave emanating equally in all directions". To construct a general solution from this, one should note that the Helmholtz equation is linear and hence a superposition of solutions is permitted. Now by considering that an arbitrary wave shape may be considered as a collection of point sources whose strength is given by the amplitude and relative phase of the wave at that point. The field, at any point in space, simply a sum of spherical waves. The only shortcoming of this argument is that it ignores the fact that the wave has a preferred direction of propagation. In the more rigorous theory this is accounted for by the inclusion of an inclination or obliquity factor.

A spherical wave in free space can be expressed in terms of the solution of (12.6) in spherical coordinates. Since the only variation of v is in the r direction, $\nabla^2 v$ becomes

$$\nabla^2 v = \frac{1}{r}\frac{d^2(rv)}{dr^2}$$

so that (12.6) reduces to

$$\frac{d^2(rv)}{dr^2} + k^2(rv) = 0 \tag{12.7}$$

The general solution of (12.7) is

$$rv = A\exp[jkr] + B\exp[-jkr]$$

or

$$v = Aexp[jkr]/r + Bexp[-jkr]/r \qquad (12.8)$$

where A and B are constants, and k is the wavenumber.

Using $exp[-j\omega t]$ for the time-dependent part (ω is the angular velocity), the first term of (12.8) represents a wave diverging from the origin and the second, a wave converging toward the origin.

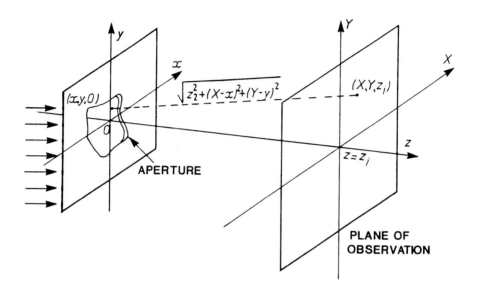

Figure 12.1 Geometry for calculating the diffraction pattern U(X,Y,z) on the screen produced by the input source u(x,y,0).

Using the rectangular coordinates as shown in Figure 12.1, Huygens' principle applied to the diverging wavefront leads to the approximate diffraction integral formulation:

$$U(X,Y) = \frac{1}{j\lambda} \int\int_{-\infty}^{\infty} u(x,y)exp[jkr]/r \; dxdy \qquad (12.9)$$

Where $u(x,y)$ is the field in the aperture from which new wavefronts are emanating, thereby leading to the integral expression for the field $U(X,Y)$ at the position (X,Y) on the observation screen and distance z away. Please note that this formula is an oversimplified diffraction formula. For instance, an obliquity factor is missing and the expression is a scalar one. The term $1/j\lambda$ makes sure that the equation is dimensionally correct.

In order to evaluate equation (12.9), the assumption is made that z^2 is much greater than $(X-x)^2+(Y-y)^2$. Then by using the binomial expansion:

$$r = \sqrt{z^2 + (X-x)^2 + (Y-y)^2} = z\sqrt{1 + \frac{(X-x)^2 + (Y-y)^2}{z^2}} \tag{12.10}$$

$$= z + \frac{(X-x)^2 + (Y-y)^2}{2z} - \frac{[(X-x)^2 + (Y-y)^2]^2}{8z^3}$$

or

$$r = z + \frac{x^2 + y^2}{2z} - \frac{Xx + Yy}{z} + \frac{X^2 + Y^2}{2z} - \frac{[(X-x)^2 + (y-y)^2]^2}{8z^3} \tag{12.11}$$

The region where only the first three terms are included is called the "far field" or "Fraunhofer" region. The region where the first four terms are included is called the "near field" or "Fresnel" region. It should be noted that in the region very close to the aperture neither of these approximations are valid because the binomial expansion may no longer be used.

It is the value of the fourth term in equation (12.11) that determines whether the Fresnel or Fraunhofer region represents the area where one obtains the Fourier transform of the input function, and a lot of theory is readily available on the properties of Fourier transforms.

To get an impression where the Fraunhofer region lies for a small area with x and y dimensions of 6 mm and an illumination wavelength of 600 nm (HeNe laser), it appears that this region is about 60 m away from the input. Obviously, if one intends to build an instrument which can produce a Fourier transform, then one is confronted with some impractical dimensions. However, it is possible to observe the Fraunhofer pattern within the Fresnel region by use of a positive or converging lens which cancels the fourth term of equation (12.11). As will be shown in section 12.4.3, the thickness of a converging lens produces a phase delay which cancels the offending term as long as the observation plane is placed in the focal plane of the lens. A heuristic explanation for this property of a lens is that all parallel lines entering a lens cross in the focal plane of the lens. Now looking at the position of the Fraunhofer region it is clear that this plane is ideally placed at infinity, i.e. the plane where parallel lines cross.

12.3.3 Approximations in the Fraunhofer Region

Two-dimensional Fraunhofer diffraction formula

As mentioned, the Fraunhofer approximation makes use of only the first three terms of (12.11). When these three terms are inserted into (12.9), the field on the observation screen is

$$U(X,Y) = \frac{1}{j\lambda z} \exp\left[jk\left\{z + \frac{x^2+y^2}{2z}\right\}\right] \cdot \int\int_{-\infty}^{\infty} u(x,y)$$

(12.12)

$$x \exp\left[-j2\pi\left(\frac{Xx}{\lambda z} + \frac{Yy}{\lambda z}\right)\right] \; dxdy$$

Where the approximation $r \approx z$ is used in the denominator of (12.12). It is obvious that (12.12) represents a two-dimensional Fourier transform, so the diffraction pattern can be expressed in Fourier transform notation as follows.

$$G(f_x, f_y) = \int\int_{-\infty}^{\infty} u(x,y) \exp[-2j\pi(xf_x + yf_y)] \, dxdy$$

(12.13)

$$= F[u(x,y)]_{f_x, f_y}$$

with $f_x = X/\lambda z$ and $f_y = Y/\lambda z$.

One-dimensional Fraunhofer diffraction formula

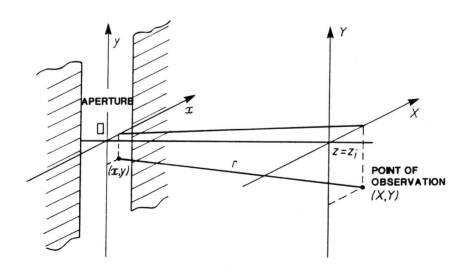

Figure 12.2 Geometry used for calculating the one-dimensional diffraction pattern.

233

Figure 12.2 shows a narrow, illuminated aperture where the input function $u(x,y)$ is only a function of the x ordinate, i.e.:

$$u(x,y) = u(x) \tag{12.14}$$

and hence the Fresnel-Kirchhoff approximation, equation (12.9), becomes

$$U(X,Y) = \frac{1}{j\lambda z} \int_{-\infty}^{\infty} \left[u(x) \left(\int_{-\infty}^{\infty} \exp[jkr]/r \, dy \right) \right] dx \tag{12.15}$$

where

$$r = \sqrt{z^2 + (X-x)^2 + (Y-y)^2}$$

In order to transform (12.15) in a more manageable approximate expression, firstly the integral I in the y direction will have to be evaluated, where

$$I = \int_{-\infty}^{\infty} \exp[jkr]/r \, dy \tag{12.16}$$

Denoting ρ as the projection of r onto the plane $y=0$ (see Figure 12.2), where

$$\rho = \sqrt{z^2 + (X-x)^2} \tag{12.17}$$

then r can be expressed by

$$r = \sqrt{\rho^2 + (Y-y)^2} \tag{12.18}$$

Introducing a change of variable

$$y - Y = \rho \sinh(t) \tag{12.19}$$

leads to the following expression for r;

$$r = \rho \cosh(t) \tag{12.20}$$

Furthermore,

$$dY = \rho \cosh(t) dt \tag{12.21}$$

This results in an expression for the integral I

$$I = \int_{-\infty}^{\infty} \exp[jk\rho\cosh(t)]dt = j\pi H_0^{(1)}(k\rho) \qquad (12.22)$$

where $H_0^{(1)}(k\rho)$ is the zeroth-order Hankel function of the first kind, now referred to by H_0. This transforms equation (12.22) into the one dimensional Fresnel Kirchhoff formula:

$$U(X) = \frac{\pi}{\lambda} \int_{-\infty}^{\infty} u(x)H_0^{(1)}(k\rho) \qquad (12.23)$$

Under certain conditions, the integration in (12.23) can be simplified. When $k \gg 1$ the Hankel function can be approximated as

$$H_0(k\rho) \approx \frac{1}{\pi}\sqrt{\frac{\lambda}{\rho}} \exp[j(k\rho-\pi/4)] \qquad (12.24)$$

and when $z^2 \gg (X-x)^2$, ρ can be written as

$$\rho = z+(X-x)^2/2z \qquad (12.25)$$

Inserting equations (12.24) and (12.25) into (12.23) yields

$$U(X) = \frac{\exp[jk(z-\pi/4]}{\sqrt{\lambda z}} \int_{-\infty}^{\infty} u(x)\exp[jk(X-x)^2/2z]dx \qquad (12.26)$$

This equation is the formula for the one-dimensional Fresnel approximation. When $\lambda z \gg X^2$, the formula for the one-dimensional Fraunhofer approximation is obtained

$$U(X) = \frac{1}{\sqrt{\lambda z}} \exp[j(kz+kX^2/2z-\pi/4)].F[u(x)]_{f=x/\lambda z} \qquad (12.27)$$

Mathematically, the Fraunhofer approximation looks very attractive. However it has an enormous drawback, namely, it can only be observed at a great distance away from the input, thereby rendering it useless for basing an instrument upon. Fortunately, the Fresnel approximation can be used in conjunction with a converging lens, as will be shown in section 12.4.3.

Approximation in the Fresnel region

The Fresnel approximation is obtained by inserting the first four terms of equation (12.11) into (12.9). Two types of expressions for the Fresnel approximation can be obtained depending on whether or not $(X-x)^2 + (Y-y)^2$ is expanded; one is in the form of a convolution and the other as a Fourier transform.

1. Convolution expression for the Fresnel region
If $(X-x)^2 + (Y-y)^2$ is inserted into equation (12.9) without expansion for r,

$$U(X,Y) = \frac{e^{jkz}}{j\lambda z} \cdot \int\int_{-\infty}^{\infty} u(x,y)$$

$$\times \quad exp[jk((X-x)^2 + (Y-y)^2)/2z]dxdy \qquad (12.28)$$

is obtained. This expression represents a convolution, and the diffraction pattern can be written using the convolution symbol, $\otimes$, as follows:

$$U(X,Y) = u(X,Y) \otimes f_z(X,Y) \qquad (12.29)$$

where

$$f_z(X,Y) = \frac{1}{j\lambda z} exp[jk(z+(X^2+Y^2)/2z]$$

Here $f_z(X,Y)$ is of the same form as the approximation obtained by binomially expanding r in the expression $exp[jkr]/r$ for a point located at the origin.

2. Fourier transform expression for the Fresnel region
If $(X-x)^2+(Y-y)^2$ of equation (12.11) is expanded and used in (12.9), one obtains

$$U(X,Y) = \frac{1}{j\lambda z} exp[jk(z+(X^2+Y^2)/2z] \cdot \int\int_{-\infty}^{\infty} u(x,y)$$

$$\times \quad exp[jk(x^2+y^2)/2z - j2\pi((Xx+Yy)/\lambda z)]dxdy \qquad (12.30)$$

This resembles the Fourier transform formula, it can be rewritten as

$$U(X,Y) = \frac{1}{j\lambda z} exp[jk(z+(X^2+Y^2)/2z] . F[u(x,y)$$

$$\times \quad exp[jk(x^2+y^2)/2z]]$$

$$= exp[jk(X^2+Y^2)/2z] . F[u(x,y) . f_z(x,y)]_{f_x,f_y} \qquad (12.31)$$

with $f_x = X/\lambda z$ and $f_y = Y/\lambda z$.
  Both expressions (12.29) and (12.31) give the same answer, but often one method is easier for computation. Generally speaking, it is more convenient to use the convolution

expression when $U(X,Y)$ has to be further transformed, because the Fourier transform of a convolution is a multiplication of the Fourier transforms.

12.4 FOURIER TRANSFORM PROPERTIES OF LENSES

The normal functions that lenses perform include the convergence or divergence of light beams, and the formation of real and virtual images. An additional interesting property of lenses is that they can perform 1-D or 2-D Fourier transforms. This property will now be examined by analyzing the design and performance of a plano-convex lens. A special feature of the optical Fourier transform is that the position of the transformed image is not influenced by the location of the input image.

12.4.1 Design of a Plano-convex Lens

The contour of an ideal plano-convex lens is made in such a way that light from a point source becomes a parallel beam after passing through the lens.

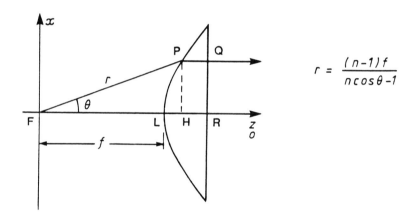

$$r = \frac{(n-1)f}{n\cos\theta - 1}$$

Figure 12.3 Design of the curvature of a plano-convex lens.

This means that it is designed so that the optical paths F - P - Q and F - L - R in Figure 12.3 are equal. Let H be the point of projection from P to the optical axis. Since $\overline{PQ}=\overline{HR}$, the optical paths will be equal if $\overline{FP} = \overline{FH}$. The optical paths are expressed as follows

Optical path of $\overline{FP} = r$

Optical path of $\overline{FH} = f + n(r\cos\theta - f)$

Where n is the refractive index of the glass, and f is the focal length $\overline{FL}$. When the optical path of $\overline{FP}$ is set equal to that of $\overline{FH}$, the formula for the contour of the convex surface of the lens is found to be

$$r = (n-1)f / (n\cos\theta - 1) \qquad (12.32)$$

Next, equation (12.32) is converted into rectangular coordinates. For now, consider only the $y=0$ plane. Taking the origin as the focus of the lens, the coordinates of the point P are

$$r = \sqrt{x^2 + z^2}$$
$$\cos\theta = (z / \sqrt{x^2 + z^2}) \qquad (12.33)$$

Inserting equation (12.33) into (12.32) yields

$$\frac{(z-c)^2}{a^2} - \frac{x^2}{b^2} = 1 \qquad (12.34)$$

where $a=f/(n+1)$, $b=f\sqrt{[(n-1)/(n+1)]}$ and $c=nf/(n+1)$. It can be seen from equation (12.34) that the contour of this lens is a hyperbola. These lenses are normally classified as aspherics. Much effort has gone into the production of these lenses. At present they are produced by computer controlled diamond turning machines, whereas normal (spherical) lenses are produced by grinding.

12.4.2 Wave Properties of Lenses

In section 12.4.1., the lens was designed with the aid of geometrical optics. However, the lenses form part of a wave optical system. The phase redistribution of a wave transmitted through a lens will now be examined. As shown in Figure 12.4, the incident beam is incident parallel to the lens z-axis.

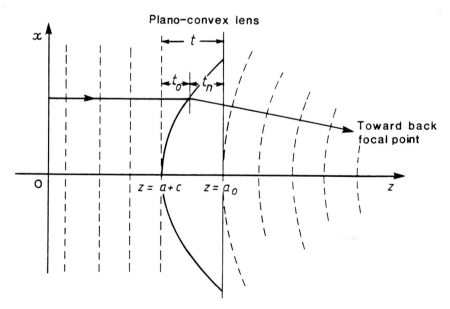

Figure 12.4 Phase distribution of a beam passing through a plano-convex lens.

When the lens is thin, equation (12.35) can be simplified. The larger b is, the smaller the variation of z with respect to the z-axis becomes, which leads to the thin lens condition

$$(x/b)^2 \ll 1 \tag{12.36}$$

When equation (12.36) is satisfied, an approximate formula for equation (12.35) is obtained using the binomial expansion

$$z \approx a + c + ax^2/2b^2 \tag{12.37}$$

If the plane surface of the lens is at $z = a_0$, the thickness t_n of the glass at x is

$$t_n = a_0 - (a + c + ax^2/2b^2) \tag{12.38}$$

and the phase of the plane wave passed through the lens at x is

$$\phi(x) = k[a_0 - a - c) - t_n] + nkt_n \tag{12.39}$$

again, inserting the values for a, b, and c from equation (12.35), the phase ϕ becomes

$$\phi(x) = \phi_0 - kx^2/2f \tag{12.40}$$

where $\phi_0 = kn(a_0 - f)$. Since an analogous relationship holds in the y direction, and the phase change can be written as:

$$\phi(x,y) = \phi_0 - k(x^2 + y^2)/2f = \phi_0 - kr^2/2f \tag{12.41}$$

where r is the distance from the lens axis to the point (x,y).

In conclusion, the lens creates a phase distribution whereby it advances with the square of the radius from the optical axis.

12.4.3 Fourier Transformation by a Lens

As mentioned previously, the optical Fourier transform is one of the most frequently used lens function, other than image formation which can be seen as two cascaded Fourier transformations. The results of the Fourier transform depends critically on the relative positions of the lens and the output screen and less critically on the position of the input object.

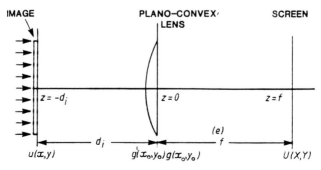

Figure 12.5 Fourier transform by a convex lens; the case when the input u(x,y) located in front of the lens and the screen is at the back of the focal plane.

Figure 12.5 shows the general situation where the input $u(x,y)$ is placed a distance d_i in front of the lens, which is located at $z=0$, and the output is observed in the Fresnel region at $z=e$. The lens has a focal length of f. For ease of analysis, the propagation distance is separated into three sections: propagation from the object to the lens, transmission through the lens and propagation from the lens to the output plane. Assuming that the light distribution just before the lens is $g'(x_0,y_0)$, then the light distribution $g(x_0,y_0)$ just behind the lens can be expressed by

$$g(x_0,y_0)=g'(x_0,y_0)\exp[j\phi_0-jk(x_0^2+y_0^2)/2f] \tag{12.42}$$

The constant phase factor ϕ_0 is not important for analysis and can be assumed to be zero. After propagation to the screen at $z=e$, the distribution $U(X,Y)$ is, according to the Fourier transform formulation for the Fresnel approximation:

$$U(X,Y) = (1/j\lambda e)\exp[jk(e+(X^2+Y^2)/2e]$$
$$\times F[g(x_0,y_0)\exp[jk(x_0^2+y_0^2)/2e]]_{f_{ox},f_{oy}} \tag{12.43}$$

with $f_{ox}=X/\lambda f$ and $f_{oy}=Y/\lambda f$ as the spatial frequencies in the Fourier domain.

Entering the expression in equation (12.42) into (12.43) and equating e to f, i.e. the observation screen is placed in the back focal plane of the lens, yields

$$U(X,Y) = (1/j\lambda f)\exp[jk(f+(X^2+Y^2)/2f]$$
$$\times F[g'(x_0,y_0)]_{f_{ox},f_{oy}} \tag{12.44}$$

The field distribution $g'(x_0,y_0)$, produced by the input $u(x,y)$ after propagation over the distance δ, can be expressed by the convolution formulation for the Fresnel approximation, i.e.

$$g'(x_0,y_0) = u(x_0,y_0)\otimes f_\delta(x_0,y_0) \tag{12.45}$$

where

$$f_\delta(x_0,y_0) = (1/j\lambda\delta)\exp[jk(\delta+(x_0^2+y_0^2)/2d] \tag{12.46}$$

Substitution of equation (12.45) into (12.44) and using the Fourier transform properties for products and convolutions, yields:

$$U(X,Y) = (1/j\lambda f)\exp[jk(f+(X^2+Y^2)/2f]$$
$$\times F[u(x_0,y_0)]\cdot F[f_\delta(x_0,y_0)]_{f_{ox},f_{oy}} \tag{12.47}$$

Performing the Fourier transform $F[f_\delta(x_0,y_0)]$ results:

$$F[f_\delta(x_o,y_o)] = \int\int_{-\infty}^{\infty}(1/j\lambda\delta)exp[jk(\delta+(x_o^2+y_o^2)/2\delta)]$$

$$\times\ exp[-j2\pi(x_oX+y_oY)/\lambda f]dx_o dy_o$$

$$=\ exp[jk\delta-j\pi\lambda\delta(X^2+Y^2)/\lambda^2 f^2] \tag{12.48}$$

$$=\ exp[jk\delta-jk\delta(X^2+Y^2)/2f^2]$$

This result leads to the general expression for $U(X,Y)$

$$U(X,Y)\ =\ (1/j\lambda f)exp[jk(f+\delta+(X^2+Y^2)(1-\delta/f)/2f)]$$
$$\times\ F[u(x_o,y_o)].F[f_\delta(x_o,y_o)]_{f_{ox},f_{oy}} \tag{12.49}$$

Now equating δ to f, i.e. the input is placed in the front focal plane of the transform lens, eliminates the spatial dependent part of the phase function, and $U(X,Y)$ is proportional to the exact Fourier transform of $u(x,y)$

$$U(X,Y)\ =\ (1/j\lambda f)exp[j2kf].F[u(x_o,y_o)]f_{ox},f_{oy} \tag{12.50}$$

In short, the amplitude distribution of light falling on a screen or detector, placed in the focal plane of a lens, is proportional to the Fourier transform of the input light distribution multiplied by a phase factor $\phi(X^2+Y^2)$. This factor can be eliminated by placing the input in the front focal plane of the transform lens.

In the following discussion it is assumed that the maximum value of the coordinates in the Fourier plane are small with respect to the focal length of the transform lens and that the input is located in or near the front focal plane of this lens. Under these conditions the variable phase function can be ignored, so that a Fourier transform of the input will be obtained.

12.5 GENERATION OF THE INPUT LIGHT DISTRIBUTION

The input light distribution $u(x,y)$ can be produced by modulating a coherent plane wavefront by reflection or transmission. The modulation can consist of an amplitude or phase modulation or a combination of both. An amplitude modulation would be ideal. However, in the case of surface height analysis of machined components, it is impossible to produce an amplitude modulation which is proportional to the surface height distribution. The only possibility left is to produce a phase modulation which is proportional to the surface height distribution $h(x,y)$, namely:

$$u(x,y)\ =\ c_1 exp[jc_2 h(x,y)] \tag{12.51}$$

Where c_1 and c_2 are constants. When $c_2 h(x,y) \ll 1$, then the exponential expression can be replaced by the first two terms of the binomial expansion and

$$u(x,y)\ \approx\ c_1(1+jc_2 h(x,y)) \tag{12.52}$$

from which it follows that, except for one constant term, $u(x,y)$ is proportional to $h(x,y)$, and the optical transform will be the desired transform of the surface height. From this transform various surface finish parameters can be extracted. The constant term in equation (12.52) produces a Dirac delta function at zero spatial frequency. Optically this light intensity peak is known as the specular reflectance or as the zero diffraction order.

There are two means of obtaining a phase modulated input field. One is by making transparent replicas of the surface height, distribution, whereby one surface of the replica matches the original surface height distribution.[7] The other method is by reflecting an incoming plane wavefront from the actual surface itself.

12.5.1 Replica Method

When a transparent replica is made of a two dimensional surface $h(x,y)$, then the resulting thickness variation will be $t-h(x,y)$, see Figure 12.6, where t is the average thickness of the replica[7].

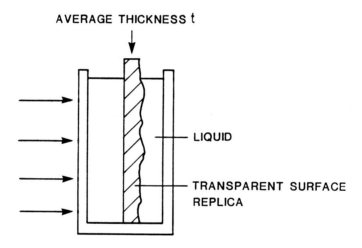

Figure 12.6 Transparent surface height replica, submerged in a liquid.

This replica can be immersed in a glass container filled with a liquid. The optical path through this container will be $c_n+(n_1-n_2)h(x,y)$, where c_n is a constant, n_1 and n_2 are the refractive indices of replica and liquid, respectively. When a plane wavefront is passed through this container, then the phase modulated input field $u(x,y)$ can be expressed by:

$$u(x,y) = c \cdot \exp[j2\pi(n_1-n_2)h(x,y)/\lambda] \qquad (12.53)$$

where c is a new constant.

By selecting the right combination of refractive indices, it is possible to make the exponent in equation (12.53) much smaller than unity, and the approximation in equation (12.52) can be employed. This method can cope with very rough surfaces, however, it has the disadvantage that a replica has to be produced first. This rules this technique out for in-process measurements of surface finish of machined components.

12.5.2 Reflection method

When a plane wavefront is incident on a surface $h(x,y)$ at an angle of incidence θ_i, then the reflected wavefront undergoes a phase modulation expressed by[8]

$$\theta = 4\pi h(x,y)\cos(\theta_i)/\lambda \qquad (12.54)$$

(a)

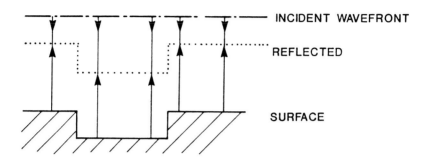

(b)

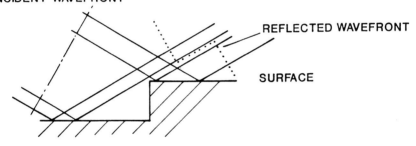

Figure 12.7 Phase modulation caused by reflecting a wavefront from a surface, for the case of (a) normal incidence, (b) off-normal incidence.

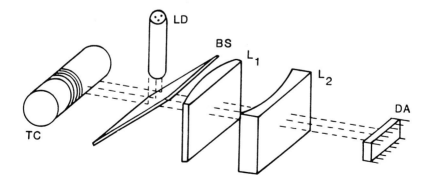

Figure 12.8 Practical transform system, consisting of a diode laser LD, beamsplitter BS, transform lens system L_1 and L_2 and a photo diode array DA. TC is a sample under test.

The Figures 12.7(a) & (b) show this phase modulation for a simple surface height model for both normal and off-normal incidence. The reflected phase modulated distribution, in a plane very close to and parallel to the mean of the surface can be expressed by:

$$u(x,y) = c.exp[j4\pi h(x,y)cos(\theta_i)/\lambda] \tag{12.55}$$

The possibility also exists here to adjust the value of the exponent by changing the incidence angle.

Figure 12.8 shows the lay out of a practical system for obtaining the Fourier transform of surface heights by reflection. Normal incidence is exploited in order to construct a compact unit. The light source is a collimated diode laser and the detector is a photo-diode array which contains 512 photo-diodes at 25 μm spacing.

This system lends itself ideally for in-process measurements, since there is no contact between the sensor and the workpiece. However, here the maximum surface height deviations are limited to about $\lambda/10$, in order to keep the exponent of equation (12.55) small enough.

12.6 RELATION BETWEEN OPTICAL FIELD & LIGHT INTENSITY

Direct observation of the optical (electrical) field is normally not possible, because of the extremely high frequencies of visible and infrared light waves (3×10^{14} Hz for $\lambda=1$ μm). The only detectable quantity is the light intensity $I(X,Y)$, which is the time average of the amount of energy which, in unit time, crosses a unit area normal to the direction of energy flow[6]. For optical fields, this is proportional to the time average of the square of the electrical field,

$$I(X,Y)=\lim_{T\to\infty} \frac{1}{T} \int_0^T E^2(X,Y,t)dt \tag{12.56}$$

where $E(X,Y,t)$ is the real part of $U(X,Y)exp[j\omega t]$, with $exp[j\omega t]$ the time dependent part of the optical field which is normally suppressed. $E(X,Y,t)$ can be expressed by:

$$E(X,Y,t)=(U(X,Y)exp[j\omega t]+U^*(X,Y)exp[-j\omega t])/2 \tag{12.57}$$

where $U^*(X,Y)$ is the complex conjugate of $U(X,Y)$. Insertion into equation (12.56) yields:

$$I(X,Y,t) = \lim_{T\to\infty} \frac{1}{4T} \int_0^T (U^2(X,Y)exp[j2wt]+U^{*2}(X,Y)$$

$$\times exp[-j2wt]+2U(X,Y)U^*(X,Y))dt \tag{12.58}$$

The time average of the first two time dependent terms equal zero, and so the intensity can be expressed by:

$$I(X,Y) = U(X,Y).U^*(X,Y)/2 = |U(X,Y)|^2/2 \tag{12.59}$$

So the light intensity $I(X,Y)$ in the observation point (X,Y) is proportional to the squared absolute value of the optical field $U(X,Y)$.

12.7 RELATION BETWEEN SCATTERED LIGHT INTENSITY & SURFACE FINISH PARAMETERS

It is assumed that a plane wavefront is normally incident on an imperfectly smooth flat metal surface, whose surface deviations from the mean equals $h(x,y)$. According to equation (12.54), the phase of the reflected wavefront, very close to the surface, is related by geometrical optics to the surface height by:

$$\phi(x,y) = 2kh(x,y) \tag{12.60}$$

This will produce an optical distribution $U(X,Y)$ in the focal plane of a transform lens. This distribution can be expressed by:

$$U(X,Y)=c\int\int_{-\infty}^{\infty} exp[j2kh(x,y).exp[-jk(Xx+Yy)/f]dxdy \tag{12.61}$$

where f is the focal length of the transform lens and c is a constant.

Since the surface finish parameters are normally defined for surface profiles, rather than rastered surface maps, it is convenient to reduce the two-dimensional Fraunhofer integral to a one-dimensional one. In reality this can be achieved by placing the detector such that $Y=0$ and having a surface height which is a function of only the x-ordinate. This may seem to reduce the usefulness of the following results, but it should be noted that the majority of engineering surfaces actually have an unidirectional lay as produced by turning and cylindrical grinding. Under these assumptions equation (12.61) can be simplified to:

$$U(X) = c\int_{-\infty}^{\infty} exp[j2kh(x)]exp[-j2kXx/f]dx \tag{12.62}$$

The light intensity $I(X)$, in the diffraction plane, is now expressed by

$$I(X) = C|U(X)|^2 \tag{12.63}$$

which, by using the well known Wiener-Khintchine theorem can be rewritten as

$$I(X)) = \int\int(exp[j2kh(x+s)] \tag{12.64}$$

$$\times\ exp[-j2kh(x)])dx.exp[-jkXs/f]ds$$

With the aid of the equations (12.62) to (12.64), various relationships can be derived which link the light intensity of the diffraction pattern to the RMS surface roughness height R_q and slope Δq.

12.7.1 Specular Reflectance and RMS Surface Height

Using equations (12.62) & (12.63) and assuming that the surface is perfectly flat, so $h(x)=0$, then the specular intensity, $I(R_q)=0$, reflected in the diffraction plane at X=0 can be expressed by

$$I(R_q=0)=C\left|c\int_0^L exp[j2k0]exp[-jk0x/f]dx\right|^2 = Cc^2 \qquad (12.65)$$

This equals all the reflected light. If the surface is not perfectly smooth, then some of the light will be scattered outside the specular direction (X=0). However, because of the law of conservation of energy, the total amount of light in the diffraction plane will still equal $I(R_q=0)$.

Gaussian surface height distribution

Assuming that the surface height distribution $h(x)$ is Gaussian with a RMS value R_q, then the height distribution $p(h)$ can be expressed by

$$p(h) = exp[-(h/R_q)^2/(2\sqrt{2\pi} R_q)] \qquad (12.66)$$

From this probability function (12.66), it follows that the light distribution at X=0 equals

$$U(0)=c\int_0^L exp[j2kj(x)]dx=c\int_{-\infty}^{\infty}exp[j2kh]$$
$$\times exp[-(h/R_q)^2/2]/\sqrt{2\pi} R_q)dh \qquad (12.67)$$

Using the equality[9]

$$\int_{-\infty}^{\infty}exp[jtx]exp[-x^2/2]\sqrt{2}dx = exp[-t^2/2] \qquad (12.68)$$

leads to the following expression for $U(0)$

$$U(0) = c.exp[-8\pi R_q/\lambda)^2] \qquad (12.69)$$

Entering this result in equation (12.63) yields for the specular reflected intensity $I_r(R_q)$ in the presence of random surface roughness R_q

$$I_r(R_q) = Cc^2exp[-(4\pi R_q/\lambda)^2] \qquad (12.70)$$

From equations (12.66) and (12.70), it follows that for a surface with a Gaussian height distribution and under normally incident light, the specular reflected intensity is

$$I_r(R_q) = I(R_q=0)\exp[-(4\pi R_q/\lambda)^2]$$ (12.71)

where $I(R_q=0)$ can be obtained by integrating the light intensity over the total diffraction plane.

Cosinusoidal surface height distribution

For a surface with a height profile expressed by $h(x)=\sqrt{2}R_q\cos(2\pi x/D)$, where D is the period length, the specular beam in the diffraction plane equals

$$U(0) = c\int_{-\infty}^{\infty}\exp[j4\pi j\sqrt{2}\,R_q\cos(2\pi x/D)/\lambda]dx = cJ_0(4\pi\sqrt{2}\,R_q/\lambda)$$ (12.72)

where $J_0(y)$ is the Bessel function of the first kind of integral zero order and with argument y. The specular reflected intensity $I_p(R_q)$ in the presence of periodic surface height equals, using equation (12.63)

$$I_p(R_q) = Cc^2J_0^2(4\pi\sqrt{2}\,R_q/\lambda) = I(R_q=0)J_0^2(4\pi\sqrt{2}\,R_q/\lambda)$$ (12.73)

where use has been made of the fact that $J_0(R_q=0)=1$. The intensity of the specular reflected beam has also been calculated for other periodic surface profiles with triangular and parabolic cusp shapes.[4] It has been found that for all these periodic surface profiles, as long as $R_q<0.07\lambda$, the specular intensity can still be represented by equation (12.73).

12.7.2 General Relationship between R_q and Diffraction Intensity Pattern

Comparing the expressions for the specular reflectance for periodic and random surface lay profiles[10], with $R_q<0.07\lambda$, shows that both expressions are very close. Statisticians will not be surprised about this result and would call it a result of the "central limit" theorem. So the surface roughness might as well be obtained by equation (12.71), or

$$R_q/\lambda = \sqrt{\log_e[I(R_q=0)/I(R_q)]/4\pi}$$ (12.74)

where $I(R_q)$ is the specular and $I(R_q=0)$ is the total integrated light intensity in the diffraction plane.

247

12.7.3 Standard Deviation of Scattered Light and RMS Surface Slope

The standard deviation σ of the scattered light intensity, $I(X)$, in the diffraction plane is defined by

$$\sigma^2 = \int_{-\infty}^{\infty} X^2 \, I(X) \, dX \Big/ \int_{-\infty}^{\infty} I(X) \, dx = M_2 / M_0 \tag{12.75}$$

where M_0 and M_2 are the zeroth and second moment of the diffraction pattern, respectively. M_0 represents also the total amount of the light intensity in the diffraction plane.

By using relation (12.64), a relationship between σ and Δ_q can be derived as follows. Application of an inverse Fourier transform on (12.64) yields:

$$k/f \int_{-\infty}^{\infty} I(X) \exp[\, j k X s / f\,] \, dX$$

$$\tag{12.76}$$

$$= c/L \int_{-L/2}^{L/2} \exp[\, j 2 k h (x+s)\,] \exp[\, -j 2 k h (x)\,] \, dx$$

substituting $s = 0$ and using (12.75) yields

$$k/f \int_{-\infty}^{\infty} I(X) \, dX = k M_0 / f = c \tag{12.77}$$

Differentiation of equation (12.76) twice with respect to s leads to

$$-(k/f)^3 \int_{-\infty}^{\infty} X^2 \, I(X) \, dX = c/L \int_{-L/2}^{L/2} \exp[\, -j 2 k h (x)\,]$$

$$\tag{12.78}$$

$$\times \exp[\, j 2 k h (x+s)\,] \cdot [\, (\, j 2 k \frac{d}{ds} h (x+s)\,)^2 + j 2 k \frac{d^2}{ds^2} h (x+s)\,] \, dx$$

Subsequently replacing d^n/ds^n by d^n/dx^n and substituting $s = 0$ results in

$$-(k/f)^3 \int_{-\infty}^{\infty} X^2 I(X) dX = -(k/f)^3 M_2 = c/L \int_{-L/2}^{L/2} [-4k^2$$

(12.79)

$$\times \quad (\frac{d}{dx} h(x))^2 + j2k \frac{d^2}{dx^2} h(x)] dx$$

Assuming that $h(x)$ has everywhere finite first derivatives and L approaches infinity, then the imaginary part in the righthand side of equation (12.79) will equal zero, because:

$$1/L \int_{-L/2}^{L/2} \frac{d^2}{dx^2} h(x) dx = 1/L \int_{-L/2}^{L/2} d(\frac{d}{dx} h(x))$$

(12.80)

$$= [h(L/2) - h(-L/2)]/L = 0$$

Now using (12.77), equation (12.79) can be reduced to

$$(k/f)^3 M_2 = k/f \ M_0 \ 4k^2 \cdot 1/L \int_{-L/2}^{L/2} [\frac{d}{dx} h(x)]^2 dx$$

(12.81)

$$= 4kM_0 k^2 \Delta_q^2 / f$$

from which follows

$$\Delta_q = \sqrt{M_2/M_0} /2f = \sigma/2f$$

(12.82)

Equation (12.82) relates the RMS surface slope to the standard deviation of the scattered light intensity in the diffraction plane.

249

12.8 PRACTICAL IMPLEMENTATIONS OF DIFFRACTION THEORY FOR SURFACE FINISH ASSESSMENT

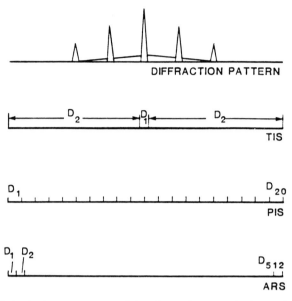

Figure 12.9 Principal arrangement of the photo-detectors in the various systems.

Three techniques have been developed into surface finish sensors, which are based upon light scattering or diffraction. The essential differences between these sensors are the number of photo-detectors used for recording the light intensity distribution in the diffraction plane. Figure 12.9 shows schematically the principal lay-out of the sensors together with a sketch of a diffraction pattern produced by a periodic surface.

12.8.1 Total Integrated Scatter (TIS)

This method was originally used by Bennett and Porteus[2] to analyse surfaces having $R_q < 0.1\lambda$. It is based upon the relation between the specular reflected intensity $I(R_q)$ and R_q as given in equation (12.75). This instrument contains essentially two detectors, where photo-diode D_1 collects the specular beam and photo-diode D_2 the diffusely scattered light intensity. Their actual system includes an aluminized collecting Coblentz sphere, which focuses the diffusely scattered light onto detector D_2. Since there are only two detectors involved, this technique can be very fast with simple electronic circuitry to obtain R_q. The upper surface roughness limit of 0.1λ is caused by the fact that for rougher surfaces the specular reflected beam effectively disappears. The lower limit is about 0.001λ, which is governed by the electronic noise in the detectors. This instrument needs accurate setting up in order to prevent the specular beam from reaching detector D_2. This renders this instrument unsuitable for in-process measurement and has to be used in a laboratory environment. Surface slope information cannot be obtained because the information for evaluation of equation (12.82) is missing.

12.8.2 Angularly Resolved Scatter (ARS)

In principle, the entire angular light intensity distribution contains a lot more information about the surface topography than can be obtained from TIS. Measurement of the angular light distribution can yield the RMS slope parameter and machine tool condition information. In order to obtain this information, a great number of photo-detectors have to be placed in the far-field of the instrument lens. The warwick system[4] contains an array of 512 photo-detectors. In order to be able to resolve the diffraction pattern, the size of the individual detectors have to be smaller than the minimum line width of the diffraction orders. To keep the size of the instrument small, photo-diode arrays have to be employed. The upper surface roughness limit is again 0.1λ for the same reason as in TIS. The lower limit is about 0.01λ because of electronic noise and the limited dynamic range of photo-diode arrays. However, the dynamic range can be increased by suitable software in the microprocessor which controls the instrument. Alignment is not a big problem, because the software can be devised to detect the position of the specular reflected beam. This information can actually be used to measure the form of the object. The acquisition of information is much slower than in TIS, because of the large number of photo-diodes.

12.8.3 Partial Integrated Scatter (PIS)

Instruments based upon this method contains less photo-detectors than the ARS based system. The Rodenstock sensor[11], for example, has 20 photo-detectors. Since they have to span the complete angular light distribution, the photo-diodes will not be able to resolve the angular distribution well enough to separate the specular light intensity. Therefore it is impossible to obtain the R_q value as described in equation (12.75), and only the Δ_q parameter can be evaluated. Again there are no severe alignment problems, and by monitoring the position of the central movement of the reflected light distribution, form information of the object can be obtained. Since this instrument contains only a limited number of detectors, the acquisition speed is higher than for an ARS system.

12.8.4 Comparison of Three Methods Based upon Diffraction

Table 12.1 lists the properties of the three techniques. A detailed comparison between the performance of the ARS and PIS systems is made because both are capable of in-process measurements, whilst the TIS method is really restricted to laboratory usage.

Table 12.1 Comparison between three light scattering techniques.

Method	Δ_q	Relative Range[1]	R_q	Range (λ)	Relative Speed[2]	In-process
PIS	yes	medium	no	----------	high	yes
ARS	yes	large	yes	0.01 - 0.1	medium	yes
TIS	no	-----	yes	0.001 - 0.1	very high	no

1. Lower limit is governed by the resolution of the diode array.

2. The speed is inversely proportional to the number of detectors.

This comparison table is based upon the capability to obtain the R_q parameter for sinusoidal surface profiles. The R_q value of these profiles range from 0.05λ to 10λ, the period length D=100 μm and the illumination wavelength is 800 nm, as produced by diode lasers. Furthermore, it is assumed that the detectors span the full angular distribution of the theoretical diffraction pattern produced by a sinusoid with an amplitude of 2.26 μm (R_q=2λ). The ARS system contains 512 and the PIS system 20 photo-detectors. The theoretical light intensity pattern I (X) is expressed by:

$$I(X) = J_m^2(4\pi\sqrt{2R_q}/\lambda)\delta(X+f.\tan(m\pi/d)) \tag{12.83}$$

where f is the focal length of the instrument lens, m the diffraction order and δ the Dirac delta function. Some of the patterns as they would be perceived by the instruments are shown in Figure 12.10. The graph in Figure 12.11 shows their theoretically obtained Δ_q values versus the analytical value equalling $2\sqrt{2}\pi R_q/D$. It shows clearly the inability of the PIS system to resolve the diffraction pattern well enough for very smooth surface profiles. It is also interesting to ask what happens when the diffraction patterns exceeds the length of the detectors. This shows that one has to be careful when using these instruments and that a graphical display of the light intensity distribution is important when judging the validity of the results.

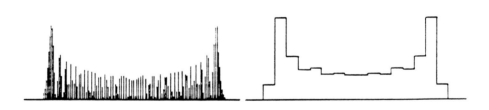

Figure 12.10 Diffraction pattern as perceived by an ARS (left) and a PIS (right) instrument.

12.9 CONCLUSIONS

Using scalar diffraction theory, two relationships have been derived which link two recognized surface finish parameters to the far-field light intensity distribution caused by light reflecting from rough surfaces. The surface finish parameters in question are the RMS height (R_q) and slope (Δ_q).

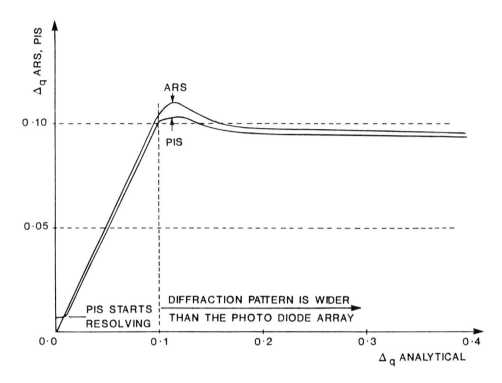

Figure 12.11 Comparison of Δ_q values as obtained with ARS and PIS methods versus the analytical value.

REFERENCES

1. Davies, H. (1984) The reflection of electromagnetic waves from a rough surface. *Proc. Elect. Engrs.*, **101**, 209-214
2. Bennett, H.E., and Porteus, J.O. (1967) Relation between surface roughness and specular reflectance at normal incidence. *J. Opt. Soc. Am.*, **51**, 123-129
3. Church, E.L., Jenkinson, H.A., and Zavada, J.M. (1977) Measurement of the finish of diamond turned metal surfaces by differential light scattering. *Opt. Eng.*, **16**, 360-374
4. Rakels, J.H. (1986) The use of Bessel functions to extend the range of optical diffraction techniques for in-process surface measurements of high precision turned parts. *J. Phys. E: Sci. Instrum.*, **19**, 76-79.
5. Stratton, J.A. (1941) *Electromagnetic Theory.* New York: McGraw-Hill.
6. Hecht, E., and Zajac, A. (1982) *Optics.* New York: Addison-Wesley Publ. Comp.
7. Whitehouse, D.J., and Jungles, J. (1974) Some modern methods of evaluating surfaces. *Proc. Int. Conf. on Prod. Engng.* Tokyo.
8. Rakels, J.H. (1986) Beugung - Ein seit langem bekanntes optisches Phaenomen, verwendet als modernes Messinstrument. *Int. Conf. on Modern Production and Production Metrology.* T.U. Vienna, 2-4 April 1986
9. Oberhettinger, F. (1978) *Fourier transforms of distributions and their inverses.* New York: Academic Press

10. Rakels, J.H. (1988) Recognized surface finish parameters obtained from diffraction patterns of rough surfaces. *SPIE*, **1009**, 119-125
11. Brodmann, R., Gast, T. and Turn, G. (1984) An optical instrument for measuring the surface roughness in production control. *Annals of the CIRP*, **33/1**, 403-406
12. Rakels, J.H. (1988) In-process surface finish measurement of high quality components. In *Proceedings of the Int. Congress for Ultraprecision Techn.* (Aachen, 1988). Springer Verlag.

Chapter 13

NANOPARTICLE VISUALIZATION FOR PARTICLE IMAGE VELOCIMETRY AT TRANSONIC SPEEDS

PETER J. BRYANSTON-CROSS

The technique of Particle Image Velocimetry (PIV) provides an instantaneous whole field visualization of a flow field. This is achieved by projecting two pulses of light of nanosecond duration into the plane of a flow field. The particles in the plane are imaged by a conventional photographic technique. By measuring the distance between any pair of particle images, the velocity of the particle and hence the fluid can be determined. The ability to track individual particles is dependent upon several factors; the particle size, the speed and turbulence of the flow, the intensity and wavelength of the light source used and the method of recording the information. The application of PIV at transonic flow speeds has been investigated in three regimes. First, when there is sufficient light being scattered from individual particles to form high resolution photographic images. Secondly, when the particles are small enough to follow the flow at transonic speeds. Thirdly, when the simplest and most direct presentation and reduction of the particle image data can be used to determine the velocity of the flow field. In this chapter, a review is made of work recently carried out at the Massechussettes Institute of Technology to study nanometre particle flows using an optical technique.

13.1 INTRODUCTION

Particle Image Velocimetry (PIV) is a technique that was first described approximately ten years ago.[1] The original concept was to create a plane of light in the flow field with a ruby pulse laser. A double pulse of laser light then forms a double image of particles within the flow field. Having developed the particle images on a photographic film, it was then reilluminated with a focussed continuous wave laser. When the laser light passes through a pair of particle images, Young's interference fringes are formed. The spacing of these fringes is proportional to the speed of the particle and its direction defines the particles trajectory. An extensive review of the many parameters involved in the application of PIV have been made elsewhere.[2]

Fringe analysis developed historically to overcome the large amount of computer processing power needed to evaluate the flow data with the level of turbulence often present. Image processing considered impossible a few years ago is now common place. The result of which has made the direct digital data processing of PIV images soluble using "brute force" computational approaches. The method of data reduction applied in this chapter is similar to that developed at both the Wright Patterson Laboratory by Goss[3] and at NASA by Wernet[4] for direct video measurements of PIV using a Silicon Intensified Tube (SIT) camera.

Two main considerations when applying the technique of PIV at transonic speeds are the resolution of the film used and the behaviour of very small particles in both scattering the incident light and following the flow field. A great deal of work exists on the scattering of light by sub-micron particles. The Mie theory is described in detail in Born & Wolf[5] and Van de Hulst.[6] More recently an excellent review of different seeding techniques was published by Melling[7] and by Patrick.[8] A detailed review of the different combinations of film and laser applied to PIV is given by Pickering and Halliwell.[9] Adrain[10] states that there is a limiting balance between the amount of light scattered from a particle and the speed and resolution of the film required to image it.

13.2 EXPERIMENTAL

The experimental arrangement was designed to demonstrate the feasibility of PIV at transonic speeds without the complications of running the MIT turbine facility. The essential feature was to provide a heated filtered high pressure air supply, as shown in Figure 13.1. The air is then accelerated through a convergent nozzle to produce a transonic flow. The size of the nozzle was 10 mm in diameter.

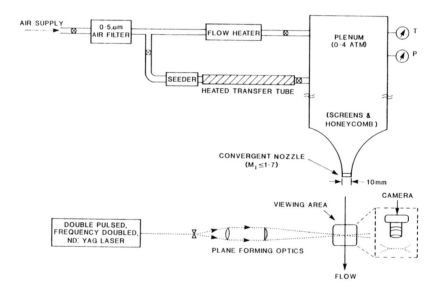

Figure 13.1 Particle image velocimetry (PIV) experimental set-up.

The Lumonics Nd:Yag laser was positioned so that, using a combination of a negative and a positive spherical lens and a positive cylindrical lens, a flat sheet of laser light was produced. The laser sheet which passed through the centre of the nozzle exit flow was 20 mm wide and 0.3 mm thick. The laser power output was normally 50 mJ/pulse. Below this, the signal to noise ratio of the film became poor. The use of lower energies has a direct impact on the application of PIV. A lower energy not only reduces the background surface scattering, but also reduces the problem of passing the beam through windows and into confined regions, such as turbomachinery blade rows.

13.2.1 Laser System

The Nd:Yag (Lumonics series) produces two 0.1 J pulses of 10 ns duration with a time interval variable over the range 50 ns to 100 μs at a wavelength of 532 nm. The double pulse firing system was developed specifically for the needs of this experiment. Previous attempts to perform PIV for sub-micron particle seeding have been made using a ruby pulse laser. There are several reasons why the ruby system is unsuitable, but in the main, most fast panchromatic films are not sensitive to the 695.4 nm ruby wavelength. Several properties of the Nd:Yag laser helped in making the PIV imaging tests successful.

1. The solid state Nd:Yag pulse has a well defined Gaussian beam profile, which can be focused to produce a laser sheet of light 0.3 mm thick.

2. A not insignificant point made by Kompenhans & Reichman[11] is that a Nd:Yag laser can be run in an "open lase mode" at a repetition rate of 10 Hz; making the alignment of the optical system a simple task. With imaging focal plane depths which are of the same thickness as the light sheet, a significant error can be generated by an out-of-focus image. When, for example a UV laser was used, the focal plane of the camera was found by focusing on "burn images" made on a card placed in the light sheet plane. A calculation of the depth of field for the optical system used is given by Adrain and Ya.[10]

13.2.2 Camera System

The camera system and type of film have been rigourously tested to evaluate the images observed. The choice of a 35 mm format SLR camera was made because Kodak only produce TMAX film in 35 mm and 120 film formats. The 90 mm Olympus Macro lens was considered to be the most suitable available lens. An Olympus 4T camera and 90 mm macro lens combined with a variable extension tube were found to produce high quality results.

The type of film used was TMAX 3200 and its properties are described elsewhere.[12] Other speeds and types of film were evaluated but TMAX currently represents the fastest black and white with a resolution of 100 line pairs/mm and an ISO of 100,000. The grain size at this magnification was seen to be of the order of 1 μm, with the smallest definable object being approximately 5 μm in size. Higher resolution slower films were evaluated. TMAX 400, was found to have a finer grain, but also a higher fog level. Kodak Pan film showed no trace of particle images. Agfachrome 1000 ASA slide film produced a high level of grain noise.

13.2.3 Seeding Materials Used

Initial experiments have been performed using a TSI particle seeder. To date only one type of particle has been used in the seeder, that of 0.6 μm styrene. An evaluation of the methods for producing styrene particles is described by Kodak.[13] A further NASA reference discusses different methods for launching the particles into the airflow. A critical part of the evaluation was to use an independent Mie scattering particle instrument. Tests were performed at different flow speeds and different particle sizes demonstrated that:

1. The carrier fluid, either alcohol or water, with which the styrene was in solution, was fully evaporated. This was easily validated, by filling the particle seeder with either water or alcohol. In both cases no particle images were found, confirming that the carrier droplets had completely evaporated.

2. The particles may have been agglomerating. The question of particle size and the ability to generate individual sub-wavelength particles is critical to the success of PIV at transonic speeds. The technique is hard to validate; because it is not possible to directly image a nanometre diameter particle optically. The images formed on the photographic film were, due to an optical distortion and lack of resolving power in the lens normally 30 μm in diameter. Thus an exhaustive series of tests have been performed to gain confidence and understanding of the particle images formed. An image showing particle agglomeration is shown in Figure 13.2. The light sheet was passed through a weak aqueous solution of the particles. From Scanning Electron Microscopy (SEM) pictures taken from the solution, the particles were found unlikely to agglomerate, Figure 13.3. Images captured on film of the particles suspended in water show an image size consistent with the smaller particle images found in air. A particle sizer, i.e. a particle measuring system which sampled and sized the particles using Mie scattering, confirmed the presence of single sub-micron non-agglomerated particles at low flow speeds. A number of scans taken of different seeding materials are shown in Figure 13.4. Images of the particles were taken at different f numbers of the camera, the pictures show the particle images decreasing with size, again suggesting that mainly single particles are present and that some film saturation was taking place. A further check has been made in that when smaller particles are used then the images captured contain correspondingly less energy. Finally, the light balance calculation shows a good agreement between the scattered light and that collected (see appendix). The reason for this exhaustive verification of particle size is that at the point where the particle size is the same as the wavelength of light, the scattering behaviour of the particle moves into the Mie region. Not only is Mie theory[4], notoriously difficult to calculate but the scattering efficiency of the particle falls with a fourth power law. There are also only a few publications,[8,9] where PIV has been attempted at sub-wavelength particle sizes. It is just at the point where most light scattering calculations have stopped,[8] that the information needed to calculate the light balance becomes critical, as shown in Figure 13.5; hence a pragmatic experimental approach has been taken in the determination of the relationship between the size of the particle in the flow and the size of the image form on the photographic film.

3. The light scattered by the particle and collected by the optical system may have saturated the photographic response of the film. The use of Kodak's TMAX series films was used after initial trails with Polaroid type 57 film which has an ASA/ISO rating of 3000. The Polaroid film has low resolution and a poor dynamic range but did confirm that an the energy being scattered was of the order of 10-13 J for a 670 nm diameter particle. The TMAX range of films 100, 400 and 3200 can all be "push" processed by both long development times and by heating the specialized developer. A typical particle pair is shown in Figure 13.6.

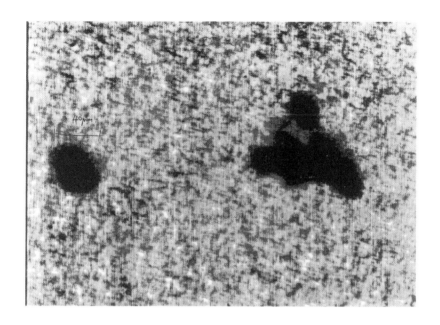

Figure 13.2 Plot showing agglomeration of styrene.

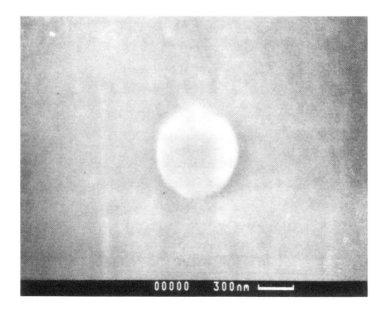

Figure 13.3 Electron microscopy image of a 500 mm styrene particle.

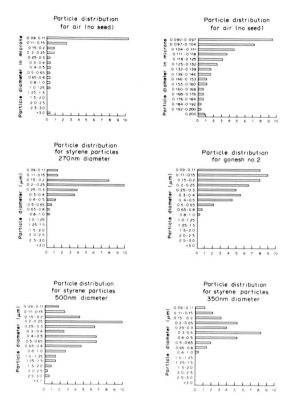

Figure 13.4 Plot showing particle distribution.

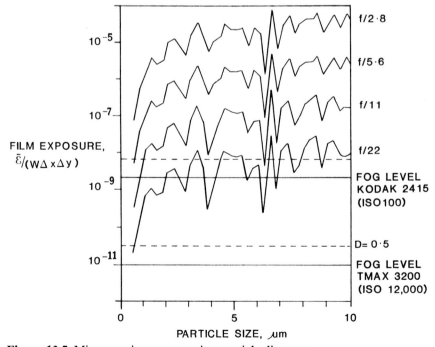

Figure 13.5 Mie scattering curve against particle diameter.

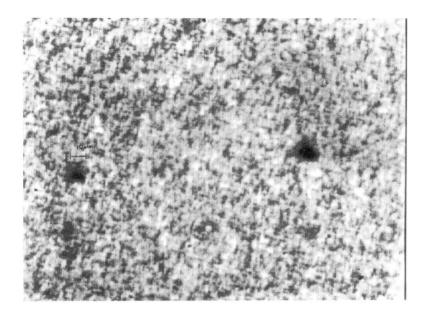

Figure 13.6 Typical PIV particle image pair.

13.3 DATA PROCESSING

There have been two approaches to PIV particle analysis. The first was adopted by Meynart[1] who used an optical fast Fourier transform to convert the particle data into essentially speed and direction data. This is an elegant solution that is often used when the particle density in the flow is high.

The method developed at MIT is a "brute force" approach where each pair of particle images has been semi-manually located. At present, the computer system has only been used to find the centre of the image and record its position. In order to move towards a fully automated system, it requires the removal of ambiguities in the data (e.g. particle direction). The two colour approach devised by Goss and co-workers[3] removes many of these particle tracking problems. This method derives its name from the fact that a single laser diode is used to pump two solid-state lasers of different wavelengths. In this way, the first laser pulse can be green and the second blue. By using a colour sensitive photographic film to record the particle tracks, it is possible to identify the direction of the particle without ambiguity. The method used for processing the data was essentially that used at MIT. The two significant differences being that a colour filter was used to digitize the particle tracks from each colour separately and the system processed the data automatically. A second approach would be to use three pulses, or two pulses with the first being longer than the second.

The PIV image can be viewed directly on 35 mm photographic negatives, each image has approximately 200 to 500 particle pairs. Several systems have been researched and evolved for the processing of PIV data. For the purpose of this evaluation exercise a direct and simple video system has been designed, as shown in Figure 13.7.

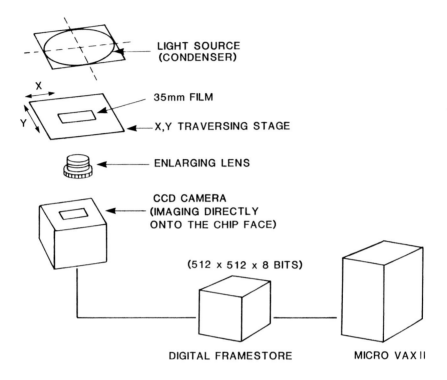

Figure 13.7 Charge Coupled Device (CCD) camera data reduction system.

Essentially a standard photographic enlarger has been used to project the enlarged (×5 to ×10) image directly onto a CCD camera. The spacing of the pixels in the camera are approximately 16 µm and the image on the 35 mm film is 1.5 times scale. Thus the digital picture has a resolution of 1 to 2 µm per pixel. The film was moved using two stepper motors so that the particle pair image could be viewed on one complete screen of a TV monitor. The position of the particles were then measured and stored in a computer. At an enlarger magnification of five, the 35 mm film frame is equivalent to approximately 600 individual TV frames. To process one PIV whole field image requires the analysis of all these 600 frames. Currently the time needed to do this represents the slowest part of the technique.

In the cases presented the film has been processed in a semi-manual manner, in that an operator has moved the film viewing one TV frame at a time. The operator moves the screen cursor to the first of a particle pair. The computer then calculated the optical centre of the particle. Next the operator moved the cursor onto the second particle in the pair. The computer was then used to find the centre of the point, record their position and compute the angle and speed of the particle.

The following optical consideration were made during the construction of the analysis system:

1. By moving the photographic 35 mm negative, the system remained optically inline. This minimized any optical distortions which may be introduced in the analysing system.

2. The addition of a third pulse in the single colour Nd:Yag laser system would remove many of the particle ambiguities.

3. The use of two different laser wavelengths that require colour sensitive films to discriminate the pulses, reduces the Signal to Noise (S/N) ratio as compared to polychromatic films by as much as 80%. This is a critical factor in the sub-micron particle imaging region. The two colour method does however remove particle tracking directional ambiguities.

4. The CCD camera used in the particle tracking exercise had a typical resolution of 512 by 512 pixels. It is recommended that a Kodak camera which has a resolution of 1000 by 1000 be used. This will reduce the number of individual TV frames which need to be scanned and make particle pairs which have a typical spacing of 0.3 mm between them easier to identify.

This type of system allowed initially a very simple image diagnostic tool. It was possible to view directly the images of the particles at a micron resolution level. As such it has been possible to assess the S/N level of the films and to calculate the image size.

Figure 13.8 Raw PIV image made of a transonic flow.

It has been found experimentally that there is a relationship between large particle images and large particles. With a S/N ratio of 2:1, it is possible to threshold digitally the image and remove the random grain noise. Having completed this, a particle is located manually. The centre of the particle image is then calculated from the weighted mean. An example of a unprocessed PIV 35 mm photographic image is shown in Figure 13.8. In this image it is possible to identify the particle image pairs, the spacing between which changes with their speed. The data is then recorded one TV frame at a time, the digital image shown in Figure 13.6 is an example of one such frame, the data can then be replotted as a vector field.

The best results had images with between 300 to 500 particles in a 20 mm by 20 mm area. The preferential size of the particle image was between 15 to 40 pixels in diameter. Smaller images fell below the general grain noise of the image, while larger images reduced the measurement accuracy of the system. Larger images also tended to associated with larger particles which would be unlikely to follow the flow. The ideal laser pulse separation was found to be that which produces an image distance between particles of 300 pixels. When an imaging system of 1.5:1 was used on the Olympus camera and magnification of ×10 in the enlargement/projection stage this represented a movement of the particle between exposures on the 35 mm film of 200 μm. At 700 ms^{-1} this movement corresponded to a pulse separation of 300 ns.

13.4 LUMONICS ND:YAG LASER SYSTEM

The Lumonics laser has a double pulse feature. It is the first ever such system to be used in a frequency doubled Nd:Yag configuration. The system has in essence been developed along the same principle as that used for ruby pulse holographic systems. The need to produce a double pulse with a separation in the region of 300 to 1000 ns is critical for PIV measurements of high speed flows; otherwise the particle pair data obtained become uncorrelated with the flow event.

The laser system has an unexpected bonus in that it was possible to produce very short stable double and triple pulses in the 50 to 500 ns range. The pulse duration of the Nd:Yag laser used was 10 ns which is within the spatial/temporal movement window for the particle. A longer pulse duration would have spread the energy of the pulse from a spot to a stream and reduced the light back-scattered off the particle. Given the pulse energy of the Nd:Yag laser is of the same order as the dye laser the results obtained were substantially better, both in terms of the light return and the particle image definition. Thus 670 nm particles could be visualized using TMAX film in air. The next objective was to produce a high speed flow close to that experienced by a turbine cascade but without the complexity and running costs of the GTL annular cascade. To do this a transonic nozzle jet flow was constructed with a filtered and heated pressurized air supply. This was completed and the styrene injected into a settling chamber before the nozzle, as shown in Figure 13.1. The first PIV images obtained at high speed yielded some worrying images. When viewed under a microscope the particle images were seen to be very large, 0.1 mm in diameter and in the process of clumping, Figure 13.2. However from this point in the PIV evaluation the laser reliability and optical suitability for the project was proven. The major question was the size of the seeding particles.

SEM images have been made from samples of the styrene which show single spherical 670 nm diameter particles. In this case, the particles had not agglomerated as previously observed in Figure 13.3.

13.5 RESULTS

The experimental results that have been obtained employed two particle materials, namely:

1. Styrene (polystyrene) with a particle diameter of 260, 350, 500 or 670 nm.

2. No. 2 Gonish incense which had a measured particle diameter distribution of 200 nm to 400 nm.

Images of the styrene have been taken with relative ease over the whole range of sizes tested. Up to 450 ms^{-1} a particle density distribution of 2 to 3 particles mm^{-2} was found to produce the best results. This gave a count of 1,000 particle pair images for each 35 mm photographic film frame. For a plane of light was 0.2 mm wide, this corresponds to a seeding density of 10 to 15 particles mm^{-3}.

Styrene was used because particles can be produced which are both spherical and uniformly sized. As a material it has a very high quantum efficiency, which makes its light scattering properties much better than other products such as oil or titanium dioxide. Its melting point, however, is only 100°C. Incense is on the other hand a burning product and although not having the same quantum efficiency can be used in higher temperature environments, as described by Patrick.[8]

Difficulty has been found with the seeding delivery, mainly with the incense burner, but also with the styrene seeder. This was due to the need to inject the seeding material into a region of high stagnation pressure. Firstly, if the particles were to be injected into a stagnation pressure of, for example 62 psi, then the seeder and all the pipework needed to be held at a higher pressure, typically 100 psi. It was found that the TSI seeder's seal began to fail at this level of pressure. Secondly, the seeding mass flow rate and pressure difference has to be strong enough to overcome the stagnation pressure in the settling tank. This was achieved by choking the outlet flow of the seeder as it entered the stagnation region.

Results have been obtained up to 680 m.s^{-1} (M_a=3.0) with styrene and up to 100 ms with Gonish seeding. Tmax 3200 film was used throughout the experiments over an aperture range of f/8 to f/22. The calculated scattered light balances with the film sensitivity for these images.

The following flow cases have been investigated. The object of the work being to show that first, the particles were following the flow; secondly that there were enough particles/image to provide a representative measurement of the flow; and thirdly that the rudimentary digitization process developed could be used to produce a whole field visualization for both external high speed aerodynamics and turbomachinery.

The results are shown in Figures 13.8 to 13.14, and correspond to two different optical diagnostics and a flow calculation. At stagnation pressures greater than 73 psi a normal shock disk of approximately 2 mm diameter was formed. The normal shock disk can be seen clearly in the shadowgraph picture made at this stagnation pressure, Figure 13.9. A diagrammatic description of the flow is given in Figure 13.10. The PIV results for the same plot, Figure 13.8, have been converted into velocity vectors in Figure 13.11, show that the 500 nm diameter particles follow the flow as predicted by Melling.[7] The particles can be seen to experience a rapid fall in velocity as they pass through the normal shock disk. The PIV image shown in Figure 13.8 represents one 35 mm photographic frame. The velocity vectors shown in Figure 13.11 were all taken from the same frame. The result shown has been repeated several times.

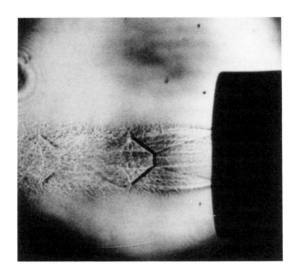

Figure 13.9 Schlieren image of nozzle flow.

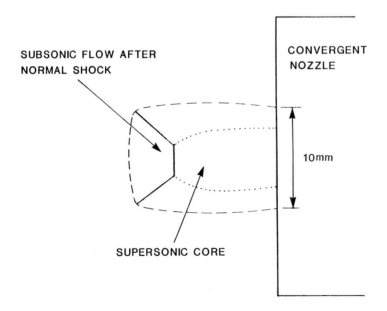

Figure 13.10 Schematic representation of the nozzle flow.

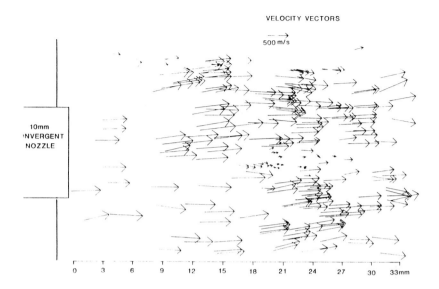

Figure 13.11 PIV data.

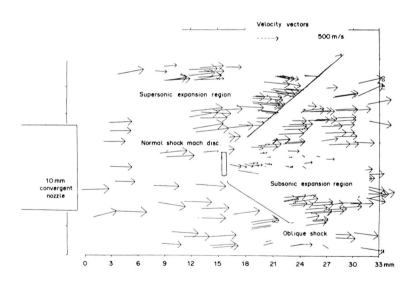

Figure 13.12 Vector representation of PIV data shown in Figure 13.11.

Figure 13.12 shows a calculation made by Giles at MIT. The Giles code used in this transonic flow comparison, is a basic mathematical algorithm which models the flow. It is an adaptation of Roe's,[14] flux-difference-splitting algorithm as implemented by Rai.[15]

It incorporates explicit flux-difference-splitting in the cross stream direction. An implicit central difference of the viscous fluxes in the cross stream direction is also employed. The advantage of this approach is that the flux-difference-splitting automatically yields very sharp shock waves (Mach disks) while implicitly treating the cross-flow speeds convergence, thus reducing the run time requirements. The calculation has used the dimensions of the nozzle entry and its stagnation condition as the boundary conditions for the numerical solution. To date the Mach disk does not form a stable numerical solution and the result obtained was achieved by stopping the programme after a number of iterations.

However, three pieces of information can be combined to this experiment to give certainty to both the measurement and the calculation.

1. The shock pattern observed in the Shadowgraph image shows the same shock positions as seen in the PIV results.

2. The shock patterns measured match both in position and velocity with those predicted by the Giles code.

3. The Giles code predicts a subsonic streamline divergence directly after the normal shock disk. This divergence creates a velocity reduction in the flow. The particle velocities measured using PIV directly after the Mach disk show velocities slower than could be expected for a flow which has experienced a normal shock but consistent with a normal shock followed by a subsonic streamline divergence. A plot along the centre-line of the nozzle for the Giles code is given in Figures 13.13 and 13.14 for the equivalent PIV measurement.

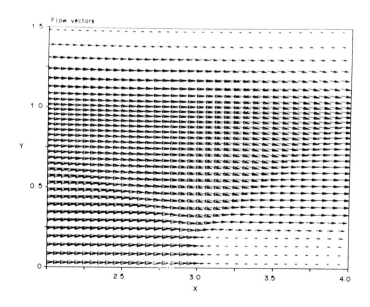

Figure 13.13 Giles calculation of the nozzle flow.

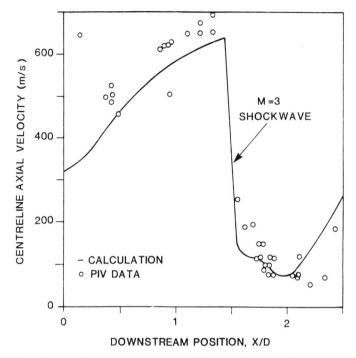

Figure 13.14 Comparison between a plot of the nozzle centre-line velocity profile as calculated by the Giles programme and measured using PIV.

Figure 13.15 Low speed flow field image made using incense smoke.

In Figure 13.15, a low speed flow is shown, very heavily seeded with incense smoke exiting from the 10 mm nozzle. Effectively, this is the same seeding rate as in the transonic cases, but with a much lower airflow to particle mixture.

13.6 CONCLUSIONS

In order to complete this work a new type of Nd:Yag double pulse laser has been developed. This laser was found to be highly stable and produce a double pulse each of duration 10 ns with a time interval of 50 to 1000 ns. It is essential to use a "green" Nd:Yag frequency doubled laser (523 nm) rather than a "red" ruby laser (694.3 nm). This is because panchromatic films of the photographic speed of Tmax are not sensitive to the wavelengths of light beyond 650 nm. No advantage has been shown in using a eximer laser operating at ultra-violet wavelengths (i.e. 300 to 350 nm).

It has been possible to record nanoparticle images onto 35 mm film using a conventional 35 mm camera, macro-lens and high speed film. The images have been captured, digitized and processed within a computer using a converted 35 mm enlarger and a CCD camera .

This is the first time the results of such a measurement have been published, and the first time the instantaneous velocity map within a supersonic nozzle of the dimensions of 10 mm has been mapped out. Moreover, a comparison of the results with a numerical model gives excellent agreement.

To summarize, an evaluation has successfully been made of the nanometric application of Particle Image Velocimetry (PIV) to turbomachinery. From which it is apparent that it is possible to record sub-micron diameter particles. The lower limit on the particle size has been found to be 270 nm when using Tmax 3200. Furthermore, double particle images have been formed showing 500 nm styrene particles travelling at speeds greater than 700 ms^{-1} in which there was no evidence of particle agglomeration. Therefore, PIV does not have any technological impediment to prevent it becoming a fully developed optical diagnostic - eventually replacing current anemometer systems in turbomachines.

REFERENCES

1. Meynart, R. (1980) Equal velocity fringes in a Raleigh-Bernard flow by a speckle method. *Appl. Opt.*, **19**, 1385-6
2. Riethmuller, R.W. (1988) Particle Image Displacement Velocimetry. March 21-25 1988, von Karmen Institute for Fluid Dynamics lecture series
3. Goss, L.P., Post, M.E., Trump, D.D., Sarka, B., McArther, C.D. and Dunning, G.E. (1989) A novel technique for blade-to-blade velocity measurements in a turbine cascade. In *Proceedings of the AIAA/ASME/SAE/ASEE 25th Joint Propulsion Conference* (Monterey, July 1989)
4. Wernet, M.P., and Edwards, R.V. (1989) A vector scanning processing technique for pulse laser velocimetry. In *Proceedings of the 13th ICIASF* (Gottingen, FRG, September 1989)
5. Borne, M. and Wolfe, E. (1959) *Principles of Optics,* pp. 654-5. Oxford: Pergammon Press
6. van de Hulst, H.C. (1957) *Light Scattering by Small Particles.* New York: Dover Publications

7. Melling, A. (1986) Seeding gas flows for laser anemometry. In *Proceedings of AGARD Conference on Advanced Instrumentation for Aero Engine Components.* (Philadelphia, May 1986), paper No. 15

8. Patrick, W.P. and Paterson, R.W. (1981) Seeding techniques for laser doppler velocimetry measurements in strongly accelerated nozzle flow fields. *AIAA*, **81-1198**, pp16

9. Pickering, C., and Halliwell, N. (1985) Particle Image Velocimetry: a new measurement technique. *Inst. Phys. Conf. Ser.* No.**77**. Also Sessions 4 paper presented at VI Conf. Photon Correlation and other Techniques in Fluid Mechanics, 1985

10. Adrian, R., and Yao, C. (1985) Pulse laser technique application to liquid and gaseous flows and the scattering power of seed materials. *Applied Optics*, **24**, 44-52

11. Kompenhans, J. and Reichman, J. (1986) Particle Imaging Velocimetry in a Low Turbulent Wind Tunnel and Other Flow Facilities. In *Proceedings of AGARD Conference on Advanced Instrumentation for Aero Engine Components.* (Philadelphia, May 1986), paper No. 35

12. Kodak publication (1989) Tech Bits, No. 3

13. Hunter, W.W. and Nichols, C. (1985) Wind Tunnel Seeding for Laser Velocimeters. *Nasa Conference Publication* No.**2359**. Presented at a workshop at NASA Langley Research Centre, (Hampton Virginia, March 1985)

14. Roe, P.L. (1981) Approximate Riemann solvers parameter vectors and difference schemes. *J. Comput. Phys.*, **43**, 357-372

16. Rai, M. (1989) Unsteady 3-D Navier-Stokes simulations of turbine rotor stator interaction. *J. propul. Power*, **5**, 305-311

APPENDIX. Calculation of Light Balance

The light balance calculation for this work has been made in a pragmatic manner. In that, it is known that the experiment being attempted was at the limit of several of the components being used: the films used have variable resolution with respect to the processing being applied; lens aberrations were apparent and were found to be dependent on both the particle size and the number; although Goss[3] has completed work on larger 50 μm particles, no work existed which showed that sub-micron particles could be imaged in this manner essential for a successful PIV result at transonic speeds.

The light balance calculation for this work was essentially treated in stages:

1. Light output from Nd:Yag laser is 50 mJ per pulse

2. Size of light sheet generated is 40 mm wide by 0.3 mm thick, i.e. the area is 12 mm². Combining 1. and 2. gives the power density of the Nd:Yag beam as 4000 Jm^{-2}

3. Cross-sectional area of particle is $\pi(2.5 \times 10^{-7})^2$ for a 500 nm diameter particle, i.e. 2×10^{-13} m²

4. Scattering ratio. Assuming that the particle could scatter light in a Lambertian manner, there would exist a spherical shell of irradiance of which a small area would be collected by the imaging Macro lens.
   i) The area of irradiance is $4\pi R^2$, i.e. $4\pi(60 \times 10^{-3})^2$ m²
   ii) Area of collection is $\pi/4.($ focal length/f/no) by macro lens $= \pi/4.(5 \times 10^{-3})$ m²
   iii) Ratio of areas is $(5/60)/16 = 4 \times 10^{-4}$.

5. Mie scattering characteristics. However, as shown in van de Hulst,[6] the light scattering properties of sub-micron particles scatter in a curve described by the Mie theory. It would be expected that the amount of light scattered orthogonally will be two orders of magnitude less than that forward or back scattered.

Thus a crude calculation of the light collected by the macro lens is given by the product of the power density, cross-sectional area of the particle, ratio of areas and Mie scattering angle dependence (Ca. 1×10^{-2}) is 16×10^{-16} J

The macro lens provides a final resolution limit to the system. It defines the size of the particle image formed on the photographic film. If the lens were diffraction limited, the spot size would be that of the Abbe limit of resolution as described elsewhere.[7] However, both the measured particle results shown in Figure 13.3 and the calibration made from a USAF test target show a limiting resolution of 15 to 20 μm.

For a 500 nm particle a typical image is measured as 30 μm. Thus, the energy density on the surface of the film is estimated by the ratio of the energy collected by the lens to the area of the image on the film, Ca. $16 \times 10^{-16}/(\pi(15 \times 10^{-6})^2)$ is 2×10^{-6} J m^{-2}

The second part of the light balance calculation is the sensitivity of the film used. The best estimate of Tmax[12] suggests a film sensitivity of 5×10^{-7} to 2×10^{-6} J m^{-1}. The sensitivity is dependent on the amount of push processing applied to the film. Kodak also point out that push processing lowers the effective line resolution of the film.

Chapter 14

HIGH PRECISION SURFACE PROFILOMETRY: FROM STYLUS TO STM

DEREK G. CHETWYND & STUART T. SMITH

An understanding of the geometry of the small scale structure of surfaces is essential if their mechanical, electrical, optical, thermal or, sometimes, chemical properties are to be controlled. Here the techniques of measuring surface geometry by profiling (as opposed to those giving a parametric description) to nanometric resolutions are examined and compared. Particular emphasis is given to the nature of the probe interaction with the surface rather than to the general design of conventional instruments. However some consideration of the precision of differing designs is included. Contact stylus methods still dominate most applications and are examined in some detail. A brief review of non-contacting optical methods follows. For nanotechnological applications, scanning tip microscopies are becoming well established and a major section discusses their operation and explores their potential.

14.1 INTRODUCTION

Measuring the topography, or geometrical form, of a surface is often a key stage in understanding, and ultimately exploiting, its properties. On a relatively large scale, a deviation from design shape might prevent reasonable operation, as with an inaccurate or warped bearing slideway. On a smaller scale, surface roughness can directly affect tribological, electrical and thermal contact and optical performance of surfaces of generally correct geometry. Here, roughness is considered to be the local variation from form impressed upon it during its production. For most conventional machining processes on metals, roughness appears with asperity heights of the order of micrometres, or somewhat larger. Some mass-produced precision items such as, ball bearings, video-tape player spindles and camera lenses, have roughness of no more than 0.1 μm and enter the realms of nanotechnology. Smaller still, some micro-electronic and micro-mechanical devices, biotechnological and other organic thin film structures require geometrical descriptions literally to a fraction of an atom.

In many cases it is possible, and sometimes ideal, to describe a surface in terms of an averaged parameter measured from its behaviour. For example, the quality of an optical surface could be expressed in terms of the degree of light scattered. These are not being considered here. The common feature of all the methods that will be considered is that they can provide a profile of the surface, that is the shape in cross section of the interface between a material and its surroundings. "Profilometry" is taken to cover all systems that can do this, whether or not it is the only thing that they can do. In particular, no specific distinction is made between systems that can map a small surface area in three dimensions and those that can provide only a single profile track, effectively in two dimensions. A further restriction on this discussion is that it will deal only with modern electronic (semi-automatic) instruments. A review of conventional surface roughness instruments and many other approaches is given by Thomas.[1]

The accurate description of geometrical form implies a measuring instrument that has high vertical sensitivity, ultimately to small fractions of a nanometre and good lateral resolution. Note, here, the widely used convention that the nominal plane of the local surface is assumed to be horizontal. Surface structures have rather similar general geometries over many orders of scale[2] and, as with the natural landscape, the average slopes of the major features tend to be small. Thus a lateral resolution that is considerably poorer than the vertical resolution may be tolerable in many applications. This is a good thing for lateral resolution is much the more difficult to provide in practice. Nevertheless, it is important that small features are not smeared out by inadequate resolution and usually some form of microprobe is scanned over the surface. The classical method of microtopographic measurement involves the contact of a diamond stylus tip. Contact with delicate surfaces is best avoided if possible and many optical instruments have been developed in recent years. Some of these provide high resolution topographical information and qualify for our definition of profilometers, indeed they are sometimes called "optical styli". The recent development of the practical scanning tunnelling microscope, and variants applying different physical principles in a similar manner, has provided a totally new measurement regime of direct application to nanotopography. These three approaches will be discussed below, the last in most detail since its metrological implications are less well explored elsewhere. In all cases the nature of the probe to surface fidelity is emphasized rather than the general design of practical instruments. However, it is appropriate first to review a few critical features of precision instrument design and microdisplacement sensor technology.

14.2 INSTRUMENT DESIGN CONCEPTS

14.2.1 Loops and Stability

Recording a surface topography requires that local height changes of nanometre or smaller scale are satisfactorily monitored. The probe scan and also, therefore, the instrument reference frame and its traverse mechanisms, must be dimensionally stable to this accuracy over at least the time taken for a measurement. The traverse path must either follow its ideal form to this accuracy or be repeatable and accurate enough not to move the sensor out of its range if computer compensation can be applied.[3] Some knowledge of how this can be achieved is necessary if the operation of profilometers is to be critically assessed.

As in many areas of system design, precision mechanical structures consist essentially of loops. Two types of loop are significant here. The underlying measurement principle of all topography instruments is that of linear displacement. To assess a displacement, one side of a transducer must be fixed to the local frame of reference and the other to the object that is moving. Thus there is a metrology loop from the probe, through the sensor, the frame and back to the workpiece. Any change in the dimensions of that loop will register at the sensor and be interpreted as surface topography. Newton's third law demonstrates that if the various parts of an instrument structure are to remain in static equilibrium the forces on every section must balance. Thus there will be force loops around the instrument to carry the effects of weight, of the drive to the traverse and perhaps other significant loads. A force loop is, by definition, stressed and so to some extent distorted. Distortion of a metrology loop from its expected shape is a major source of error. A basic rule of precision design is that the force loops and metrology loops should be isolated from each other. This is impossible to achieve completely, but it remains a sound principle on which to build. Some high precision instruments have two side-by-side structures, one to carry the drive forces and workpiece weight, while a second metrology frame carries the gauging and supports only its own weight.

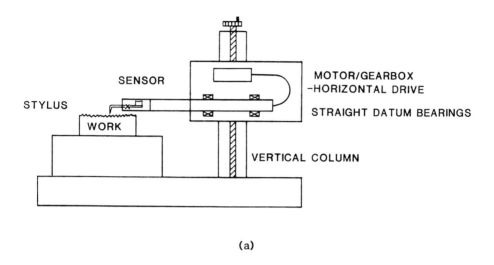

(a)

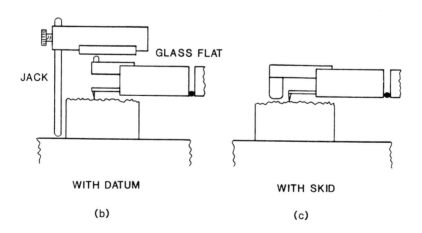

Figure 14.1 Schematics of a typical surface finish measuring instrument illustrating its metrology loops: (a) full instrument, (b) external datum, (c) skid datum.

Figure 14.1(a) illustrates a typical design for a surface roughness instrument. The stylus and sensor is cantilevered from a drive unit which contains a straight line datum bearing. The drive unit is suspended from the column of the worktable upon which the workpiece rests. The instrument metrology loop is visually obvious and provides a basis upon which the instrument can be critically assessed. Most obviously, it is very large compared to the displacements that will be measured. It is therefore vulnerable to thermal distortion and low frequency mechanical vibration. The traverse drive force loop is identical to the metrology loop so that, for instance, some differences of measurements would be expected if the traverse were reversed. More significantly, the metrology loop carries much of the instrument weight and, because the traverse is within the loop, the moments on the bearings will vary during a scan. Some variable loop distortion is inevitable. In defence of the designers, it should be stressed that they were aware of these points. The instrument is for general purpose use and must accommodate a wide variety of workpieces while being easy to use. It was acceptable to compromise on the loop design in order to get a convenient operational shape. The analysis does demonstrate that this type of instrument is not suitable for measurements of the highest precision. Figure 14.1 (b) and (c) show standard ways by which the instrument is made less sensitive to vibration and so usable for finer measurement. A separate datum surface can be used, supported directly from the worktable with a hinge relaxing the constraint to the drive axis. Clearly, the purpose of this is both to reduce the size of the metrology loop and to separate partly the metrology and force loops. If only local roughness deviation is to be measured, a skid can be used through which the "static" side of the sensor is referenced immediately to the workpiece surface.

The metrology loop is now almost as small as possible and the only force fluctuation within it will be from the sliding friction of the skid and stylus, presumed to be small. Its disadvantage is that the skid interacts with long wavelength surface structure as a non-linear filter and so distorts the profile to some extent.

The principles of kinematic design are important to the design of metrology loops. These date back certainly to the time of Lord Kelvin and are discussed in many older books on metrology. Furse[4] provides a good summary in the present context. The main idea is that an object in space can be described by exactly six coordinates, three translations and three rotations. It has six degrees of freedom. Contacting the body at one point prevents it moving in one direction and so removes one freedom (or provides one constraint). Removing a rotational freedom requires two points of contact but one of them will act as a translational constraint. Thus exactly six contacts, suitably placed, will, in the presence of a compliant closure force, completely constrain the body from movement. With less than six points, complete constraint is impossible. With more than six, there is overconstraint or redundancy. Unless by chance the position of extra contacts conform exactly to the body or forces distort the system, only a subset of six contacts will be active. This is why the closure force which prevents the body lifting away from the constraint points must be compliant, preferably a body force such as weight. A single degree of freedom mechanism must by similar arguments have exactly five constraint points and so on.

Of course, no real system can have true point contacts and even using small contacts can cause high local stresses. Instruments, where forces can often be kept quite small, can exploit the ideas almost exactly. For example, Figure 14.2 shows an example of a relocation platform. The specimen can be removed and later replaced in almost exactly the same position. Balls touching planes simulate the point contacts. The arrangement of three balls engaging three non-parallel vee-grooves will be familiar to users of surveying equipment. Another common arrangement uses three balls in a trihedral hole, in a groove and on a plane and is often called a "Kelvin clamp". In both cases there is a theoretically unique position at which the two parts of the clamp will match.

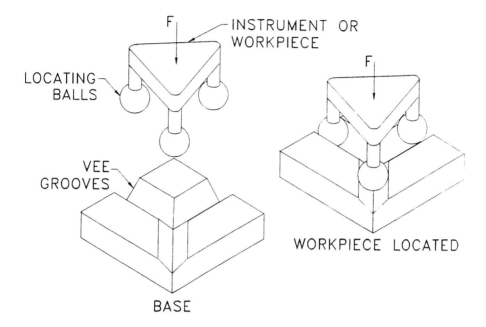

Figure 14.2 A typical geometrical arrangement for a kinematically designed relocation mount.

Errors in manufacture may cause that location to vary from the design position, but do not affect the quality of the location. Also, for example, the spacing of the balls will change through thermal expansion and so the location position may shift slightly, but a true location is maintained without strain being induced. This absence of induced strain, whether thermal or mechanical in origin, makes strict adherence to kinematic principles very attractive to the instrument designer. Real devices are imperfect and it is impossible either to bring the contacts together without sliding, so there is some residual friction force, or to totally prevent contamination at the contact points. Even so, it is normally easy to achieve consistent relocation to better than 1 µm (and 10 arcseconds) and 0.1 µm is possible with care. For almost strain-free permanent construction kinematic locations may be freely mounted and then adhesive applied to lock them into their settled position.

Kinematic design principles, applied to each component of an instrument structure, ensure that a readily assembled and minimally stressed form is realized. It should have good static stability, but there is nothing in the design method that ensures the geometrical accuracy, or even close repeatability, of a traverse bearing. The contacts will always provide the correct constraint by finding their own fit as they move position. In practice, every interface is a potential source of positioning error through the entrapment of dirt, corrosion and so on. The use of mechanisms rather than sliding bearings for some types of traverse can provide more consistent motion, although often only an approximation to the ideal motion can be achieved. However, the mechanism itself has several joints, with bearings, and so the advantage of consistency might be lost. This explains the attractiveness of elastic, or flexure, mechanisms for instrument design.

The use of flexures to provide short range movement again has its origins in distant history but has been brought to a high point of performance in modern instruments.[5,6] Figure 14.3 shows the simplest way of obtaining a nearly straight motion over a short distance. The two legs, C and D, which support the platform, A, above the base are distorted elastically and symmetrically by a centrally applied actuation force, F (see e.g. Smith[7] for information on actuators). Providing they remain within their elastic region, the legs should behave identically at every identical actuation. In practice, both material hysteresis and microslip at the clamping points degrade the ideal behaviour. Monolithic designs overcome some of these difficulties and provide some other advantages, but often involve rather intricate manufacture.[8] For example, by making the whole mechanism from one block of material deliberate over-constraint can be introduced to provide elastic averaging of errors due to manufacturing tolerances without induced stresses.

Obtaining stability of the loops in precision instruments implies close attention to both their geometrical layout and their material. Typical sources of error in metrology loops are sagging due to self-weight or impressed forces and thermally induced distortion. Thermal effects generally relate to linear expansion and may be caused by environmental drift or local heat sources within the instrument. Generally bending of the loop structure leads to larger errors than simple expansion which preserves their shape, because it offers a lever-like source of magnification of a small displacement. Thus a temperature difference across the structure may matter more than a uniform change. Bending moments and temperature gradients are minimized by using structures symmetrical about the line of action of the measurement. This is just another variant of the well-known Abbe principle: that for minimum errors from residual misalignment the sensor axis should lie on the line of action of the probe. Materials selection can follow several strategies, all based on the comparison of property ratios.[9]

Important properties include dimensional stability, elastic modulus, thermal expansion and thermal conductivity. For example, two strategies for overcoming thermal gradient related distortion would be to use a high conductivity material to ensure rapid equilibration or to use a low expansion material so that the effect of the gradient is negligible. Modern materials of interest for precision instruments include molybdenum, fused silica or quartz, ceramics, "zerodur" and similar ultra-low expansion glass ceramics, and single crystal silicon. Each offers some properties of particular interest with a good overall spread or some which are extremely good and others poor. There is rarely a single correct answer to the problem of materials selection. For relatively routine work, steel, invar and aluminium continue to be much used.

14.2.2 Displacement Sensors

With the exception of the newly developed scanning tip microscopies, which will be discussed later, virtually all practical microtopography sensors depend on one of three basic transduction principles. A much wider range of methods can be used to measure small displacements,[10] but most cannot provide high lateral resolution and there is little point in deviating from the use of a few proven and adequate methods if an intermediate stylus is needed.

(a)

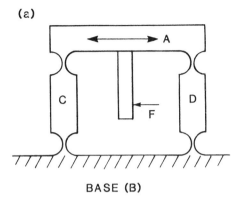

BASE (B)

(b)

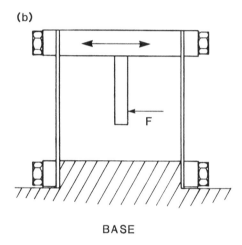

BASE

Figure 14.3 Simple forms of elastic linear translation mechanisms: (a) monolithic construction using notch hinges, (b) fabricated construction using leaf springs.

Inductive gauges are the most popular choice in high precision metrology. Two closely related forms are used, the Linear Variable Differential Transformer (LVDT) and the Linear Variable Differential Inductor (LVDI). As shown in Figure 14.4, both sense the position of a ferromagnetic core (or slug), S, relative to a pair of symmetical coils, A and B, by means of the changing flux linkage. The LVDT is a true transformer in which the two (sensing) secondary coils lie symmetrically with respect to a primary, P. An excitation signal is supplied to the primary, typically a few volts at a few kHz. The secondaries are wired in antiphase so that if the core is central their induced signals cancel. As the core moves the net signal increases, so the output is a carrier signal from the primary amplitude modulated by the displacement signal. The LVDI does not have a primary but the two coils are wired in-phase and used as half of an AC bridge circuit. The output signal is similar to that of an LVDT. Simple rectification and filtering can provide a displacement signal, but better performance is obtained by phase-sensitive, or synchronous, detection.

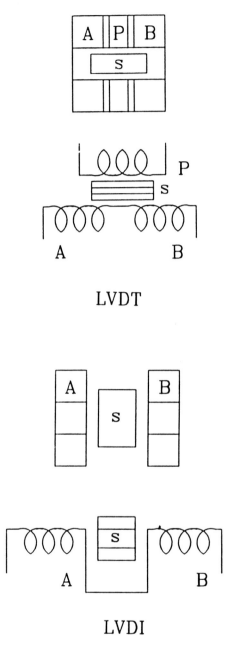

LVDT

LVDI

Figure 14.4 Linear inductive displacement sensors.

Capacitive displacement sensors exploit variations of the parallel plate capacitor. Two parallel plates have a capacitance, C, given by

$$C = \varepsilon A / d \tag{14.1}$$

where A is the area of overlap, d the spacing and ε the permitivity of the gap. Sliding one plate by the other to vary the area can provide a large range and good linearity.

Varying the gap gives a non-linear characteristic. By differentiating equation (14.1) with respect to d, the sensitivity (change in output for a small change of input) is seen to increase as the nominal gap decreases. Thus for very short range measurements extreme sensitivity can be obtained by using a small gap. The devices are AC excited and used in a bridge circuit or directly coupled to a charge amplifier. High performance depends particularly on very high input impedance at the detector which integrated circuits can now provide at moderate cost. Static and low frequency displacement can be measured with noise levels as low as 1 pm $Hz^{-0.5}$ in practical configurations.[11] Interestingly, charge amplifier driven capacitive gauges used for ultrasonic signal detection are also sensitive to small fractions of a nanometre, but this time in the MHz frequency band.[12] Capacitive gauges can be used as non-contact displacement sensors on conducting surfaces, but the plate size needed to provide a reasonable signal is too large to provide the lateral resolution needed for topography. One possible exception is a fringe field probe[13] that exploits the electric field distortion at the edge of a thin plate, an effect most designers of capacitive gauging spend much effort to eliminate.

Optical measurement is usually performed by interferometry and, while many special interferometers have been invented, the basic concept of them all is shown by the Michelson arrangement, Figure 14.5. An adequately coherent light source, usually a laser, is split to provide a reference beam and a measurement beam. Each is returned by a mirror to the beam splitter and two coincident return beams interfere at a detector. If the two path lengths differ by an exact multiple of the wavelength, λ, there will be constructive interference and a large signal. If one mirror moves by 0.25λ, the path difference changes by 0.5λ and destructive interference occurs. Displacement is measured by counting the passing fringes and interpolating. The basic fringe spacing of 0.5λ corresponds to around 0.25 µm, fringe division by 100 is fairly routine (1000 is state of the art) so nanometric resolution is available. Polarization is commonly used to improve the beam splitting and to avoid interference before the detector. Corner cube or roof prisms are preferred to plane mirrors because they are less sensitive to parasitic rotations.

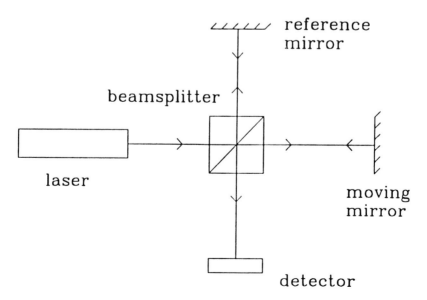

Figure 14.5 Basic configuration for a length measuring Michelson interferometer.

14.3 CONTACT STYLUS INSTRUMENTS

14.3.1 Instrument Configurations

Stylus based surface finish measuring instruments nearly all conform to the general arrangement shown in Figure 14.1(a), although details vary considerably. This reflects the need for general purpose instruments to cope with bulky workpieces and to be able to measure within recessed features and relatively small diameter holes. The cantilevered probe is essential for the latter purpose and generally easier to use than an external datum. Other topographic measurements such as straightness and roundness use similar approaches involving a datum motion from a translational or rotational bearing, respectively. It is equally valid to move either the workpiece or the sensor, since it is the relative motion that creates the datum. Often there are pragmatic reasons for choosing one form over the other, for example to avoid heavy workpieces loading the precision bearings, but, especially in roundness measurement, there are also some more subtle effects.[14] Form measuring instruments do not need such high bandwidths as roughness instruments. A more massive probe assembly is then tolerable and that, in turn, makes it easier to mount it in a better position relative to the datum bearings, even if it is further from the workpiece surface.

A few workshop roughness instruments, in which robustness takes precedence over ultimate precision, use piezo-electric sensing elements. They tend also to have relatively poor bearing precision in the drives since they are intended for use with a skid reference, Figure 14.1(c). Discussion here is concerned with higher precision instrument types. In these an inductive sensor is almost always used. Their drives generally have maximum traverses of 50 to 150 mm with deviations from best straight line of a few tenths of a micrometre. Errors are smoothly varying, so better straightness is obtained over shorter traverses. The sensor acts through a lever to the stylus, mainly to provide free space near the stylus tip. Typically the stylus is 50 mm from a knife-edge pivot and the ferrite slug of the sensor a further 25 mm along a thin tubular arm. It is remarkable that a demagnifying lever is used in such high sensitivity applications! Stylus conditions should conform to a suitable National or International Standard. The stylus is almost invariably of diamond, with the most common shapes being either conical with a 2 μm radius tip or 90° pyramids with the tip truncated to a 2 μm square. Earlier instruments tended to have a "chisel" tip, about 2 μm in the traverse direction, but wider in the transverse direction. 10 μm tips are sometimes used. The nominal static force of contact onto the surface should be 0.7 mN and this should not change appreciably throughout the range of the sensor. The lever arrangement to the stylus causes an arcuate movement that reduces the linearity at larger displacements. Consequently the active range is limited to around 100 μm (or around ×500 in the notation of vertical magnification commonly used in this field). Realizable discrimination depends generally on electrical and mechanical noise at the input and is of the order of 0.1% of full scale at the selected magnification. The larger instruments are difficult to use at magnifications above ×50,000 (perhaps ×200,000 with a skid system) because of vibration sensitivity in the large metrology loops. Typical traverse speeds are from perhaps 10 μm s^{-1} to 1 mm s^{-1}.

Straightness and roundness instruments tend to have broadly similar sensor systems, somewhat scaled up in size. Styli are usually spherical or radiused chisels of around 1 mm radius and made of ruby, sapphire or tungsten carbide. Because of their larger size, both range and ultimate magnification tend to be a factor of ten or so lower than for roughness instruments.

The Rank Taylor Hobson Form Talysurf[15] is still the only commercial instrument that combines fully the function of straightness and roughness measurement. Its basic layout is similar to a conventional roughness instrument, but the inductive sensor is replaced by a miniature laser interferometer. This is of the Michelson type, using polarized beams to obtain quadrature fringe signals so that the direction of motion can be determined. Because the interferometer is an incremental counting device, the normal constraints on range to resolution of linear sensors do not apply. A range of 8 mm with a resolution of 2.5 nm is available so that curved surfaces may be simultaneously checked for shape and fine surface textures. Interchangeable styli are again mounted on a pivot arm. With its large range, the arcuate motion produces significant errors in the lateral positioning of the stylus from its expected position during some traverses. The scan position is monitored by a second interferometer and a correction term computed according to the signal recorded, a considerable computational task.

For measurements of the highest sensitivity, more specialized instrument geometries are used. They are sometimes called step height instruments since a common use is for the measurement of thin film and microelectronic devices. For this task a relatively blunt stylus, perhaps a 10 μm flat, is used at low contact force. The sensor is placed in line with the stylus to satisfy the Abbe principle and the metrology loop reduced in size compared to general purpose instruments. Only small, or at least thin, workpieces can be accommodated but stability is much improved. Fine topography can be measured with chisel styli of about 0.1 μm in the traverse direction at contact forces of around 40 μN. The Rank Taylor Hobson Talystep remains one of the most precise commercial instruments even though its basic design dates from the 1960's. It has only a 1 mm traverse range provided by a single, very deep flexure hinge. The motion is slightly arcuate in the plane of the workpiece surface, but very consistent vertically. Vertical magnifications of up to ×2,000,000 are available as standard. The sensor is an LVDI and the stylus arm is supported on ligaments as a simple parallel motion elastic mechanism. The lack of capacitive sensors in instruments of this type might seem surprising and, indeed, some are now appearing. However, until recently the electronics needed for high sensitivity measurements cost considerably more for capacitive systems than for inductive ones and the performance of the sensors is not much different. Capacitive systems are probably still more expensive and instrument makers have no strong incentive to change a perfectly acceptable system.

14.3.2 Limits of Stylus Technology

There are two aspects to the question of the smallest features observable by stylus instruments, those concerned with the interaction of the stylus with the surface and those concerned with the rest of the sensor system. Of the useful sensor principles, optical interferometry has a discrimination determined by the wavelength of the light and the degree of fringe division that can be reliably achieved. Around 0.5 nm seems to be the best that can be expected of current technology. Both capacitive and inductive sensors have a theoretically infinitesimal response, with the practical limit arising from noise associated with their implementation. Under favourable conditions, the LVDI sensors used in Talystep have been shown repeatedly to respond to slowly changing displacements at the 10 pm level.[16] There is no reason to suppose that this is a real limit, it was merely the discrimination limit of the particular test system that was used. Capacitive gauges are being studied by the same methods and appear to be equally good.

The ultimate sensitivity of the sensor is unlikely to be achieved in a real instrument, partly because the bandwidth needed will admit more noise and partly because the instrument metrology loop and traverse will have an influence. Theoretical modelling of inductive and capacitive gauges using parameters typical of surface metrology instruments suggests that for a 25 Hz bandwidth the input noise should be around 0.1 to 0.2 nm RMS, while practical assessments find this estimate to be slightly conservative.[17,18] Thermal drift during single traverses can be significant with such high sensitivities even in a temperature controlled room. The best Talystep transducers (not the complete instrument) have a net thermal expansion of around 50 nm K^{-1} because the metrology loop cannot be single material[16], so mK levels of drift cannot always be neglected. The traverse bearings also become a major source of uncertainty over traverses too long for convenient, stable elastic mechanisms. Work originating at the National Physical Laboratory (NPL) has sought to address two problems through the use of ultra-low expansion glass ceramics.[19,20] The loop net expansion coefficient of the instrument can be reduced by a factor of around 250 to a few nm K^{-1}. Simultaneously, the use of polymeric bearings on the optically worked slideway surfaces provides a traverse of 40 mm with a straightness deviation of about 15 nm and scatter on repeated tracks of 1.5 nm RMS. Shorter traverses give somewhat better performance. These instruments appear to reflect the current state of the art.

The physical contact of the stylus on the surface presents some problems. Local stressing is an obvious source both of measurement error and workpiece damage. The Hertz model of contact can be used to estimate these effects. Simplifying the situation to that of a sphere pressing against a plane and taking the elastic modulus of a diamond stylus to be so high that its distortion is negligible compared to that of the workpiece, the maximum stress is

$$P_{max} = \frac{3F^{1/3}}{2\pi}\left(\frac{4}{3RG}\right)^{2/3} \tag{14.2}$$

and the indentation depth

$$z = \left(\frac{1}{R}\right)^{1/3}\left(\frac{3GF}{4}\right)^{2/3} \tag{14.3}$$

where F is the contact force, R the tip radius and the compliance $G = (1-\eta^2)/E$ where E is the elastic modulus and η Poisson's ratio. This implies that a 1 mN force on a 2 μm tip against a steel surface will indent it by about 20 nm if the process is elastic and provide a peak stress of over 10 GPa. This is well over ten times the bulk strength, yet little if any damage to the surface is observed in practice. The surface layer is much stronger than the bulk. On soft materials such as copper, a scratch left by a standard stylus may be seen by eye, but scanning electron microscopy often shows this to be an interpolation in the brain. The damage usually occurs only at the tips of asperities and seems caused more by the lateral traverse of the stylus than by its normal loading.

Providing the surface is not unacceptably damaged, elastic deflection does not necessarily introduce a significant error. The profile may still be followed faithfully with a constant indentation. There is some evidence that this happens on very compliant surfaces.[1] The dynamic forces of the traverse are much smaller than the static load. A reasonable estimate for a conventional instrument traversing over a ground surface at 1 mm s^{-1} suggests that the maximum angular acceleration is around 1 rad s^{-2}. For typical stylus arm inertias this requires a tip force about 1 μN. Thus in normal operation dynamic forces can be totally neglected compared to the static loads. Another source of error is that the mountings of a real stylus have finite transverse stiffness, so that its track may meander slightly as it deflects around asperities rather than climbs over them. Again, Thomas[1] considers this point.

The overall conclusion is that, although they are subject to several potential errors none of which is easy to assess in practice, stylus instruments are remarkably good and genuinely deliver sub-nanometre vertical resolution. A remaining difficulty cannot be overcome. The stylus must present a significant area of contact in order to avoid totally unreasonably stress and a very fine tip lacks mechanical robustness and soon becomes larger in practice. The lateral resolution is fundamentally limited because of this. 0.1 μm tips are the smallest that have been made and used reliably.

14.4 OPTICAL TECHNIQUES

Optical techniques of surface profiling are taken here to imply that the surface is interrogated directly by a beam of light. Such methods are attractive since they are non-contacting and many variations on a few basic themes have been developed in the last decade. As the purpose here is to explore the ultimate capabilities of optical devices only a few archetypal profiling instruments will be discussed. A number of papers at a recent conference[21] provide a good recent survey of available methods. For microtopography assessment, all the systems operate in the general form of a microscope, with a large numerical aperture objective used to provide a small spot size. This gives good discrimination but tends to limit the maximum displacement readily achieved. Generally, vertical ranges are broadly comparable with the highest sensitivity LVDI sensors.

One approach to profiling with a microscope is to scan it over the surface while maintaining focus and using the displacement needed to hold focus as a measure of the surface profile. The first attempts[22] to do this practically moved the microscope body to maintain the focus condition detected by a Foucault edge at its back focus. Neither its resolution nor its operating speed were adequate for other than some specialized applications. Little happened in this area until the compact disc player created a large market for laser diode optics that could track a distorted disc. The technology has been successfully adapted to precision measurement[23] and appears to be capable of detecting sub-nanometre height differences. Figure 14.6 shows schematically the basic concept around which such systems are built. A low cost version has been reported which still resolves to better than 100 nm.[24] In all these variations, lateral discrimination is of the order of the spot size and so depends on the source and the objective aperture. At best spots of around 1 to 2 μm are used, giving lateral discrimination similar to standard stylus instruments.

The second main class of optical profilers uses interferometry. Essentially they are more sophisticated versions of Figure 14.5 with one mirror replaced by a microscope objective and the surface. The basic scheme is shown in Figure 14.7. To ensure good stability in the (optical) metrology loops the beams use a common path as far as is possible. Also detection generally involves a phase-stepping method in which successive small shifts in the optical path are introduced at each measurement position.[25] This improves greatly the quality of fringe division that can be achieved, that is, it improves vertical discrimination. Several commercial systems have been produced involving the reference mirror located in the objective and with the interference field detected on CCD linescan or full area photodiode arrays (see, for example,[26]).

FROM INSTRUMENTATION TO NANOTECHNOLOGY

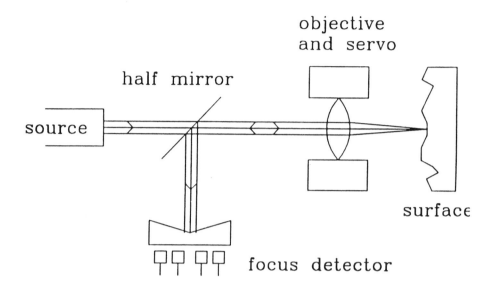

Figure 14.6 Essential features of optical stylus instruments that exploit constant focus conditions.

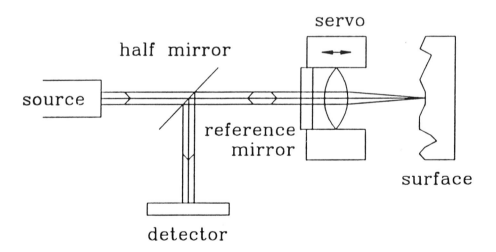

Figure 14.7 Typical layout of the major features of an interference microscope for surface profiling with phase-stepping drive and near common path optics.

All such systems rely on a heavy computational effort to collate and decode the data into height information. Another virtually completely common path method uses 90° polarization between the two interferometer beams together with a birefringent objective.[27] The focal length for the two beams is somewhat different, so that when one is accurately in focus, the other is defocussed to produce a spot of 10 to 20 μm around the focussed one. This acts as a reference beam which integrates local roughness to provide the average surface position in an optical equivalent of the stylus and skid. It has the similar advantage of a very small metrology loop and disadvantage of limited response to longer wavelengths.

Following the discussion in section 14.2, the vertical discrimination of any of the interferometer systems is set by the degree of fringe division that can be achieved. Conventional length interferometry still has difficulty in providing precision better than 1 nm.[28] The closely controlled conditions of the profilers seem to deliver rather more, although precise figures are difficult to obtain. Certainly the computer output from the systems can discriminate to below 0.1 nm and output profiles appear plausible, but whether they are accurate is less clear. The Downs common path system[27] has shown evidence of sensitivity at the 0.02 nm level. The lateral resolution is limited always by the smallest spot size on the surface and in some systems also by the spacing of the cells in the photodiode array. The best results resolve to rather less than 1 μm. The lateral range of imaging systems is restricted by the field of the objective and so reduces as higher powers are used for greater sensitivity and might be 50 to 100 μm across at top magnification. The defocussed spot reference limits that system to an upper wavelength of about 15 μm for accurate profile reproduction. One criticism of all optical instruments is that they depend on surface reflectivity. This can be a serious problem, but is less so with the types described here. Three- or four-phase interpolation allows wide variations in the amplitude of the fringe signal obtained without significant loss of accuracy. Any reflectivity above a few percent is acceptable. A less discussed but more serious issue occurs with surfaces that are not of one material. Different parts may then cause different optical phase changes at the reflection and these cannot be differentiated from profile height changes.

Finally one more general technique might be considered.[29] This relies on "frustrated total internal reflection" in which the presence of a surface very close to the internally reflecting surface of a prism captures energy from the evanescent wave. Under the correct conditions a grey-scale image of a surface structure can be observed in a modified microscope from which computer analysis can extract nanometric amplitude data. Lateral resolution is similar to other optical systems. This interesting method is perhaps better applied as a scanning probe microscopy, of which more below.

14.5 SCANNING TUNNELLING MICROSCOPY

14.5.1 Introduction and Principle of Operation

The concept of the scanning tunnelling microscope was first postulated, and named the "Topographiner", in 1972 by Young *et al.*[30] as an extension of the "field emission ultramicrometer"[31], an ultra-sensitive electrode to specimen separation transducer. A sharp tungsten tip was suspended above a conducting surface in a high vacuum and a voltage potential applied across the gap. They showed that at sufficiently high voltages the Fowler-Nordheim theory of electron emission would accurately predict the current flowing across the gap. If a constant current is passed through the emitter the applied voltage will be very sensitive to the emitter to surface spacing (hereafter called the emitter spacing). In the original instrument this voltage could be amplified and applied to a piezoelectric ceramic upon which the probe was mounted in order to maintain a constant separation. Assuming that the voltage-displacement characteristic of the piezo element is linear (see Smith[7]), the applied voltage can be used as a monitor of the surface altitude at the point of the highest asperity of the probe. Thus we have a direct analogy with the stylus technique in that a point probe is maintained at a constant distance above a surface, in one case by maintaining a constant field emission voltage and in the other by the contact forces. To obtain a profile with the Topographiner, all that is required is to move the specimen along a straight line and monitor the perpendicular displacement (the servo signal) of the probe. The essential difference between the two techniques is that the latter method is *non-contact.*

Before discussing the practical implementation of this instrument and its derivative, the Scanning Tunnelling Microscope (STM), it is important to investigate both their principles of operation and their limitations. Although the theory of Fowler-Nordheim emission is not of importance to this discussion, Young[31] was able to show that the separation sensitivity, S, of an emission tip of radius r_0 with an emitter spacing x_0 is

$$S = -\left(\frac{1}{2x_0}\right)\frac{1}{(x_0/r_0)^{0.5} - 1} \tag{14.4}$$

This equation shows that not only will the sensitivity increase with a reduction in the separation, but also, at small separation, the radius of the tip becomes less important enabling the use of very sharp tips and so very high lateral as well as vertical resolutions to be achieved. Consequently, as the dimensions of the probe and the emitter spacing are reduced the measurement resolution increases. This was realised at the time by Young who wrote[31]

"The system has the remarkable property that sensitivity increases as the electrode spacing decreases over many orders of magnitude, like a continuously expanding vernier on a mechanical instrument."

Young *et al.* also realized that as the probe approached sufficiently closely to the surface, the phenomenon of electron tunnelling would dominate over field emission. In this, the position of electrons can only be defined in terms of the probability wave of the Schrodinger equation and there will be a finite probability of them extending beyond the tip as an electron cloud having a density that decays exponentially with distance. If there is a potential difference across the emitter spacing a current will flow with an exponential dependence on the separation. The current density, J, across two perfectly flat electrodes of infinite area at an emitter spacing x_0 across which is an applied potential, V, is given by[32]

$$J = \frac{e}{2\pi h (\beta x_0)^2} \times$$

$$\left\{\bar{\phi}\exp(-A\bar{\phi}^{1/2}) - (\bar{\phi}+eV)\exp\left[-A(\bar{\phi}+eV)^{1/2}\right]\right\} \tag{14.5}$$

where $A = 4\pi\beta x_0 (2m_e)^{0.5}/h$, e and m_e are the electronic charge and mass, $\bar{\phi}$ is the average potential barrier height between the electrodes, h is Planck's constant and β is a correction factor, usually close to unity. The left and right hand sides of the expression in brackets describe the number of electrons flowing from the high potential to the low and from low to high respectively. Additionally as the voltage tends towards zero, the average barrier height is nearly equal to half of the average work function of the two materials. As the voltage increases, these conditions no longer apply and the situation becomes far too complicated to model using the assumptions used in the original derivation of this formula. At low voltage, the equation 14.5 reduces to

$$J = \left[\frac{(2m_e)^{1/2}}{x_0}\right]\left(\frac{e}{h}\right)^2 \overline{\phi}^{1/2}V \ exp\left(-A\overline{\phi}^{1/2}\right) \qquad (14.6)$$

$$J = \frac{k_1\overline{\phi}^{1/2}V}{x_0} \ exp\left(k_2 x_0 \overline{\phi}^{1/2}\right) \qquad (14.7)$$

where k_1 and k_2 are constants for a particular system. At constant separation, the junction is ohmic with the absolute value being dependent upon both the separation and the work function of the two electrodes. Again, if a constant resistance is maintained between the emitter and specimen, and if the work function (or, in reality, the barrier height) remains constant, then so too will the altitude of the probe above the surface. To see how this phenomenon may be used for surface profiling it is necessary to examine the design of the scanning tunnelling microscope.

14.5.2 Scanning Tunnelling Microscope Design

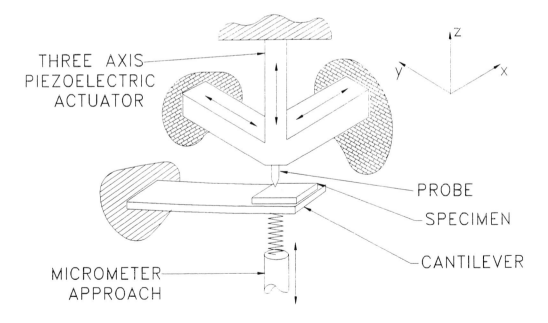

THREE AXIS PIEZOELECTRIC ACTUATOR

PROBE

SPECIMEN

CANTILEVER

MICROMETER APPROACH

Figure 14.8   Schematic representation of a typical design for a scanning tip microscope.

A schematic diagram of a typical STM system is shown in Figure 14.8. In this the probe is attached to three orthogonal piezoelectric elements that can translate the specimen in all three cartesian axes.

To operate the microscope, the specimen is first moved towards the probe until a tunnelling current is monitored (typical values for the total current and tip/specimen potential are 1 nA and 100 mV, respectively, giving an ohmic junction of 10 MΩ). The probe is then raster scanned over the specimen using the x- and y-axis piezos while the z-axis actuator is servo controlled to maintain a constant tunnelling current. To linearize the relationship between the tunnel current and the emitter spacing, the output of the current amplifier is usually fed into a logarithmic amplifier prior to its use to control the probe servo. Plotting the applied voltages to the three drives gives a three dimensional image of the specimen surface in terms of its equipotential barrier height. It must be emphasized that the image can only be interpreted as an accurate representation of the surface geometry if it is assumed that the actuators are linear and the barrier height is constant.

It is difficult to ascertain the exact altitude of the probe above the surface, but for surface finish measurement, it is only necessary that the probe is maintained at a constant altitude. In practice, a tunnelling current is first observed at a tip specimen separation of approximately 2 nm and it will be controlled to maintain a separation of less than half of this. Thus if any of the parameters were to change, causing for example an apparent change of one half of this value, then the uncertainty of the position of the surface would still be in the order of a fraction of a nanometre.

Although it was possible to obtain a tunnelling characteristic with the Topographiner, problems with electronic noise and external vibrations precluded the use of this instrument for imaging.

A decade later Binnig et al.[33] were able to achieve a stable tunnel current and measure its exponential character. They succeeded only by placing the microscope inside a vacuum chamber and mounting the complete assembly on a large concrete block to isolate it from vibration.

Further vibration isolation was provided by floating the tunnelling probe and specimen on superconducting magnets. Following the success of this experiment, a new design was constructed which has formed the basis of almost all succeeding STMs. This consisted of three orthogonal piezo drives upon which the probe was attached with the specimen mounted on a piezo driven "louse". The louse is a fine feed device that operates by electromagnetically or electrostatically clamping one end of a platform while the other end is extended. The extended end is then clamped followed by the unclamping of the opposite end and retraction of the platform. Repetition of this cycle results in the progression of the louse along a controlled path in a manner similar to the movement of a caterpillar (for discussion of drives see Smith[7]). Later on in the same year, these workers were able to publish the first ever STM images of $CaIrSn_4$ and Au surfaces in which monoatomic steps could be clearly resolved.[34,35] Further research led in the following year to the imaging of a 7×7 reconstruction on the close-packed plane of a silicon wafer.[36] In this image, not only could individual atoms be clearly resolved but also their arrangement in unit cells. This work was probably instrumental to the award, in 1986, of Nobel Prizes to both Gerd Binnig and Heinrich Rohrer. After this spectacular image was presented there followed a large global effort into both the use and design of this and other related techniques (to be briefly covered in section 14.5.4).

Since this initial work, commercial STM systems have been developed which incorporate image processing techniques that make them almost routine laboratory tools.

Surprisingly, although the annual publications on STM are now in the hundreds, there have been few improvements on the original design of Young *et al.* The major improvements have been in the increased stiffness of the probe driver (mainly due to its miniaturization) enabling its use with common laboratory isolation stages and the replacement of the louse with micrometer approach mechanisms, as shown in Figure 14.8. The most commonly used vibration isolation method for high vacuum applications is to mount the instrument onto a laminate of stainless steel plates separated by "Viton" rubber O-rings. A stack of five plates is commonly used to provide adequate isolation. For operation in air, the instrument is usually suspended from three elastic ropes. There is little point in using systems that will provide greater isolation, since electronic and mechanical noise injected through connecting cables often dominate.

The most significant improvement in the stiffness of the STM driver was to replace the three orthogonal drive elements with a single tube that has been slit to achieve the same three dimensional translation capability.[37] Commonly called a "tube scanner", this could be produced having total dimensions of a few millimetres square and capable of operating at speeds of 10 kHz or more to enable real time imaging.[38,39]

Variations in the tribological and dielectric properties of the interfaces of the louse mechanisms[40] cause a varying applied force within the measurement loop that resulted in an unpredictable movement of the specimen relative to the probe. Micrometers have been used almost exclusively for the coarse approach of the specimen in subsequent designs. Most incorporate a lever mechanism with the micrometer to give submicron resolution of movement. The most common method has been to attach a weak spring to the end of the spindle which pushes against a stronger spring upon which the specimen is mounted, Figure 14.8. The gearing of this drive is directly proportional to the ratio of the stiffnesses of the weak and strong springs whereas the stiffness of the complete assembly is equal to the sum of the two stiffnesses because springs connected in parallel behave as series stiffnesses. This type of lever system has the advantage of being linear and continuous, but the disadvantage of a very long measurement loop that encompasses all of the spring elements as well as the micrometer (early examples of this approach can be found in references[41-45]). One method for the reduction of the thermal sensitivity of these instruments is to design the system so that the relative expansion of different elements compensate each other. Using such a system, it has been shown that (after a settling period of approximately half an hour) it is possible to achieve drift rates of less than 0.1 nm per minute.[46] Another technique for the reduction of the measurement loop length is to use a wedge. This has the advantage that the microscope can be retracted from the loop upon positioning and the wedge can be simply made from thermally stable materials such as "Zerodur".[19,47-49] Using a micrometer driven wedge made from Zerodur we have found that a positional resolution of ≈1 nm can be achieved by manual adjustment in a normal laboratory environment.

Although STMs show sub-atomic lateral and vertical resolution there has been little work dealing with their use for metrological purposes. One major reason is that piezoelectric actuators have a high hysteresis, temperature dependence and, more importantly, creep. Initially, calibration was achieved by imaging an atomically flat surface and using the known lattice spacing as a ruler. This has the drawbacks that the lattice may be distorted and it is not yet fully understood to what extent the high electron current density influences the surface being measured. Using simple calculations, it is clear that both interatomic and electrical forces will be significant leading to distortion of both the probe and specimen. During the tunnelling process forces in the region of 10 nN have been measured[50] and at a lateral resolution of 1 nm this indicates a pressure of approximately 10 GPa. This is comparable to the values of the surface hardness, H_v, of even the hardest of materials (H_v for silicon is approximately 5 GPa[51]). It is also apparent from equations (14.5) to (14.7) that for a constant current the separation will be a function of both the geometry of the interface (the above equations have been derived by assuming the electrodes are of similar materials and are both perfectly flat and parallel) and the tip potential. A further complication is that if the tip polarity is positive then the electrons are being "pulled" from a very small surface area in which the concentrated fields will induce an additional resistivity across the interface. If, however, the polarity is reversed, then the electrons can disperse over the specimen surface yielding a different value for the resistivity of the interface. Although theories for these phenomena have been presented,[52-54] they have not found wide application to the interpretation of STM images.

For many applications in surface physics, chemistry or biology much can be learned even if the drives lack precision, but for metrology accurate knowledge of the tip translation in all three axes is required. Although there has been an attempt to linearize the piezo actuators[55] there have been very few attempts to achieve an absolute position reference for the translation stages. An STM has been developed at the National Research Laboratory of Metrology (NRLM) in Japan in which the three orthogonal drives utilize simple notch type linear springs, Figure 14.3, driven by piezoelectric actuators. The relative displacement of these is then monitored using a laser interferometer to remove the drive non-linearities.[56] An alternative to the piezoelectric drive has been developed in which a linear electromagnetic force actuator driving an elastic spring mechanism has been employed.[57] There remains a lot of work to be carried out in this area.

14.5.3 Milestones in the Use of STM

The initial work of Binnig *et al.* in which it was shown that atomic structure could be directly investigated was rapidly expanded to include the examination and even modification of surface features at the atomic level. There have been so many new applications of this technique in surface science that a detailed discussion of all of these advances would be both too long and too diffuse (for a review of STM instrumentation see Kuk and Silverman[58]). As a consequence only a few spectacular and interesting examples have been chosen for this discussion, with apologies to those we have omitted.

From equations (14.5) and (14.7) it can be seen that during the tunnelling process there is an exchange of electrons both from the tip to specimen and vice versa. In theory, any two waves of equal frequency should create a standing wave pattern and this is indeed predicted for the case of tunnelling. This effect is enhanced when the gap voltage is increased and although many researchers had reported this phenomenon the first real evidence, including a comparison with theory, was presented in 1985.[59] Periodicities of the order 0.2 nm could be observed in the electron density across the tunnel gap.

It was also realized during the early development of STMs that the probe is so close to the specimen that molecules are unlikely to be present or will be excluded. The instrument can continue to operate in air as long as the surface does not grow an insulating layer of thickness greater than the tunnelling gap. This was exploited for the imaging in air of biological samples in which strands of DNA were directly measured.[60] Although providing no new information, this work made the scientific community aware of the potential for extending the use of STM to other research areas such as chemistry and biology.

In 1987 Becker *et al.*[61] showed that, by applying a large potential across the gap, it is possible to remove an atom from the surface and, by reversing the procedure, to replace it elsewhere. This was immediately followed by the presentation of results whereby the tip was used as a micromechanical tool to cut small grooves[62] and to pick up and place biological samples for examination.[63] Two years later researchers were able to obtain the first images of single stranded nucleic acids.[64]

To finish, it is interesting to note that in April of this year (1990), two research workers were able to accurately manipulate xenon atoms on a nickel surface to spell out the letters IBM.[65] This is the smallest data storage medium yet. Later in this same year convincing images showing atomic scale detail of single strands of DNA were reported.[66] Atomic scale manipulation and examination are now available to science.

14.5.4 Alternative Scanning Tip Sensors

The STM operates only with conducting surfaces. This is a real problem if atomic resolution measurement is wanted for normal applications on insulators or in a working environment with surface contamination and the formation of insulating surface layers. Binnig and his co-workers solved it by the invention of the Atomic Force Microscope (AFM).[67]

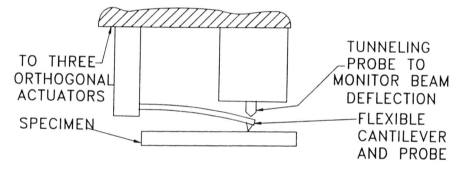

Figure 14.9 Schematic representation of the atomic force proximity sensor using a tunnelling probe to monitor beam displacement.

In this, the probe is attached to a cantilever beam which, due to the interatomic forces, will bend due to Van der Waals and Lennard-Jones forces as it is brought into proximity with the surface. The deflection of the beam was monitored by a tunnelling probe, signal from which was used to servo the complete cantilever and probe to maintain a constant force, see Figure 14.9. The majority of subsequent designs have used optical methods for monitoring the deflection of the cantilever beam. In essence, the AFM is directly related to conventional stylus techniques, but with forces in the region of nanonewtons which is six orders of magnitude lower. Atomic resolution images have been achieved using this technique, but it is much more difficult[68] and, in general, the AFM is considered to have a lower lateral resolution than the STM (the question is not fully resolved). AFM imaging has again been extensively developed and commercial systems are now available. Using this technique, researchers have been able to image insulators such as cleaved mica and polymers and to image in real time biological processes such as the clotting of a protein.[69]

Contemporary with the development of the AFM was the utilization of the capacitance between the probe and surface as a proximity sensor. This resulted in the development of the Scanning Capacitance Microscope.[70,71] In this the probe is controlled to maintain a constant dielectric value between the tip and specimen to allow the examination of insulating layers and their associated electronic properties.

Two other techniques for the microscopic examination of surfaces were developed by the same group of workers. One uses resonant phenomena as a monitor of surface proximity. In this, a standard AFM cantilever is excited into resonance by modulating the probe drive signal and the amplitude of the resultant oscillation used as a measure of the proximity of the surface.[72] This has the advantage that insulating surfaces can be detected at a probe stand-off of 15 nm or more. The other technique utilizes a micro-thermocouple in which the conduction of heat between the tip and specimen is used to maintain a constant separation. A lateral resolution of 100 nm and vertical resolution of 3 nm has been achieved.[73]

Finally, two recent probe microscopes have been developed for the imaging of soft surfaces. The first of these is the Scanning Ion Conductance Microscope (SICM).[74] In this a micropipette is filled with an electrolyte which is then immersed in a reservoir of electrolyte and advanced towards the surface to be examined. As the space between probe and sample is decreased so too is the gap through which ions can flow. The conductance between the pipette and electrolyte is monitored and used as the surface proximity signal for scanning. The surface of a micropore filter has been mapped by this means. The other probe consists of a fine lens of sub-micron diameter at the end of an optical fibre. Upon excitation by a photon beam, this will create a bright spot due to the near field phenomenon. If it is brought into proximity with a surface, there will be a reflection in which the intensity varies rapidly with separation. Again this can be used to maintain a constant surface altitude to investigate the optical properties of the surface. It has also been observed that there is a localised increase in the intensity of reflection due to a localized increase in the density of electron charge (plasmons) and this recent discovery is now being actively investigated.[75]

14.6 SUMMARY

All the instruments reviewed in this chapter operate on the principle that a fine probe is drawn across a surface and the subsequent path of the follower monitored as a measure of the surface profile. The variety of techniques for sensing the proximity of the surface is so diverse that a direct comparison is not practicable. Indeed, as different techniques rely on different physical principles for detecting the "surface", it is rash to assume that they measure the same features. Although left out of this discussion because of its relatively poor vertical discrimination, the scanning electron microscope provides a further example of an instrument that provides high quality information that cannot always be correlated well with others. Increasingly, we may need to talk of the mechanical surface, the optical surface, the electronic surface and so on. Nonetheless, a few general observations about profilometry are appropriate.

An interesting method for the comparison of different families of profilometer instruments is to present their vertical and horizontal range and resolution on a log-log plot, see Figure 14.10. The vertical axis ranges from 0.01 nm as a limit to the vertical resolution up to 10 mm as a limit to the maximum range of the probe. The horizontal axis plots from the surface wavelength that can be resolved by the complete instrument to the maximum allowed by its traverse. Although most of the transducers are continuous, the resolution of both probe and stylus will be limited by its geometry, by the surface geometry and by the design and operating conditions of specific instruments. For STM it has been shown that subatomic lateral resolution can be achieved with very smooth surfaces but that as the size of features increases these details are lost due to the integrating effect of the overall probe profile. This is also true for stylus techniques, but at a larger radius of curvature, around 1 μm for a sharp tip and more in many applications. As a consequence, at low surface wavelength and amplitude, the locus of probe microscopy techniques approaches the limits of the plot. Stylus techniques, however, are limited to a maximum lateral resolution of approximately 100 nm. Present results from AFM instruments suggest that vertical and horizontal resolution as good as the STM is not readily achievable and this has been acknowledge by the dashed line in the lower wavelength region.

As the size of features increases, probe microscopes tend to reach a limit of measurement range due almost entirely to the maximum extension that is practicable with current piezoelectric actuators (the maximum scan range available with commercial systems is 100 μm in both axes). Another limitation to tunnelling probe performance is the fact that the tip must be servo controlled to remain within approximately 2 nm of the surface. As piezo actuators are stacked up, or bimorphs are used, their stiffness is reduced. This introduces problems of vibration isolation which will eventually result in a loss of the signal and/or the probe contacting and damaging the surface. In theory, stylus instruments are only limited by the length of the traverse mechanism. However, as the traverse is increased in size the measurement loop becomes unacceptably large, resulting in unacceptable thermal drifts and an increased susceptibility to vibration. Maintaining adequate precision in the datum also becomes more difficult. There have been attempts to reduce these effects by the use of ceramic materials for the instrument components with a maximum traverse range of around 100 mm being the present limit.

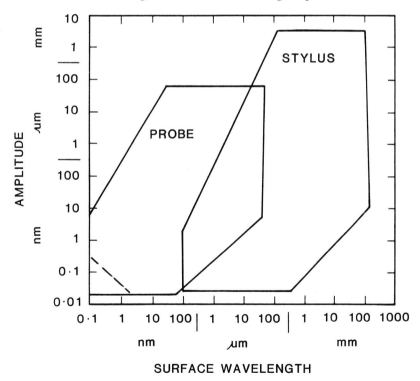

Figure 14.10 Potential performance characteristics for surface profilometry of stylus instruments and probe microscopes plotted in amplitude-wavelength space. The probably limited resolution of the AFM is indicated by the dashed line.

The locus of performance for optical surface profilometers has been omitted because limitating values for range and resolution, although different in nature, reveal in a locus that is similar to stylus methods.

The important feature of the amplitude wavelength plot is that it indicates the range of measurements that can be covered by various instruments. In particular, it is apparent that modern probe microscopes are ideal tools for imaging and identification of small features, usually in the sub-micrometre region.

They are complemented by stylus instruments which are ideally suited for measurements ranging from sub-micrometre to millimetre sized surface features. The overlap between these regimes provides the necessary continuum and a means of comparison and cross-calibration within a limited bandwidth. This will be of fundamental importance in the future if reliable interpretation of measured profiles is to be achieved.

Ideally, both types of instrument would be capable of completely filling the plot of Figure 14.10 to give complete information over all wavelengths of interest, so providing two complementary measurements to aid the verification of the data. It is to this end that many researchers into engineering metrology and instrumentation will aiming over the next few years in an attempt literally to broaden the range of nanotechnology.

REFERENCES

1. Thomas, T.R. (1982) Stylus instruments and Other measurement techniques. In *Rough Surfaces*, edited by T.R. Thomas, pp. 12-69. London: Longman.

2. Sayles, R.S. and Thomas T.R. (1978) Surface topography as a nonstationary random process. *Nature*, **271**, 431-434.

3. Chetwynd, D.G. (1990) Algorithms for computer aided precision metrology. In *From Instrumentation to Nanotechnology*, edited by J.W. Gardner and H.T. Hingle, Ch. 5. London: Gordon and Breach

4. Furse, J.E. (1981) Kinematic design of fine mechanism in instruments. *J. Phys. E: Sci. Instrum.*, **14**, 264-271. Reprinted in *Instrument Science and Technology* edited by B.E. Jones (1983), vol. 2, pp. 11-20. Bristol: Adam Hilger

5. Jones, R.V. (1962) Some uses of elasticity in instrument design. *J. Sci. Instrum.*, **39**, 193-203

6. Sydenham, P.H. (1984) Elastic design of fine mechanism in instruments. *J. Phys. E: Sci. Instrum.*, **17**, 922-930

7. Smith, S.T. (1990) Nanoactuators for controlled displacements. In *From Instrumentation to Nanotechnology*, edited by J.W. Gardner and H.T. Hingle, Ch. 15. London: Gordon and Breach

8. Smith, S.T., Chetwynd, D.G. and Bowen, D.K. (1987) Design and assessment of monolithic high precision translation mechanisms. *J. Phys. E: Sci. Instrum.*, **20**, 977-983

9. Chetwynd, D.G. (1989) Materials classification for fine mechanics. *Precision Engineering*, **11**, 203-209

10. Garratt, J.D. and Whitehouse, D.J. (1979) Displacement transducers below 50 mm - a comparative survey. *Annals of the CIRP*, **28**, 511-518

11. Hicks, T.R., Reay, N.K. and Atherton, P.D. (1984) The application of capacitance micrometry to the control of Fabry-Perot etalons. *J. Phys. E: Sci. Instrum.*, **17**, 49-55

12. Kim, K.Y. and Sachse, W.H. (1986) Self aligning capacitive transducer for the detection of broadband ultrasonic displacement signals. *Rev. Sci. Instrum.*, **57**, 264-267

13. Garbini, J.L., Jorgensen, J.E., Downs, R.A. and Kow, S.P. (1988) Fringe-field capacitive profilometry. *Surface Topography*, **1**, 131-142

14. Chetwynd, D.G. (1987) High-precision measurement of small balls. *J. Phys. E: Sci. Instrum.*, **20**, 1179-1187

15. Garratt, J.D. (1982) A new stylus instrument with a wide dynamic range for use in surface metrology. *Precision Engineering*, **4**, 145-151

16. Bowen, D.K., Chetwynd, D.G. and Schwarzenberger, D.R. (1990) Sub-nanometre displacements calibration using x-ray interferometry. *Meas. Sci. Technol.*, **1**, 107-119

17. Whitehouse, D.J., Bowen, D.K., Chetwynd, D.G. and Davies, S.T. (1988) Nano-calibration for stylus based surface measurement. *J. Phys. E: Sci. Instum.*, **21**, 46-51

18. Bowen, D.K., Chetwynd, D.G. and Davies, S.T. (1985) Calibration of surface roughness transducers at angstrom levels using x-ray interferometry. *Proc. SPIE*, **732**, 412-419

19. Lindsey, K., Smith, S.T. and Robbie, C.J. (1988) Sub-nanometre surface texture and profile measurement with NANOSURF 2. *Annals of the CIRP*, **37**, 519-522

20. Garratt, J.D. and Bottomley, S.C. (1990) Technology transfer in the development of a nanotopographic instrument. *Nanotechnology*, **1**, 38-43

21. Various Authors (1988) in *Surface Measurement and Characterisation, SPIE*, **1009**

22. Dupuy, O. (1968) High precision optical profilometer for the study of microgeometrical surface defects. *Proc. Instn. Mech. Engrs.*, **182**, Part 3K, 255-259

23. Kohno, T., Ozawa, N., Miyamoto, K. and Musha, T. (1988) High precision optical surface sensor. *Appl. Opt.*, **27**, 103-108

24. Sayles, R.S., Wayte, R.C., Tweedale, P.J. and Briscoe, B.J. (1988) The design, construction and commissioning of an inexpensive prototype laser optical profilometer. *Surface Topography*, **1**, 219-227

25. Creath, K. and Wyant, J.C. (1988) Measurement of ultraprecision components using non-contact interferometry based instrumentation. In *Ultraprecision in Manufacturing Engineering*, edited by M. Weck and R. Hartel, pp. 287-302. Berlin: Springer Verlag

26. Lange, S.R. and Bhushan, B. (1988) Use of two- and three-dimensional noncontact surface profiler for tribology applications. *Surface Topography*, **1**, 205-217

27. Downs, M.J., McGivern, W.H. and Ferguson, H.J. (1985) Optical system for measuring the profiles of super-smooth surfaces. *Precision Engineering*, **7**, 211-215

28. Downs, M.J. (1990) A proposed design for an optical interferometer with sub-nanometric resolution. *Nanotechnology*, **1**, 27-30

29. Guerra, J.M. (1988) Photon tunnelling microscopy. *Proc. SPIE*, **1009**, 254-263

30. Young, R.D., Ward, J. and Scire, F. (1972) The Topographiner: An instrument for measuring surface microtopography. *Rev. Sci. Instrum.*, **43**(7), 999-1011

31. Young, R.D. (1966) Field emission ultramicrometer. *Rev. Sci. Instrum.*, **37**(3), 275-278

32. Simmons, J.G. (1963) Generalised formula for the electric tunnel effect between similar electrodes separated by a thin insulating film. *J. Appl. Phys.*, **34**(6), 1793-1803

33. Binnig, G., Rohrer, H., Gerber, Ch. and Weibel, E. (1982) Tunnelling through a controllable vacuum gap, *Appl. Phys. Lett.*, **40**(12), 178-180

34. Binnig, G., and Rohrer, H. (1982) Scanning tunnelling microscopy. *Helv. Phys. Acta*, **55**, 726-735

35. Binnig, G., Rohrer, H., Gerber, Ch. and Weibel, E. (1982) Surface studies by scanning tunnelling microscopy. *Phys. Rev. Letts.*, **49**(1), 57-60

36. Binnig, G., Rohrer, H., Gerber, Ch. and Weibel, E. (1983) 7x7 reconstruction on Si(111) resolved in real space. *Phys. Rev. Letts.*, **50**(2), 120-123

37. Binnig, G. and Smith, D.P.E. (1986) Single-tube three-dimensional scanner for scanning tunnelling microscopy. *Rev. Sci. Instrum.*, **57**, 1688-1689

38. Bryant, A., Smith, D.P.E., and Quate, C.F. (1986) Imaging in real time with the tunnelling microscope. *Appl. Phys. Lett.*, **48**(13), 832-834

39. Smith, D.P.E., Binnig, G., and Quate, C.F. (1986) Atomic point-contact imaging. *Appl. Phys. Letts.*, **49**(18), 1166-1168

40. Greenwood, J.A.,and Williamson, J.B.P. (1966) Contact of nominally flat surfaces. *Proc. Roy. Soc. Lond.*, **A295**, 300-319

41. Coombs, J.H. and Pethica, J.B. (1985) Properties of vacuum tunnel currents: Anomalous barrier heights. *IBM/STM Workshop, Oberlech, Austria*, July 1-5

42. Van der Walle, G.F.A., Gerritsen, J.W., van Kempen, H. and Wyder, P. (1985) High stability scanning tunnelling microscope. *Rev. Sci. Instrum.*, **56**(8), 1573-1576

43. Drake, B., Sonnenfeld, R., Schneir, J., Hansma, P.K., Slough, G., Coleman, R.V. (1986) Tunnelling microscope for operation in air or fluids. *Rev. Sci. Instrum.*, **57**(3), 441-445

44. Gerber, Ch., Binnig, G., Fuchs, H., Marti, O. and Rohrer, H. (1986) Scanning tunnelling microscope combined with a scanning electron microscope. *Rev. Sci. Instrum.*, **57**(2), 221-224

45. Smith, D.P.E. and Binnig, G. (1986) Ultrasmall scanning tunnelling microscope for use in liquid helium storage dewar. *Rev. Sci. Instrum.*, **57**(10), 2630-2631.

46. Jericho, M. H., Dahn, D. C., and Blackford, B. L. (1987) Scanning tunnelling microscope with micrometer approach and thermal compensation. *Rev. Sci. Instrum.*, **58**(8), 1349-1352

47. Heinzelmann, H., Grutter, P., Meyer, E., Hidber, H., Rosenthaler, L., Ringger, M, and Guntherodt, H.-J. (1987) Design of an atomic force microscope and first results. *Surface Science*, **189/190**, 29-35

48. Parker, J. L., Christenson, H. K., and Ninham, B. W. (1989) Device for measuring the force and separation between two surfaces down to molecular separations. *Rev. Sci. Instrum.*, **60**(10), 3135-3138

49. Haase, O., Borbonus, P., Muralt, P., Koch, R., and Rieder, K.H. (1990) A novel ultrahigh vacuum scanning tunnelling microscope for surface science studies. *Rev. Sci. Instrum.*, **61**(5), 1480-1483

50. Durig, U., Gimzewski, J.K., and Pohl, D.W. (1986) Experimental observation of the forces acting during scanning tunnelling microscopy. *Phys. Rev. Letts.*, **57**(19), 2403-2406

51. Pethica, J.B., Hutchings, R. and Oliver, W.C. (1983) Hardness measurement at penetration depths as small as 20 nm. *Phil. Mag.*, **48**(4), 593-606

52. Tersoff, J., and Hamann, D.R. (1985) Theory of the scanning tunnelling microscope. *Phys. Rev. B*, **31**(2), 805-813

53. Feuchtwang, T.E., Cutler, P.H., and Miskovsky, N.M. (1983) A theory of vacuum tunnelling microscopy. *Phys. Letts.*, **99A**(4), 167-171

54. Garcia, N., Ocal, C., and Flores, F. (1983) Model theory for scanning tunnelling microscopy: Applications to Au(110)(1x2) surfaces. *Phys. Rev. Letts.*, **50**(25), 2002-2005

55. Kaizuka, H. (1989) Application of capacitor insertion method to scanning tunnelling microscopes. *Rev. Sci. Instrum.*, **60**(10), 3119-3122

56. Tsuda, N., and Yamada, H. (1989) STM, AFM and interferometers in NRLM. *Bulletin of NRLM*, **38**(4), 403-407

57. Smith, S.T., and Liu, X. (1991) An electromagnetically driven linear spring for tunnelling profilometry. *Nanotechnology*, **2**, to appear

58. Kuk, Y. and Silverman, P.J. (1989) Scanning tunnelling microscope instrumentation. *Rev. Sci. Instrum.*, **60**(2), 165-180

59. Becker, R.S., Golovchenko, J.A. and Swartzentruber, B.S. (1985) Electron interferometry at crystal surfaces. *Phys. Rev. Letts.*, **55**(9), 987-990.

60. Baro, A.M., Miranda, R., Alaman, J., Garcia, N., Binnig, G., Rohrer, H., Gerber, Ch. and Carrascosa, J.L. (1985) Determination of surface topography of biological specimens at high resolution by scanning tunnelling microscopy. *Nature*, **315**, 253-254.

61. Becker, R.S, Golovchenko, J.A. and Swartzentruber, B.S. (1987) Atomic scale surface modifications using a tunnelling microscope. *Nature*, **325**, 419-421

62. McCord, M.A. and Pease, R.F.W. (1987) Scanning tunnelling microscope as a mechanical tool. *Appl. Phys. Letts.*, **50**(10), 569-570

63. Foster, J.S., Frommer, J.E., and Arnett, P.C. (1988) Molecular manipulation using a tunnelling microscope. *Nature*, **321**, 324-326

64. Dunlap, D.D., and Bustamante, C (1989) Images of single-stranded nucleic acids by scanning tunnelling microscopy. *Nature*, **342**, 204-206

65. Eigler, D.M., and Schweizer, E.K. (1990) Positioning single atoms with a scanning tunnelling microscope. *Nature*, **344**, 524-526

66. Driscoll, R.J., Youngquist, M.G., and Baldeschwiegler, J.D. (1990) Atomic-scale imaging of DNA using scanning tunnelling microscopy. *Nature*, **346**, 294-296

67. Binnig, G., Quate, C.F. and Gerber, Ch. (1986) Atomic force microscope. *Phys. Rev. Letts.*, **56**(9), 930-933

68. Binnig, G., Gerber, Ch., Stoll, E., Albrecht, T.R., and Quate, C.F. (1987) Atomic resolution with the atomic force microscope. *Surface Science*, **189/190**, 1-6

69. Drake, B., Prater, C.B., Weisenhorn, A.L., Gould, S.A.C., Albrecht, T.R., Quate, C.F., Cannell, D.S., Hansma, H.G., and Hansma, P.K. (1989) Imaging crystals, polymers and processes in water with the atomic force microscope. *Science*, **243**, 1586-1589

70. Matey, J.R. and Blanc, J. (1985) Scanning capacitance microscopy. *J. Appl. Phys.*, **57**(5), 1437-1444

71. Bugg, C.D. and King, P.J. (1988) Scanning capacitance microscopy. *J. Phys. E; Sci. Instrum.*, **21**, 147-151

72. Martin, Y., Williams, C.C. and Wickramasinghe, H. K. (1987) Atomic force microscope - force mapping and profiling on a 10 nm scale. *J. Appl. Phys.*, **61**(10), 4723-4729

73. Williams, C.C., and Wickramasinghe, H.K. (1986) Scanning thermal profiler. *Appl. Phys. Letts.*, **49**(23), 1587-1589

74. Hansma, P.K., Drake, B., Marti, O., Gould, S.A.C., and Prater, C.B. (1989) The scanning ion conductance microscope. *Science*, **243**, 641-643

75. Pohl, D.W., Fischer, U.Ch., and Durig, U.T. (1988) Scanning near field optical microscopy (SNOM). *Microscopy*, **152**(3), 853-861

Chapter 15

NANOACTUATORS FOR CONTROLLED DISPLACEMENTS

STUART. T. SMITH

In this chapter a variety of actuator mechanisms that are capable of operating at the nanometre level are reviewed. Each mechanism is discussed in terms of its relative merits such as dynamic response, repeatability, size, linearity and cost. The types of actuator covered in this discussion are: piezoelectric, mechanical micrometers, friction drives, magnetostriction, magnetoelasticity, shape memory alloys and bimetallic strips, electromagnetic, electrostatic, hydraulic and the Poisson's ratio drive. These techniques are summarized at the end of the chapter to clarify individual advantages and to highlight the reasons for choosing between different mechanisms.

15.1 INTRODUCTION

This chapter briefly reviews some of the techniques that may be used for primary displacement actuators that are both continuous and can be operated at the nanometre level. By primary actuator it is implied that the translation device is not being driven by another mechanism via a lever or gearing system or some other form of coupling. In practice it is very difficult to adhere to this definition due to the fact that, for example, an electromagnetic actuator operates by generating a force. This must, in turn, be transformed into a displacement by pushing against a spring which, due to its stiffness, will determine the displacement. Thus a lever mechanism has been incorporated in the actuator at the outset. However, there also remains the problem of outlining the limitations of each individual technique. Taking the above example, the speed of response of a magnetic field to that of a changing electric field is equal to the velocity of light. This cannot be exploited in practice due to the inertia of elements that must be driven by the actuator. To obtain a value for the response, it is useful to look at related and well developed technologies (in this case Hi-Fi loud speaker design). Although not exactly a scientific approach, until more research work is carried out in this field, it is felt that the values that have been chosen are representative of what may be possible with presently available technology. Additionally, because of the very wide scope of this chapter, it is likely that some important attributes of these mechanisms have been overlooked and oversimplified. The author will be grateful to receive any new information relating to the contents of this chapter as well as the results of present and future research projects.

When specifying an actuator for applications in nanotechnology, the decision is usually based upon the three following criteria:

1. Stiffness- will determine both the dynamic response and the magnitude of the transmissible forces.

2. Positional accuracy- limits repeatability and is often enhanced by implementing feedback control. Although there are exceptions to the rule, this usually results in an increase in the cost of such a device.

3. Range- at a resolution of one nanometre, the range tends to be rather restricted in terms of normal engineering applications. Generally, costs can be considered roughly proportional to the range/accuracy ratio for a given resolution. On average a value of 100 is reasonable, 10,000 is getting rather expensive and 10^8 is beyond present capabilities below lengths of 1 m.[1] There are often physical restrictions on the range of some of these actuators such as magnetic saturation, yield stress etc.

15.2 TYPES OF ACTUATOR

In this section a range of mechanisms will be presented that have the potential for applications in smooth, continuous actuators of sub-micrometre or even sub-nanometre resolution.

15.2.1 Piezoelectric Drives

Of the twenty one crystal classes that do not have a centre of symmetry twenty of these can experience a dimensional change upon application of an electrical potential gradient (electric field) and such materials are known as piezoelectric. This lack of a centre of symmetry is a necessary precursor for this phenomena. Ten of these classes will also generate a surface charge upon heating and these are known as pyroelectric materials. Although *all pyroelectrics are piezoelectric*, the reverse is not always true due to the fundamentally different reasons for the occurrence of these effects. In a pyroelectric material, the net dipole is changed by a change in the magnitude of polarization when heated. In piezoelectric materials that are not pyroelectric, the dipoles which are invariably altered in the presence of heat are so arranged that they compensate each other resulting in no net change in dipole moment. However, upon application of a stress some dipole moments in specific directions are increased more than the others and a net dipole change is induced. Additionally, some materials can be polarized by the application of a sufficiently strong electric field. By analogy with ferromagnetism these are called ferroelectric although both do not necessarily contain iron. *All ferroelectrics are both piezoelectric and pyroelectric* and constitute all of the commonly used actuator materials because, like magnetic materials, they can be easily and permanently polarized (commonly referred to as poling) at or near to their Curie temperature (the temperature at which permanent magnetization is lost). This also enables these materials to be first sintered as a ceramic and then polarized in a desired direction. Poling, however does not result in a perfect polarization in these polycrystalline materials but will result in the creation of a large number of domains in which there is an overall polarization direction. If one imagines an ideal material in which there are N parallel molecules of length δ having a charge of +q and -q at each end, then the magnitude of the total polarization is given by

$$P_x = Nq\delta_x \qquad (15.1)$$

The only way to change this value is to physically squeeze the poles together by the application of an applied stress or to displace the charges with an applied field. Under uniaxial conditions the total displacement current, D, is given as a linear combination of the two phenomena, i.e

$$D_x = \varepsilon E_x + dT_x \qquad (15.2)$$

where ε is the materials permittivity at constant stress (Fm^{-1}), d is the piezoelectric strain constant $(CN^{-1}$ or $mV^{-1})$, T is the applied normal stress (Nm^{-2}) and E is the field strength (Vm^{-1}). In fact this treatment is very much an oversimplification and in practice these effects are heavily anisotropic with a more complete analysis necessitating the use of tensor analysis. Excellent introductory reviews can be found in the papers of Jaffe and Berlincourt[2] and Gallego-Juarez[3]. Conversely, the strain, S, for a given applied field or stress is given by the equation

$$S_x = d^T E_x + s T_x \qquad (15.3)$$

where d^T is transposed piezoelectric strain matrix (mV^{-1}) and s is the elastic compliance at constant applied field $(m^2 N^{-1})$.

Again values for these constants and other inter-related properties for a wide range of materials can be found in the paper of Jaffe and Berlincourt.[2] It should be stressed that not only are most of these materials pyroelectric and thus very sensitive to temperature but there is a finite field above which these will depole. For the very popular lead zirconate titanate (PZT) material at 25°C this will be in the region 0.7-1.0 kVmm^{-1}.

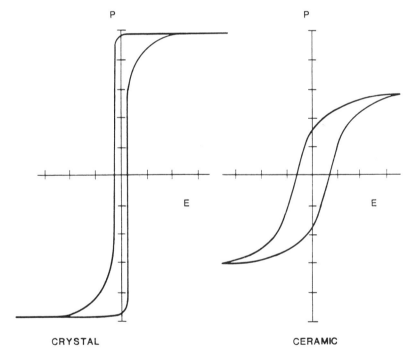

Figure 15.1 Typical hysteresis curve for single crystalline and ceramic barium titanate.

From the above simple analysis it can be seen that, in the absence of an applied stress, there will be a net displacement current required to obtain a proportionate strain. It is usual to use voltage drives for this type of actuator and, as such, the actual displacement will depend upon the size, distribution and stability of the electric domains within these polycrystalline ceramics as well as other influences such as homogeneity of the material.

The resulting relationship between applied voltage and displacement will exhibit considerable hysteresis that can only be reduced by feedback control or a charge drive. A typical hysteresis curve for both single crystalline and ceramic barium titanate is shown in Figure 15.1. The latter approach was first proposed by Newcomb and Flinn[4], who suggested using a charge amplifier as the driving element. This idea was further simplified by Kaizuka and Siu[5], Kaizuka[6], by inserting a compensating capacitor in series with the piezoelectric and using a voltage drive. Using this technique it is shown that the percentage hysteresis is reduced by up to one fifth with, however, a similar reduction in the sensitivity. This is outweighed by far in that the apparent creep is reduced by a similar amount. This is caused by the gradual changing of domains analogous to Barkhaussen jumps, and will quite commonly be in the region of 30% over a period of 2-20 minutes[7], which precludes these devices for metrological application in open loop operation.[8] The magnitude of these effects is very difficult to assess and control and will vary from material to material and even from batch to batch. Another disadvantage with these materials is the cross-coupling which not only results in a lateral expansion of approximate magnitude 0.25-0.65 but often leads to a bending motion that can result in angular deflections of up to 20° although this is usually less than a few percent of the total motion.

To overcome the nonlinearities associated with this actuator, closed loop control strategies have been developed and such devices are now commercially available. Using capacitive feedback techniques Hicks et al.[9], were able to control Fabry Perot etalons (parallel optical flats for use in spectrometry) at a dynamic gap separation maintained constant at the 1 $pmHz^{-1/2}$ level and commercial devices are now available which operate a range/resolution ratio of 10,000:1 (Queensgate Instruments Ltd, Ascot, UK & Lambda Photometrics, Harpenden, UK) . The feedback transducers in these applications include capacitance and strain gauges. Generally, a displacement of approximately 10 μm is obtained at an applied field of 1000 V. Under a load W the displacement will reduce by an amount WL / EA, where L is the length of the piezo and E and A are the elastic modulus and the cross-sectional area perpendicular to the load. Thus the maximum stiffness is given by a short length and large area. The maximum practical ratio of area to length is approximately 100:1 giving a stiffness of 100E. Although this will scale, as the devices become larger so too does the voltage necessary to obtain the same field strengths. Thus these actuators are usually used for small scale manipulators of the order of a few millimetres or less in dimension. One of the key features of such devices is their speed of response. For a PZT device, the speed of sound of the material in the polar axis can be in the region which enables these to be operated at frequencies up to 1 GHz. There has recently been a gigantic surge in the use of these devices for scanning tunnelling microscopes where the tip is controlled by three orthogonal piezoelectric actuators which can collectively occupy less than two cubic millimetres.[10]

The apparent limitations on the available displacement are overcome by stacking many actuators in series or creating a bi- or multi-laminar element analogous to the bimetallic strip.[11] Three such arrangements are shown in Figure 15.2.

As the number of actuators increases both the stiffness and the dynamic performance of these devices will reduce. The bimorph of Figure 15.2b can be further enhanced by arranging the poling such that the actuator naturally bends in an "S" shape. This relieves the flexures from restriction due to the end clamps. Using such a method Matey et al.[12], were able to obtain displacements of 120 μm in all three axes with the resolution limited by electronic and mechanical noise. There is evidence that the hysteresis will increase due to creep in the bonds. This effect which theoretically reduces with temperature has been observed by, Blackford *et al.*[13], and Tritt *et al.*[14] The lowest natural frequency of the stage presented in these papers was 190 Hz.

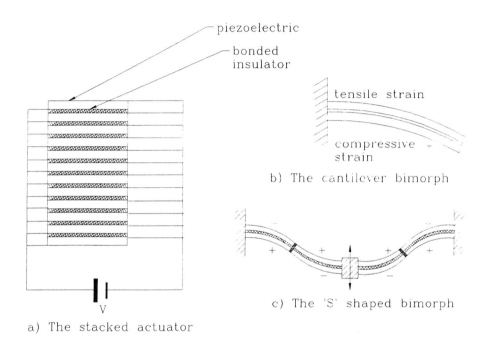

Figure 15.2. Long range piezoelectric actuators.

Finally, for even longer range applications a device called an "inchworm" has been developed. As the name suggests, this device achieves motion an a manner similar to some caterpillars. A piezoelectric actuator is mounted onto either a slideway or an optical flat with a conducting surface. Attached to the base of the actuator are two or more conducting pads separated from contact with the slideway by a thin layer of insulating material. The inchworm is actuated by first clamping one pad to the slideway by an electrostatic force and then extending the piezoelectric material. This free end is then clamped and the fixed end is released. By then contracting the piezo and repeating these steps, it is possible to obtained unlimited motion. This type of device was first proposed by Joyce and Wilson[15], and became famous as the coarse feed in the first scanning tunnelling microscope of Binnig and Rohrer[16], and was subsequently investigated by many commercial organizations. Although problems with stability of these actuators have prevented their being commercially available until Ca. 1989[17], devices are now available with a traverse of up to 100 mm with individual step sizes of 2 nm or less (Burleigh Instruments Inc. NY 144532). A less accurate but still interesting extension of this idea is the piezoelectric motor that operates on the same principle (there are now versions that operate by creating a circulating ripple).[18,19]

15.2.2 Mechanical Micrometers

These will invariably be of a hysteretic nature due to the nonlinearity of contact forces and, unless well designed, will have backlash. Because of the intricate and precise shape of components that make up a feedscrew, it is only economically feasible to construct these from metals. As a consequence the thermal properties of feedscrews tends to be rather poor. However, micrometers are available in differential form and these can be manually positioned to about 0.1 μm under low load conditions or possibly 0.01 μm with feedback.[20] The open loop accuracy is mainly restricted by pitch errors in the feedscrew. Although these can be reduced by mechanical averaging, this will invariably result in loss of stiffness. Under closed loop control positional accuracies of better than 10 nm are achievable and ruling engines have been known to be repeatable to two orders of magnitude better than this.[21-24] The stiffness of this type of mechanism is difficult to generalize. However, considering a rather squat device whereby the area of contact is approximately ten percent less than the square of the length of the feedscrew and that approximately half of the stiffness is lost at the material interface, then the stiffness of such devices is approximately 0.45E. Speed of traverse is also difficult to generalize but a high speed precision drive might typically be capable of positioning a carriage over 100 mm in a second or two.

15.2.3 Friction Drives

There are two distinct groups of friction drive, one of which relies on the clamping friction in a similar way to how the human hand grips while the other method utilizes the inertia of objects to overcome frictional forces in a manner akin to the trick of removing a table cloth without moving the cutlery. The inchworm described in the previous section is a hybrid of the this and the above actuation mechanism.

The former actuator is a derivative of the feedscrew and consists of a polished bar that is squeezed between two rollers that can be rotated with high precision. The main advantage of this mechanism is that the contact forces between driver and driven remains relatively constant and the range of traverse is virtually unlimited. Again these include most of the problems associated with feedscrew devices plus additional problems associated with rotary accuracy/resolution, performance of the bearings supporting the squeeze rollers, surface finish of the contact, alignment of drive rod axis perpendicular to the squeeze roller axes and Heathcote slip. The latter of these problems is likely to be the most difficult to reduce. It is caused by the differing relative velocities between sliding area contacts that occur in both the ball bearings supporting the squeeze rollers and at the roller/drive rod interface. These energy losses will result in a hysteretic torque displacement characteristic. If the drive is subject to any load, then an additional hysteresis due to the finite slip in the contact zone would be expected.[25] However, with feedback control these mechanisms exhibit a resolution of approximately 5 nm and appear to be a serious challenger to the feedscrew (see Drives and Controls, July/August, 1988). There is very little information on friction drives for ultra-high precision applications although many such drives have been designed and used to achieve long range motion at relatively high velocity (100 mm/s) with sub-micron resolution,[26] and devices are now commercially available covering a variety of specifications. The author would expect that these additional interfaces may result in a further reduction in the stiffness of this drive and is likely to result in a reduction of attainable stiffness by an order of magnitude or more and a rough estimate of ≈ 0.05E is not unreasonable. This compares with a stiffness of 360 Nm^{-1} given in the paper of Reeds *et al.*,[26] which for a length of 100 mm gives the stiffness per unit length as 0.04 GPa which is two orders of magnitude below this figure.

The other type of drive that utilizes friction as the displacement mechanism is the dynamic translation device. In this the moving carriage is placed on to the slideway and is held in place by frictional forces acting at the interface. The slideway is then subject to an impulsive force that will cause the carriage slip along the slide due to its inertia. The slideway is then allowed to relax slowly taking the carriage with it. Repetition of this cycle results in a net motion in the direction of the impulsive force. A typical example of such a device (taken from Pohl[27]) is shown in Figure 15.3

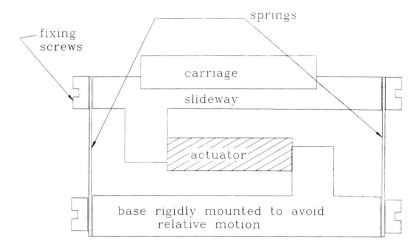

Figure 15.3. A dynamic linear spring translator employing friction as the primary actuation mechanism.

In this the carriage is mounted onto a platform of a linear spring. Attached to it is a piezoelectric translator that is, in turn, rigidly mounted onto the base. This is subsequently excited with a triangular waveform to achieve the desired displacement. Using this device, it was possible to translate loads of up to 1 kg at a velocity of 0.2 mm/s with a resolution of a few nanometres. A similar technique was employed by Smith and Elrod,[28] only in this instance an electromagnetic force actuator has been used. Using this device it was again possible to position the carriage to within a few nanometres and the traverse speed could be controlled up to 1 mms^{-1}. Using a combination of these and mounting the specimen stage on a glass flat it was also possible to achieve a two dimensional scan. Both of the above developments were instigated in an effort to reduce the problems of reliability that were being encountered at that time with the inchworm. The stiffness of such a device is clearly low and will be almost entirely dependent on the coefficient of friction and the normal applied forces.

15.2.4 Magnetostriction

This phenomena is very similar to that of the piezoelectric effect only it involves a change in dimension in the presence of a uniform magnetic field. This means that the displacement per unit field will increase with dimension making this favourable as a large scale or heavy duty actuator. As would be expected, this phenomena also has intrinsic hysteresis and will creep with time. A material demanding recent interest is an alloy of terbium, dysprosium and iron (called Terfenol-D) which has giant magnetostrictive properties.

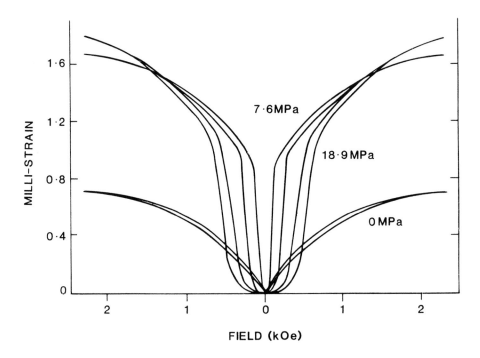

Figure 15.4 Magnetostriction of Terfenol-D

Figure 15.4 is a plot of strain against magnetic field when the material is subject to a stress of ≈ 20 MPa, a moderate field strength of 70 kA m^{-1} will result in a strain of 700 $\times$ 10^{-6}, reproduced from Clarke et al.[29] From this it can be seen that the strain is also dependent upon the applied stress and will fall off rapidly either side of this value and it is recommended that the working stress is maintained between 8 and 20 MPa.[30] The size of such an actuator is restricted primarily by the cost of the material (terbium is very expensive) and that of the field generator. A reasonable device for the control of a grinding wheel might have dimensions of 100 mm in length and an area of 20,000 mm^2 giving a stiffness of 200E and a displacement of 80 μm under a load of 400 kN. This clearly has potential as a heavy duty nanoactuator under feedback control. The speed of response of such a device is restricted mainly by the inertia of the material and for the large scale applications being considered, a typical resonant frequency of 1 kHz may be expected.[31] Yuan et al.[32] have used this type of actuator for the closed loop control of a tool feed compensator with a 125 Hz bandwidth at 10 nm resolution. However, as the size and thus displacement amplitude reduces, devices having an operating frequency of up to 300 kHz have been produced.[30]

15.2.5 Magnetoelasticity

This is frequently referred to as the ΔE effect and is related to the magnetostrictive coefficient.[33] All magnetostrictive devices will change elastic modulus in the presence of a uniform magnetic field. To utilize this as a mechanical actuator the material must first be subjected to strain via a dead weight, spring or internal stresses. These will result in a distortion of the material, the magnitude of which depends on its elastic modulus. If the material is then placed within a magnetic field the elastic modulus and thus the distortion will change. The magnitude of this distortion is dependent upon both the percentage change in elastic modulus and the magnitude of the initial distortion. A typical value for percentage change is around 2-10% and the maximum strain for many materials will be in the region of 0.1%, thus the achievable strain is approximately 10 – 50 microstrain or a change in length of 2 –10 μm for a 100 mm long device. A material of particular interest is Elinvar or Ni-Span-C which is a 42% Ni-Fe alloy. This composition results in a near zero thermoelastic coefficient near to room temperature and a thermal expansion coefficient of 3×10^{-7} K^{-1}. It does, however, undergo a change in Youngs modulus of approximately 0.1% in the presence of a high magnetic field (Inco Alloys International Ltd, Hereford, UK). Thus it is feasible to use this as a magnetoelastic device that is dimensionally sensitive to magnetic field but is relatively immune to temperature. The main advantage of this type of actuator over all others is that the displacement per unit field strength will depend upon the load condition and therefore will be **tuneable**. Although this is a rather small effect there are materials with much greater sensitivities to the electric field (in particular Terfenol-D) and up to 20% strains are possible with some Ni-Fe and Co-Fe alloys.[34] Additionally, the change in elastic modulus does is not linear and can have both positive and negative slope and contain considerable hysteresis making this more suitable as part of a closed loop system. Again the stiffness and the propensity for large scale devices of such an actuator is similar to the magnetostriction device above only, the cost of alloys is a lot less for this actuator.

15.2.6 Shape Memory Alloys and Bimetallic Strips

This family of devices requires heat as the actuating mechanism. In shape memory alloys (usually Ni-Ti) the material is bent at a high temperature and then allowed to cool whereby it will return to its original shape. Upon reheating the material will return to its deformed shape at that original temperature as if it had remembered. For a review of these materials see Schetky[35], Golestaneh.[36] Actuators using this effect are capable of very large displacements for a relatively small device size. A typical example would be a deflection of 20 mm for a cantilever beam of around 100 mm in length and a drive force of 300 Nm^{-1} with a time constant of approximately 4 seconds.[37] The heating is usually supplied by passing a current through the actuator. More recently, this type of actuator has been used to align optical fibres by pushing them into very accurate vee grooves. Using this technique it is possible to position an individual fibre to within 0.1 μm.[38] There is also interest in using this type of actuator for miniature robot arms.[39] Bimetallic strips are of similar performance the difference being that the mode of operation is differential thermal expansion of two intimately contacted thin materials. A typical example is the old vehicle indicator mechanism. Again, under controlled conditions, it is not unlikely that similar range and resolution to the shape memory alloys may be achieved.

15.2.7 Electromagnetic

There are a large number of possibilities for the design of electromagnetic actuators. These, in general, involve the use of the magnetization and demagnetization of soft magnets. The range of devices considered in this section will be restricted to only those with linear mechanisms that employ permanent magnets in the presence of weak field strengths. The most common A.C. linear displacement actuator is the audio loud speaker. In precision electromagnetic devices, a saturated permanent magnet(usually NdBFe or SaCo) is surrounded by a coil (or vice versa) and a force proportional to both the strength of the magnet and the coil current is generated between the two. To obtain a controlled displacement, the generated force is usually applied to a linear spring and it is this which appears to be the main limitation to the achievable accuracy of such a device. For a saturated magnet in a relatively weak field, the force/current characteristic has been found to be linear to within better than one part in 10,000. Using a spring manufactured from single crystal silicon it is possible to obtain open loop displacements of up to 100 nm with a resolution of 5 pm as measured using X-ray interferometry (see chapter 16). This type of actuator requires some form of stiffness against which to act and can therefore only be considered as a primary actuator in need of some secondary mechanism by which controlled displacements may be achieved. Because of this the available range and resolution are not easy to assess. In the paper of Smith and Chetwynd,[40] it is shown that the output force per unit input power is dependent upon both the permanent magnet material (in particular its intrinsic magnetization) and the conductivity of the input windings. Clearly the use of superconductors will relieve these limitations. The linearity of such a device is proportional to the fractional demagnetization and this is, in turn, closely related to the recoil permeability which should be as low as possible for maximum linearity. Demagnetization curves for the materials NdDyFeBNb, NdDyFeB and NdFeB are shown in Figure 15.5 (reproduced from Parker et al.[41]). Nonlinearities are due to demagnetization as the intrinsic magnetization follows the characteristic curve and complete demagnetization will occur when the applied field strength equals the coercive strength of the magnet. As a consequence, the recoil permeability is related to the applied field and reduces to zero at very low fields. In normal applications relatively weak fields are employed and it is the constancy of the B-H characteristic that limits the linearity of operation.

From Figure 15.5 it can be seen that the simple NdFeB material has the flattest characteristic in the presence of a low field strength with the alloys of dysprosium and niobium creating a larger slope but also considerably increasing the coercive strength. The Curie temperature is approximately 270 °C for these materials but this can be increased by the addition of cobalt to approximately 500 °C.[42]

The two different ways of utilizing an electromagnetic force actuator are the driven coil or the driven magnet. In the first case a permanent magnet is constructed to consist of a permeable loop that is broken by a small gap in which the driven coil of winding length, L, is positioned. By passing a current, I, through this coil it will experience an axial force, F_x, given by (for a detailed discussion see Hadfield[43])

$$F_x = B_x I L \tag{15.4}$$

B is the flux density in the gap and from considerations of continuity this can be approximated from the equation

$$B_x = \sqrt{\frac{\mu_o H_m B_m V_m}{V_g}} \tag{15.5}$$

where μ_o is the permeability of the gap, $H_m B_m$ is the energy product for the magnetic material, V_m and V_g are the volumes of the magnet and gap, respectively.

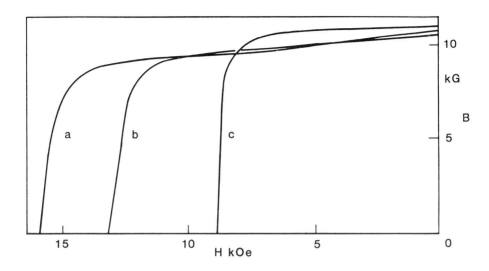

Figure 15.5 Demagnetising curves for (a) NdFeDyBNb (b) NdFeDyB and (c) NdFeB magnets.

From equation 15.5 it can be seen that the optimum condition is obtained for a magnet material having the highest energy product and the minimum gap volume. Thus, if there is no restriction on space, then it is best to use the largest volume of magnet possible. The biggest drawback with this system is due to the fact that the active element is attached to the moving stage and the energization of the winding will create a source of heat in the mechanism. One way of overcoming this problem is to use two counter-rotating windings and operate the actuator in differential mode while maintaining a constant input power. This should, in practice, result in a constant heat input that will eventually reach an equilibrium temperature and remain constant during operation. The disadvantage with this method is that the extra winding is redundant in terms of the actuating force requiring additional gap space and thus reducing the force per unit power that is obtainable from such an actuator. Because of the large mass of magnet required for this type of actuator, it is not feasible to attach the magnets to the moving platform. To utilize such a device a completely different principle of operation must be employed. In this, the permanent magnet is attached to the moving platform and this is, in turn, surrounded by a coil that is rigidly mounted onto the base of the device. The force on the magnet is then proportional to both the magnitude of the magnetic moment and the gradient of the field along the axis of magnetization. In the case of a permanent magnet material having a constant flux density with applied field (a condition nearly satisfied by the selection of a magnet of low recoil permeability and avoiding strong fields as mentioned above) the moment is equal to the intrinsic magnetization multiplied by the volume. Additionally, under these conditions, the intrinsic magnetization becomes equal to the remanence of the magnet, B_{rem}, and the resultant force on the magnet is given by

$$F_x = B_{rem} V_m \frac{H_{x1} - H_{x2}}{x1 - x2} \tag{15.6}$$

311

where H_{x1} and H_{x2} are the values of applied field at the poles of the magnet situated at positions $x1$ and $x2$ (these are usually near to the end of a permanent magnet) and F_x is the force along the x-axis which is parallel to the vector $(x1, x2)$. For a relatively short magnet in a field of gentle, gradient this can be written in differential form

$$F_x = B_{rem} V_m \frac{dH_x}{dx} \tag{15.7}$$

A simple actuator can be constructed by fixing or glueing a series of disk type magnets to a linear spring device and surrounding these by a uniformly wound circular cylindrical coil. The maximum field gradient in this instance is near to each end of the coil and thus it is favourable to place a magnet at each end with opposing poles (the gradients are equal and opposite at each end), see Figure 15.6. This type of device has the advantages that the force on the magnet is insensitive to position and there is no mechanical coupling which results in a near to zero stiffness force actuator.

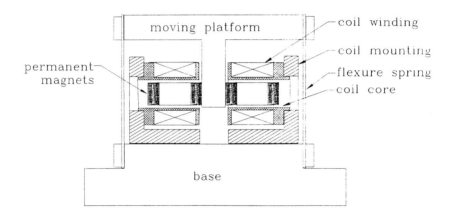

Figure 15.6. A two coil, four magnet solenoid magnet driven linear spring.

A disadvantage is that the coil should be as close to the magnets as is possible. This is hard to achieve without the magnets touching the core. A useful method for checking this is to electrically isolate the magnets from the core and then check for a short circuit as an indicator of mechanical contact. The optimum geometry of the coil for such an actuator is presented in the paper of Smith and Chetwynd,[40] while the design parameters for a coil having a uniform field gradient are presented in the paper of Pouladian-Kari.[44]

At present it is possible that the common loud speaker may be one of the most efficient mechanisms achieving displacements in the region of 500 µm over a 20 kHz bandwidth. However, the flexure would have to be considerably redesigned if dimensional stability is to be achieved. Given this restriction it is feasible that nanometre level resolution may be possible.

15.2.8 Electrostatic

As with the above actuator this is a non-contact force driven device that uses the principle that any two electrodes separated by an insulating barrier will be attracted towards one another if there is an electric potential difference between them. The magnitude of this force is given by

$$F = \frac{\varepsilon A V^2}{2x} \qquad (15.8)$$

From this it can be seen that the force is always positive and is related to the permittivity (ε) and the reciprocal of the separation (x) and also the square of the potential. This force can then be changed by varying either the polarization or the applied potential to affect a change in the separation. The stiffness of this system is related to both the square of the ratio of voltage to stiffness and the dielectric constant. For an area of 100 mm^2 a separation of 0.01 mm and a bias voltage of 100 V such a device operating in air would have a stiffness of magnitude 4.3×10^{-4} Nmm^{-1} and will generate a force of 4.3×10^{-8} N. Thus, because of the very low forces associated with this type of actuator, it is most commonly used for ultrasonic excitation or in micro miniature devices such as submillimetre sized motors,[45,46] and microelectronic resonance accelerometers and may find application in resonant probe microscopy for the combined excitation and monitoring of small probes (possibly made from silicon, diamond or titanium diboride for high frequency scanning). Although the force characteristic is nonlinear, it is an exact relationship extending over many orders of magnitude and restricted mainly by the dielectric value of the insulating barrier. A disadvantage of this type of device is that, if air is used as the separating dielectric, then this is known to vary with both temperature and humidity and thus for very precise application a closely controlled environment is required.

15.2.9 Hydraulic

In contrast to the electrostatic actuator, hydraulic drives are capable of generating very large forces per unit size of actuator. This can take the form of a servo controlled hydrostatic bearing whereby displacement is obtained by applying a differential pressure. These have been used to obtain displacements of up to 100 µm with a resolution of 1 nm.[47] These are currently being used in Japan for the fine feed mechanism in ductile grinding machines. These devices tend to be very linear and have a stiffness in the region 10^8 Nm^{-1}. Another method for the utilization of this force is to use the hydraulic fluid to pressurize a very stiff balloon. Considering that pressures of 300×10^5 Pa are readily achievable and controllable then a tube of diameter 100 mm would be subjected to a tensile force of 236 kN. If the desired displacement is 100 µm, then the tube stiffness will be ≈ 2.36 GNm^{-1} or 0.01E. For a steel cylinder of length and outside diameter 100 mm, the inside diameter would be 70 mm. It is believed that this method has not been employed, therefore it is not certain whether this mechanism would be linear (although it is to be expected that with pressures so far below the yield stress of steel things would be relatively well behaved). A major problem is the elimination of noise being transmitted through the fluid from the pumps and control valves. One method for reducing this is the introduction of accumulators in the system which will provide damping at the expense of operating speed.

15.2.10 The Poisson's Ratio Drive

The above simple model of a pressurized cylinder does not take into account the effects of Poisson's ratio in the calculation of the axial strain. Imagine the previous cylinder, only this time the outside of the cylinder is subjected to the same pressure and there is a core inside so that there are no piston type forces acting on the ends. Now the cylinder walls are subjected to a uniform stress of 30 MPa in the radial direction. According to elasticity theory, there will be an axial strain in proportion to Poisson's ratio. For the cylinder in the example above, the extension due to this pressure will be 4.7 μm assuming a Poisson's ratio of 0.33. This will also be linear and of a similar stiffness. This type of actuator is also being considered for applications in ultra-stiff machine tool technologies. The speed of response of a hydraulic actuator is going to be restricted by the volume of fluid required for actuation and the response of the regulator mechanisms. Commercial suppliers are only capable of operating at frequencies in the region of a few Hertz even though, unlike the direct hydraulic drive the Poisson's ratio drive does not, in theory, require a large volume and should be capable of operating at higher speeds than the above.

A variation on the mechanism is that of applying a controlled squeeze to a shaft in a direction perpendicular to the required motion via a clamping arrangement. Using this technique, Jones,[20] was able to position a linear spring mechanism to within 1 nm using feedback.

15.3 SUMMARY

A large number of devices have been discussed in this chapter with an attempt to briefly outline their relative attributes. The objective of this summary is to recap the main advantages and limitations and provide a quick reference for choosing between actuator types. These will be listed with a brief sentence or two describing what may be considered to be the key points of interest.

Piezoelectric: These are cheap, ideal miniature, stiff, very high speed and usually operate over a displacement range below 10 μm. Hysteresis, creep and the high sensitivity of the coupling equations with temperature limit the achievable accuracy.

Mechanical micrometers: The large range makes these most desirable for larger scale applications. Manual operation capability makes these a desirable option for relatively simple tasks. However, at the nanometre level, they require considerable dexterity and usually some form of feedback due to backlash, medium stiffness, temperature sensitivity and many other inherent nonlinearities.

Friction drives: Again similar limitations to those of the micrometer with nonlinearity reduced by the friction grip of the moving stage. With feedback or using a stick slip type drive, large traverses are possible with nanometre level resolution.

Magnetostriction: Using some of the new giant magnetostrictive alloys that are now available, these devices will give relatively large displacements depending on the size of the device. These will ideally scale to very stiff actuators for control of large forces over submillimetre ranges. Presently, the alloying elements are prohibitively expensive. This and the magnetoelastic mechanism mentioned below exhibit considerable hysteresis and creep.

Magnetoelasticity: Actuators employing this effect are tuneable and can thus operate over a variety of ranges using the same device. Again, the preferred scaling is for large actuators with high drive forces.

Shape memory alloys and bimetallic strips: These will form low stiffness but large range actuators. Applications have been mainly in micromechanical robot arms or large conventional engineering mechanisms. They work mainly in a pseudo on-off mode and will generate a heat source that will result in temperature gradients within the device. The dynamic response of this device is almost invariably very slow.

Electromagnetic: Linear, characterizable actuators can be produced with the electromagnetic drive. The actuation mechanism is usually indirect and takes the form of an electromagnetic force actuator pushing against a linear spring. Such devices are cheap and easy to implement with a fast dynamic response and can even be designed for a near zero stiffness. Obviously, unless well screened, electromagnetic interference may be problematical.

Electrostatic: Again, this is a force generating actuator which must be transformed to a displacement by a spring of some sort. Although this phenomena is not linear, this affords possibly the largest range to resolution ratio of all. Forces generated are usually small and, as such, this makes miniature actuators more favourable. It will be sensitive to the permittivity of the gap separation and limited by its dielectric strength.

Hydraulic: The hydraulic actuator is most useful as a high force and stiffness actuator for the fine motion of large masses. It will be linear to a high accuracy within a reasonable range. The main disadvantage is that a hydraulic pressure generation unit is both expensive and bulky and it is hard to prevent from transmitting vibrations to the instrumentation.

Poisson's ratio: This has similar attributes to the hydraulic drive only approximately an order of magnitude less range.

REFERENCES

1. Hocken, R. and Justice, B. (1976) Dimensional stability, *NASA report* NAS8-28662, p.48
2. Jaffe, H. and Berlincourt, D. A., (1965) Piezoelectric transducer materials, *Proc. IEEE*, **53**(10), 1372-1386
3. Galleg-Juarez, (1989) Piezoelectric ceramics and ultrasonic transducers, *J. Phys. E: Sci. Instrum.*, **22**, 804-816
4. Newcomb, C. V. and Flinn, I. (1982) Improving the linearity of piezoelectric ceramic actuators, *Electronics Letters*, **18**(11), 442-444
5. Kaizuka, H. and Siu, B. (1988) A simple way to reduce hysteresis and creep when using piezoelectric actuators, *Japan J. Appl. Phys.*, **27**(5), L773-L776
6. Kaizuka, H. (1989) Application of capacitor insertion method to scanning tunnelling microscopes, *Rev. Sci. Instrum.*, **60**(10), 3119-3122
7. Jaffe, B., Cook, W. R. and Jaffe, H. (1971) *Piezoelectric Ceramics*, Volume 3 of a series of monographs on Non-Metallic Solids. London: Academic Press, pp. 83-85
8. Drake, B., Sonnenfeld, R., Schneir, J., Hansma, P. K., Slough, G. and Coleman, R. V. (1986) Tunnelling microscope for operation in air or fluids, *Rev. Sci. Instrum.*, **53**(3), 441-445
9. Hicks, T. R., Reay, N. K. and Atherton, P. D. (1984) The application of capacitance micrometry to the control of Fabry-Perot etalons, *J. Phys. E: Sci. Instrum.*, **17**, 49-55
10. Smith, D. P. E. and Binnig, G. (1986) Ultrasmall scanning tunneling microscope for use in a liquid helium storage dewar, *Rev. Sci. Instrum.*, **57**(10), 2630-2631
11. van Randeraat, J. and Setterington, R. E., (1974) Piezoelectric Ceramics, Mullard Ltd, London.

12. Matey, J. R., Crandall, R. S. and Bryki, B. (1987) Bimorph-driven x-y-z translation stage for scanned image microscopy, *Rev. Sci. Instrum.*, **58**(4), 567-570
13. Blackford, B. L., Dahn, D. C. and Jerico, M. H. (1987) High-stability bimorph scanning tunneling microscope, *Rev. Sci. Instrum.*, **58**(8), 1343-1348
14. Tritt, T. M., Gillespie, D. J., Kamm, G. N. and Ehrlich, A. C. (1987) Response of piezoelectric bimorphs as a function of temperature, *Rev. Sci. Instrum.*, **58**(5), 780-783
15. Joyce, G. C. and Wilson, G. C. (1969) Micro-step motor, *J. Phys. E: Sci Instrum.*, **2**(2), 661-663
16. Binnig, G. and Rohrer, H. (1982) Scanning tunneling microscopy, *Helv. Phys. Acta*, **55**, 726-735
17. Mamin, H. J., Abraham, D. W., Ganz, E. and Clarke, J. (1985) Two-dimensional, remote micropositioner for a scanning tunneling microscope, *Rev. Sci. Instrum.*, **56**(11), 2168-2170
18. Tojo, T. and Sugihara, K. (1985) Piezoelectric driven turntable with high positioning accuracy, *Bull. Japan Soc. of Prec. Engg.*, **19**(2), 135-137
19. Hatsuzawa, H., Toyoda, K. and Tanimura, Y. (1986) Speed control characteristics and digital servo system of a circular travelling wave motor, *Rev. Sci. Insturm.*, **57**(11), 2886-2890
20. Hatsuzawa, T., Tanimura, Y., Yamada, H. and Toyoda, K. (1986) Piezodriven spindle for a specimen holder in the vacuum chamber of a scanning electron microscope, *Rev. Sci. Instrum.*, **57**(12), 3110-3113
21. Jones, R. V. (1988) *Instruments and Experiences*; Papers on Measurement and Instrument design, J. Wiley and Sons, NY, paper XVIII, 219-247 and paper XVII, 203-204
22. Rowlands, H. A. (1902) *The Physical Papers of H. A. Rowlands*. Baltimore: The John Hopkins Press, 691-706.
23. Merton, T. (1950) On the reproduction and ruling of diffraction gratings, *Proc. Roy. Soc.*, **A201**, 187-191
24. Hall, R. G. N. and Sayce, L. A. (1952) On the production of diffraction gratings. II: The generation of helical rulings and the preparation of plain gratings therefrom, *Proc. Roy. Soc.*, **A215**, 536-550
25. Mindlin, R.D., Deresiewicz, H. (1953) Elastic spheres in contact under varying oblique forces, *Trans ASME: J. Appl. Mech.*, **75**, 327-344
26. Stanley, V. W., Franks, A. and Lindsey, K. (1968) A simple ruling engine for diffraction gratings, *J. Phys. E: J. Sci. Instrum.*, **1**, Ser. 2, 643-645
27. Reeds, J., Hansen, S., Otto, O., Carroll, A. M., McCarthy, D. J. and Radley J. (1985) High speed precision X-Y stage, *J. Vac. Sci. Technol.*, **B3**(1), 112-116
28. Pohl, D. W. (1987) Dynamic piezoelectric translation devices, *Rev. Sci. Instrum.*, **58**(1), 54-57
29. Smith, D. P. E. and Elrod, S. A. (1985) Magnetically driven micropositioners, *Rev. Sci. Instrum.*, **56**(10), 1970-1971
30. Clark, A. E., Teter, J. P. and McMasters, O. D. (1988) Magnetostriction "jumps" in twinned $Tb_{0.3}Dy_{0.7}Fe_{1.9}$, *J. Appl. Phys.*, **63**(8), 3910-3912
31. Oswin, J. R., Edenborough, R. J. and Pitman K. (1988) Rare-earth magnetostriction and low frequency sonar transducers, IOP/IEE/IEEE day seminar on Magnetic Sensors and Amorphous Materials, 20th April 1988, reproduced in *Aerospace Dynamics*, **24**, 9-13

32. Engdahl, G. and Svensson, L. (1988) Simulation of the magnetostrictive performance of Terfenol-D in mechanical devices, *J. Appl. Phys.*, **63**(8), 3924-3926

33. Yuan, Z., Feng, Z. and Zhao, W. (1987) Quick responding magnetostrictive micro-feeder, *Proc. 6th Int. Conference on Precision Engg.*, Osaka, 593-598

34. Squire, P. T. (1990) Phenomenological model for magnetization, magnetostriction and ΔE effect in Field-annealed amorphous ribbons, *J. Magnetism and Magnetic Materials*, **87**, 299-310

35. Bates, L. F. (1963) *Modern Magnetism.* Cambridge: Cambridge University Press, Chpt. XI.

36. Schetky, L. M. (1979) Shape memory alloys, *Scientific American*, 74-82.

37. Golestaneh, A. A. (1984) Shape-memory Phenomena, *Physics Today*, **37**(**4**), 62-70

38. Yaeger, J. R. (1984) A practical shape-memory electromechanical actuator, *Mechanical Engineering*, 51-55.

39. Jebens, R., Trimmer, W. and Walker, J. (1989) Micro actuators for aligning optical fibers, *Proceedings of the second IEEE workshop on Micro-Electro Mechanical Systems*, Salt Lake City, 35-39

40. Kuribayashi, K. (1989) Millimeter-sized joint actuator using a shape memory alloy, *Sensors and Actuators*, **20**, 57-64

41. Smith, S. T. and Chetwynd, D. G. (1990) An optimized magnet-coil force actuator and its application to linear spring mechanisms, *Proc. Inst. Mech. Eng.*, **204**(C4), 243-253

42. Parker, S. F. H., Pollard, R. J., Lord, D. G. and Grundy, P. J., (1987) Precipitation in NdFeB-type materials, *Trans. IEEE*, **Mag-23**(5), 2103-2105

43. Sagawa, M., Fujimmura, S., Yamamoto, H., Matsuura, Y. and Hiraga, K. (1984) Permanent magnet materials based on the rare earth-iron-boron tetragonal compounds, *Trans. IEEE*, **Mag-20**(5), 1584-1589

44. Hadfield, D. (1962) *Permanent Magnets and Magnetism* London: J. Wiley and Sons, chapters 2, 6 & 7.

45. Pouladian-Kari, R., Parkes, D., Jones, R. M. and Benson, T. M. (1988) A multiple coil solenoid to provide an axial magnetic field with a near-linear gradient, *J. Phys. E: Sci. Instrum.*, **21**, 557-559

46. Fan, L., Tai, Y. and Muller, R. S. (1989) IC-processed electrostatic micromotors, *Sensors and Actuators*, **20**, 41-47

47. Tai, Y. and Muller, R. S., (1989) IC-processed electrostatic micromotors, *Sensors and Actuators*, **20**, 49-55

48. Miyashita, M. and Yoshioka, J. (1982) Development of ultra-precision machine tools for micro-cutting of brittle materials, *Bull. Japan Soc. of Prec. Engg.*, **16**(1), 43-50

Chapter 16

CALIBRATION OF LINEAR TRANSDUCERS BY X-RAY INTERFEROMETRY

D. KEITH BOWEN

A review of the techniques of X-ray interferometry relevant to micro-displacement metrology is given, with particular emphasis to its role in calibration. Monolithic interferometer designs are advocated because of their favourable metrology loop structures; these can achieve standard deviations as low as 5 picometres. Examples of the calibration of differential transformer, laser interferometer and capacitance gauges are given.

16.1 INTRODUCTION

Many engineering displacement transducers have resolution in the nanometre region, though this has only been exploited in recent years. Most strikingly, the last decade has seen the development of a range of scanning tip microscopies and other profiling methods that can be used reliably with atomic resolution. These include scanning tunnelling, atomic force, capacitance, thermal, magnetic and resonance beam microscopies, optical probes and stylus probes with optical or differential transformer readout. All these instruments possess reasonably high bandwidths ($\sim$ kHz) so that detailed information can be collected at a rate acceptable for regular use.

The traditional calibration gauge, a step standard, has not kept up with this development. However, during the same decade X-ray interferometry has been established as a metrological tool of picometric sensitivity (i.e. one part in 10^{12}). Its bandwidth is very small ($\sim$ Hz) with practical X-ray sources and so it does not compete with the other sensors directly. Its one major advantage over all the others is that its traceability to the length standard is both very direct and extremely well defined. It has thus become the natural tool for the calibration of displacement transducers at nanometre, and smaller, scales.

The traceability derives from the direct use of the silicon lattice parameter as a secondary standard of length. The lattice spacings of extremely pure, electronics grade single crystal silicon have been very precisely determined. An absolute determination is extremely expensive, but once it has been done for one crystal, others can be readily compared to it. In fact, such is the control of modern production that the lattice parameters can be predicted in advance to a few parts in 10^6 depending upon the manufacturing process used (float zone or Czochralski, level of doping). X-ray diffraction methods allow the comparison of the lattices of two silicon crystals to a precision of 10^{-8}. The error associated with using different X-ray interferometers to provide a local length standard for a calibration will be certainly no more than a few femtometres!

Since its invention in 1965 by Bonse and Hart[1], X-ray interferometry has been used mainly as a special instrument for experimental materials physics. The only dimensional metrology to which it has been generally applied, and then for its own characterization, is the calibration of the silicon lattice against the optical length standard, originally at the National Bureau of Standards (NBS), Deslattes[2], Deslattes and Henins[3] and later at PTB, Becker, Seyfried and Siegert[4] and elsewhere, for example Basile *et al.*[5]; the thrust of many of these studies was a better determination of the Avogadro number. Even though Hart[6] had pointed out its usefulness in metrology, its use as a sensor or calibrator was not pursued until 1983 when Chetwynd, Siddons and Bowen[7] showed that it could be used under practical conditions to calibrate inductive microdisplacement gauges.

The purpose of this chapter is to review the operation of the X-ray interferometer, to survey the developments in calibration systems since its introduction and to indicate how X-ray interferometric calibration may be applied to a wide range of practical situations.

16.2 PRINCIPLES OF X-RAY INTERFEROMETRY

If a beam of X-rays impinges on a thin crystal at the Bragg angle for lattice planes that lie nearly normal to the plane of the crystal, at an angle given by

$$n\lambda = 2d\sin\theta \tag{16.1}$$

where λ is the wavelength, n an integer, d the lattice spacing and θ the angle of incidence of the beam on the lattice planes, three beams will emerge from the other side. One is merely an attenuated direct beam which is of no interest other than as a source of noise in the detectors. The others are the "diffracted" and "forward diffracted" beams which emerge symmetrically at the Bragg angle from the same point on the crystal surface. The crystal has acted as a beam splitter, providing two wavelength selected beams diverging from each other. A second thin crystal, or "blade", can then be used to split each of these beams, providing two which continue to diverge and two which converge. When these meet they will interfere and produce standing waves of the electric field of the X-ray propagation. The two combining beams will thus "know" the spatial phase of the two blades. If a third blade is placed at this convergence point, there will be interference between this standing wave and that present in the crystal by its distribution of electrons - the atomic lattice. If the lattice planes are displaced laterally, the power switches between the two emergent (diffracted and forward-diffracted) beams of the whole system. Figure 16.1 shows the ray paths for this system which comprises all the X-ray optics of an interferometer.

If the third blade is traversed slowly in its own plane and vertically as shown in Figure 16.1, the power in both the O and H beams will vary sinusoidally, completing one cycle for each displacement through one lattice spacing. The device is symmetrical and any one blade may be traversed; if the central one is used the displacement fringes occur with a pitch of half the lattice parameter. On real devices there is always considerable power in both the O and H beams throughout with a fluctuation in intensity representing low-contrast fringes. Although the physics is quite different, there is a close formal analogy to an optical Moire fringe system with a grating of atomic pitch.

A practical device will clearly require the use of lattice planes providing a strong reflection so that good intensity output is obtained. Silicon (220) and (111) planes are both suitable. The spacings at 20 °C are 0.3135625 nm for (111) planes and 0.1920170 nm for (220) planes, Hart[8], for float-zone silicon. In each case the experimental uncertainty of the values is below 2 fm.

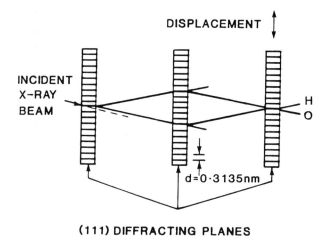

Figure 16.1 X-ray optical paths through an interferometer.

Several features of the X-ray interferometer make it particularly suited to calibration and in respect of its error analysis it is inadequate to follow the analogy with optical interferometry or moire gratings. These features are:

1. The pitch of the displacement fringes depends only upon the crystal lattice and is independent of the X-ray wavelength used to read it; the wavelength only determines the angle at which diffraction occurs.

2. Manufacturing errors reduce contrast in the fringes but generally do not affect their sinusoidal nature.

3. No errors are known that could introduce odd harmonics.

4. Defocussing is the only known source of second harmonic distortion and this is negligibly small with the beam apertures used in practice.

5. All refractive indices are within a few parts per million of unity at X-ray wavelengths, and air absorption for hard X-rays is small; these facts mean that air turbulence and other environmental fluctuations have negligibly small influence upon the operation of the system.

6. Since ray sources have low intensity compared to optical sources, it is necessary to integrate over, typically, several seconds to get a good signal to noise ratio. The compensating advantage of this is that photon counting detectors can be used which provide inherently linear intensity measurement without the square-law distortion usually found in photocurrent detectors.

7. The only noise in the output is the statistical fluctuations in the rate of photon arrivals, which is known to be governed by Poisson statistics and is calculable from the measured data.

The reliability of fringe division is therefore likely to be much higher than for optical instruments.

Silicon and germanium are the only feasible materials for making interferometers at an acceptable cost, and silicon is both cheaper and has lower X-ray absorption. Its mechanical and thermal properties are fortunately good for this purpose. It is quite hard and apparently perfectly elastic-brittle at room temperatures. If the crystal is stressed it will either fracture or return to its original form on removal of the stress. This is a highly desirable characteristic in a secondary standard: it will either maintain its calibration or, if too badly abused, be obviously broken. It has high thermal conductivity so there are unlikely to be thermally induced strains to distort the lattice and the blades are unlikely to be at significantly different temperatures. Its coefficient of linear thermal expansion is quite small, about 2×10^{-6} K^{-1}, and will influence the lattice parameter directly but overall expansion of the device has no effect on the measurement scheme. Even with a shift of 10 K from the nominal working temperature the fringe pitch error will be only about 0.006 pm.

16.3 PRACTICAL IMPLEMENTATIONS

Conceptually the production of an X-ray interferometer is simple. The three blades of identical material and orientation are produced by machining away the surface of a suitably oriented single crystal to leave them standing proud. Since one blade must translate relative to the others a slideway has to be introduced. Most research groups do this by first testing the crystal to verify correct interferometer operation, then slicing the crystal between the blades and mounting the pieces containing two and one blade onto a separate translation mechanism. There are considerable practical difficulties in executing this plan. The crystal structure in the machined blades must remain virtually perfect. The crystals must be mounted so that there is minimal strain in the blades since this will distort the lattice. The separated blade must be re-aligned to the first two with extreme parallelism and the translation stage must maintain that parallelism during operation.

Achieving low damage in the blades is a matter of experience in machining brittle materials, using high surface speed diamond grinding and etching to remove machining damage. Careful design of the crystals, incorporating strain relief cuts, can ensure that there is a long path through the crystal from the mounting points to the blades. Any strain field induced by the mounts will then have decayed to a negligible intensity at the blades (St. Venant's principle). The translation system is usually an elastic mechanism driven through reduction levers by piezoelectric actuators. Examples of this type are given by Deslattes[2] and Becker Seyfried and Siegert.[9] Both also discuss briefly the question of aligning the blades. Generally the lattices should be oriented to each other to better than 5 nrad in the most sensitive axis and perhaps 5 μrad in the least sensitive. The stages in doing this are first kinematic location, often involving cementing in position, then optical alignment and finally observations of the X-ray propagation through the blades. Detailed descriptions of systematic procedures for alignment are given in Mewes, Lehrke, Rademacher, Seyfried and Reim.[10]

The advantages of the separated blade method are that it allows flexibility of application and, in principle, long traverses; silicon crystals can be obtained in lengths up to 1 m! This can be important in lattice calibration where the optical standard fringes are around 1000 times the size of the X-ray fringes and many fringes need to be counted to achieve precision. The complexity of the re-alignment is traded off against the relatively simple working needed to produce the silicon components. For calibration purposes we prefer a monolithic construction in which an elastic translation mechanism supporting one blade is machined into the single silicon crystal. Such designs require fairly complex machining of the brittle crystal and provide modest maximum movements but alignment is relatively simple since the blades are never totally separated from each other. This approach, incorporating a magnetic drive in place of the usual piezoelectric actuator was favoured by Hart and it was with one of his monoliths that Chetwynd et al.[7] first demonstrated transducer calibration. They proposed new monolith designs which preserved many of the features of Hart's but which were optimized for metrological applications. Further development of these has occurred since.

For metrology purposes, the two design types can be compared by examining the metrology loops and force loops needed for calibrating a stylus-driven sensor, using the arguments of Chetwynd et al.[11] The metrology loop of the monolith is from the moving blade, through the translation stage to the fixed blades and is completed by the X-ray paths, which in mechanical terms can be considered a zero-length connector. The sensor loop is from the moving blade through the stylus arm and sensor to its support frame, from that frame to the fixed blades and then through the translation stage back to the fixed blade. This loop is bound to carry some of the force used to hold the sensor to the interferometer (usually gravitational). A force loop passing through the translation stage is needed to cause it to displace. Separated blade designs can be arranged to follow the precision design guide-line that this force loop has as little common path as possible with the metrology loops. However they inevitably require that there are several physical interfaces in the X-ray metrology loop which will also be part of the sensor loop. These are likely to reduce the dimensional stability of the loops at the sub-nanometre level. This potential error is significant for a micro-displacement calibrator although unimportant for most applications to which X-ray interferometry has been applied.

Monolithic interferometers optimize the metrology loops. The X-ray loop is contained totally within the single crystal and has no interfaces. The sensor loop has a minimal number of interfaces, one at the probe and one at the frame mounts to the fixed blades plus those inherent to the sensor and its frame. However the drive force loop must share path with the metrology loops since the monolith is distorted to provide the motion. By using a large monolith to provide a metrological base to which other sub-systems are referenced, it can be arranged that the common path be common to all the loops so that any distortion within it is cancelled from the calibration process. Thus a typical modern metrological interferometer has the form shown in Figure 16.2. In operation the blade motion is vertical and governed by the simple leaf spring mechanism. The monolith rests kinematically located on a base plate that contains a solenoid coil forming half of the force drive. A small saturated magnet cemented to the lower arm of the monolith sits in the neck of the coil so that a controlled deflection occurs as the current is varied. As there is no physical contact between the base and the moving stage of the monolith, vibration coupling is minimized. The sensor and its carrying frame are mounted kinematically onto the upper surface of the monolith with the probe contacting the upper arm. The magnet centroid and the contact point of the probe lie in the plane of the moving blade to minimize Abbe offset errors. This plane intersects the centre of the leaf springs to minimize parasitic deflections, Jones and Young.[12] The arms are cantilevered from the stage to increase the path length for the attenuation of strain fields induced at the drive points.

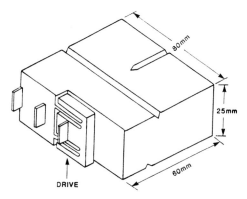

Figure 16.2 Metrological X-ray interferometer monolith.

The bulk of the monolith carries the kinematic mounts and the fixed blades. The springs are the only part of the system shared between the principal loops, assuming that the strain fields from the upper and lower face mountings are dissipated in the bulk without significant interaction. The springs also carry the probe force loop thus providing in-built compensation for probe spring rate effects. This simple feature is the source of the superiority of monolithic designs for calibration work and it is unlikely that a separated blade design would be used for this purpose unless an exceptionally large traverse were needed. Experimental evaluation of spring mechanisms similar to those used in the interferometers, with length 20 mm, breadth 5 mm and thickness 1 to 2 mm, indicates stiffnesses of 0.1 to 1.0 N m^{-1} and displacements of around 100 μm before breakage. These are quite close to theoretical expectations and were obtained while using only routine good practice when cutting the silicon. Monolithic designs can clearly provide traverses quite long enough for nanometre sensitivity calibration.

Figure 16.3 shows a monolith in relation to the other principal system components, a collimated X-ray source set at the Bragg angle to the blades and a scintillation detector with slits to select only the O beam from the output (Bowen, Chetwynd and Schwarzenberger[13]). Although we often use an active air suspension for vibration isolation, the monolithic interferometer is remarkably insensitive to vibration considering its extreme sensitivity and simple damping mats and acoustic shielding normally suffice. Thermal shielding is also provided to minimize convection currents and the experiments are performed in a temperature controlled-room. In practice we find that the system being calibrated is usually much more sensitive to both temperature fluctuations and vibration. The X-ray source is a fixed anode tube allowing up to 1.5 kW input power with a molybdenum target. Collimation provides a rectangular beam approximately 1 mm by 7 mm (its long axis would project out of the paper in Figure 16.1). Normally the output slits select only about 3 to 5 mm from this width. The Bragg angle for Mo K$_\alpha$ radiation on silicon (111) planes is close to 6.5°. Typical operating conditions for the tube are 45 kV and 30 mA. Depending on the beam width used this provides mean intensities of 1000 to 2500 counts per second in the O beam.

16.4 MONOLITH MANUFACTURE

Monoliths such as those illustrated here are produced from 100 mm (4") silicon boules with most operations using a conventional silicon/quartz slicing machine (Meyer and Burger QS3). Full details have been given elsewhere.[11,13] Manufacturing tolerances are not at all critical (Smith, Chetwynd and Bowen[14]). The main mountings are kinematic, and an accuracy of 100 μm in spring width, 10 μm in blade thickness and 1 μm in blade straightness (these last two only over the few mm^2 actually in the X-ray beam) are adequate; greater accuracy would result in better contrast and hence faster measurement but is not essential for operation.

Figure 16.3 A typical calibration system.

16.5 ALIGNMENT AND TWIST MEASUREMENT

The most critical alignment issue in setting up an interferometer is associated with relative twist of the moving blade in its own plane. A twist by a small angle α will cause X-ray Moire fringes across the output beam with a pitch

$$s = d / \alpha \qquad (16.2)$$

where d is the lattice parameter. So on (111) planes fringes of 1 mm pitch will be caused by a twist of around 0.3 μrad. These fringes are distinct in type from the displacement fringes which are a modulation of the net intensity across the whole beam as the blade traverses. However, as the blade traverses in their presence, each position across the beam passes through maximum and minimum intensities at different instants so that the change in the total intensity is small: the moiré fringes reduce the contrast of the displacement fringes. Indeed there is zero contrast if an exact number of cycles of the moire pattern lies across the aperture. In practice, successful operation can only be achieved if there is less than one half cycle of moire in the beam. For a 5 mm beam this requires that the twist is less than 30 nrad (about 0.005 arcseconds). Final adjustment for twist is by the direct observation of the moire fringes. Previously this has been a lengthy process involving photographic exposure, development and measuring but recently the process has been greatly speeded up by the use of an X-ray sensitive image detector and fringe-recognition image-processing programs (Chetwynd, Cockerton and Smith 1990).

16.6 DISPLACEMENTS CALIBRATION

The simplest method of calibrating a sensor is to drive the monolith stage, measuring its deflection against the sensor output, and to count the fringe peaks. (Note that because long counting times are needed the movement must be stepwise rather than continuous). Estimates are made of the part fringes at either end of the traverse and the combined fractional fringe count is multiplied by the lattice parameter to provide a value for the "true" displacement. Following the conventions of optical interferometry, fringe estimation would use in essence a least squares sinusoidal fit to the intensity data. This may certainly be done but the situation is not the same as the optical one. The fringe contrast is low, perhaps down to 20% or less, and estimating the contrast is itself subject to experimental error. Further, because of the low X-ray intensities, the variance on individual points from the Poisson distributed counting statistics may be significantly large. If J is the measured intensity (counts s^{-1}) then the total count over time t seconds will be Jt and the variance,

$$var(J) = J/t \tag{16.3}$$

Thus it is always possible to improve the counting precision by extending the time spent at each point. In practice this may not be desirable because extending the measurement time increases the vulnerability of the calibration to environmental drift. In the presence of these errors it is no longer clear that least squares fits are superior to simpler approximate methods and there is a good case for estimating fringes from methods based on the Stirling approximation. Bowen et al.[13] discuss this approach and consider in detail the minimisation of errors in real measurements.

Hart[6] observed that phase quadrature signals akin to those provided by optical Moire fringe transducers can be produced by an X-ray interferometer. This has attractions for metrology applications since it improves somewhat the precision of fringe interpolation and also provides directional information. It is no longer necessary to assume that the traverse moves monotonically in the direction specified. While this is not too important in simple calibrations, it places severe restrictions on more sophisticated operations. The H and O beams both contain fringe information and generally it will be in different phases. The actual phase difference depends on the exact thickness of silicon traversed by each of the interfering beams and is not easily controlled during manufacture. In none of the interferometers built by our group has the phase difference been sufficiently close to $\pi/2$ to be useful. Also the background noise is inherently higher in the H beam. Other groups have directly exploited both beams together, for example Bergamin, Cavagnero and Mana.[15]

An alternate way of providing quadrature outputs is to split the H beam and to introduce a phase plate into either of the beams converging on the third blade. Several slightly different approaches may be used, but the one that we have found most cost effective and reliable is to use the whole beam width for good counting statistics, moving the phase plate alternately into and out of the beam. It requires that the phase plate be removed and replaced with good precision and low vibration. Polymer films of the order of 100 μm thick make suitable phase plates of low absorption. The exact phase shift can be adjusted by rotating the film so that the beam passes through it at an angle.

Optical interferometry commonly uses phase stepping to provide improved fringe analysis, see Grievenkamp[16] for a general review, and similar methods can be used with the X-ray interferometer. A three phase system ideally has the phases equally spaced at $2\pi/3$ but a simple analytical form also arises if the two outer phases are symmetrically placed relative to the middle one. The shift cannot be too far from $2\pi/3$ or numerical errors start to become important. Schwarzenberger, Chetwynd and Bowen[17] showed that such a system can readily be implemented. The central phase uses the interferometer simply and the outer phases are provided by placing a phase plate into each of beams R or T (or RR and TR) in turn. Provided the plate remains parallel to the blades this will provide the symmetrical shifts. It cannot be rotated to adjust the phase so a suitable thickness must be selected. This is not critical although the closer the shift is to $2\pi/3$ the better. This approach requires three counts at each step and so is slower than the quadrature systems, but it has distinct compensating advantages: in particular, the fringe interpolation does not depend upon assumptions that the contrast and mean intensity remain constant from fringe to fringe as do the simpler systems. A suitable phase stepper consists of a small piece of 90 μm polyacetate film cemented to the needle of a microammeter. Switching preset voltages to the meter movement provides a precise, low inertia positioner.

Three phase X-ray interferometry represents the current state of the art for calibration systems for it provides high quality fringe interpolation with a minimum of assumptions about the system behaviour. With a contrast of around 20% and a mean intensity in the region 1000 to 2500 counts s^{-1}, counting for 3 to 5 s at each phase provides a 95% confidence interval of less than 5 pm. A typical calibration run might involve some hundreds of points and so take 20 to 60 minutes to perform. This is acceptable if extreme precision is required but perhaps rather slow for more routine calibrations. Thus single phase systems still have a role to play, providing 95% precisions of 10 to 20 pm according to conditions in roughly one third of the time.

Silicon is ideally elastic and so, provided that the magnet on the drive remains fully saturated, the traverse is extremely linear with drive current and has no detectable hysteresis. Once initially calibrated against the lattice, the drive may be used with high confidence as an open loop positioner to a small fraction of a fringe.[17] Thus for point to point calibrations (equivalent to step standards) rather than full linearity tests, much faster operation can be achieved by making multifringe steps knowing that the whole number of fringes skipped can be accurately determined from the drive demand signal. Interpolation is used at each end to find the fractional part of a fringe for the actual step. Such procedures are quite straightforward except that care must be taken to avoid a one fringe error being introduced if the step ends at a relative fringe phase close to a multiple of λ.

Demonstration calibrations of various surface metrology gauges and industrial laser interferometers have been reported elsewhere, for example Bowen, Chetwynd and Schwarzenberger,[13,18] and examples are quoted here without detailed discussion. Figure 16.4 shows a calibration result with a Talystep gauge and a single-phase interferometer measurement. An up-down ramp of 0.4 nm was used to check hysteresis and to exclude the thermal drift from the calibration. The fidelity with which the transducer follows the reversal of the ramp - indicating better than 12 pm sensitivity and freedom from hysteresis - is quite remarkable. The importance of thermal stability is seen when it is realized that the Talystep/interferometer calibration difference apparent here (if perfectly calibrated the Talystep signal would be identical to the coil demand line) could also be ascribed to a temperature drift of 0.005 K over 500 s. This stability requirement is primarily on the engineering transducers; as discussed above the interferometer is much more stable.

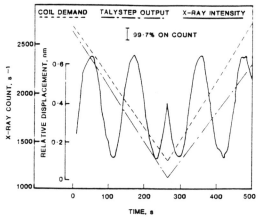

Figure 16.4 Calibration result with a Talystep gauge.

Laser interferometers are often regarded as the ultimate metrology tool, but there must always remain questions of fringe interpolation and turbulence corrections. This is illustrated in Figure 16.5, which shows a calibration run on an HP laser interferometer; in order to run both interferometers simultaneously the HP ×36 expander unit was used, and effectively we are interpolating well beyond the manufacturers' specification. This result is therefore more illustrative of the possibilities of calibrating optical instruments. The laser instrument suffers from quantization noise and turbulence apart from calibration error and it is clearly seen that the X-ray instrument can quantify these errors.

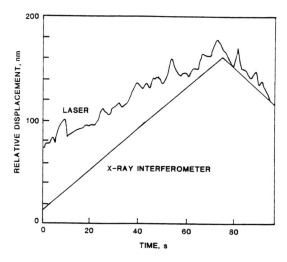

Figure 16.5 Calibration result with an over-interpolated HP laser interferometer.

Obviously there is no difficulty in principle in taking an optical reference from the moving stage and the low stylus forces of the Talystep-type gauges neither couples vibration into the blade nor applies torque to twist it to an extent which seriously reduces the contrast. With other types of sensors this may not be so clear and certainly the same conditions would not prevail during an attempt at direct calibration of a micro-actuator. A capacitive sensor may be built in series with the X-ray interferometer to provide a buffer, as is shown in Figure 16.6.

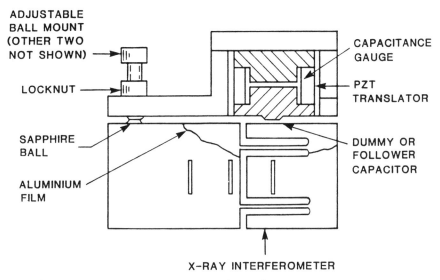

Figure 16.6 Schematic of a series capacitor calibrator.

Small parallel plate capacitors with gaps of around 10 μm interrogated by modern low-noise electronics have picometric sensitivity although they have non-linearity and, in many cases, holding an absolute calibration to that level is problematic. In the illustration an actuator is tested by maintaining a constant capacitance by tracking the actuator against the interferometer displacement and relating its drive signal to the fringe count. The capacitor is used as a nulling device so that its displacement characteristics, other than its high sensitivity, are unimportant. A modification of this scheme allows the calibration of higher contact force probes by contacting them with the upper capacitor plate so that the actuator, either piezoelectric or a coil-magnet drive, provides the probe displacement force in response to the movement of the monolith stage.

Experimental evaluation of open capacitive sensors of approximately 6 mm diameter and 10 μm gap, nominal capacitance 20 pF, with specially designed electronics (supplied by Queensgate Instruments Ltd) indicates that real calibrations can often be simplified. The small range performance is so good that the capacitor may be used as a transfer standard and not just as a nulling device. Figure 16.7 shows a typical output from the gauge when the interferometer coil drive is ramping slowly up and down with nominally 4 nm amplitude.

A bandwidth of 25 Hz has been used on the detectors to indicate total noise levels. In calibration use this bandwidth could be reduced to 1 Hz or lower. The noise is mainly from the electronics and acoustic pick-up, so this reduction of bandwidth is very effective: at 1 Hz the noise on a 1 nm ramp is compatible with the interferometer precisions discussed above.

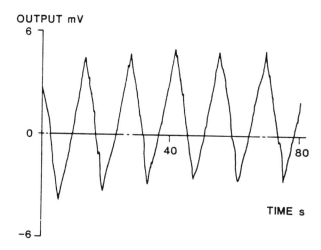

Figure 16.7 Response of 6 mm capacitive sensor to a 4 nm ramp.

The transfer calibration is carried out in one set-up so that no accuracy is lost between the two stages. First the interferometer is used to calibrate the capacitive gauge against the lattice parameter. Then the actuator, either one being calibrated or one driving another sensor, is operated with its displacement measured by the capacitive gauge. If required the capacitor can be rechecked against the lattice immediately after the main test. The traceable accuracy is slightly reduced by the extra step in the chain, but in practice this may be more than offset by the higher speed of calibrating the external device. The monolith drive can also be used as a transfer standard in some applications and the combination of these two options allows a wide range of practical problems to be tackled efficiently.

16.7 CONCLUSIONS

This chapter has shown how X-ray interferometry can be used effectively as a practical calibration tool. Monolithic designs of interferometer are preferred for this purpose because of their inherent metrological structure and because they ease the setting procedures, especially if direct video imaging to analyse the X-ray moire fringes is used. Their application is extended by the use of capacitive gauges in nulling mode to isolate subsystems and both capacitive and coil drive transfer methods to provide more rapid traceable calibrations to picometre sensitivities. These techniques make practical the use of the silicon lattice as an industrially usable, readily portable, secondary standard of length.

ACKNOWLEDGEMENTS

Dr DG Chetwynd, Dr S Cockerton, Dr ST Smith, Mrs DR Schwarzenberger and Dr. S Harb have contributed significantly to the body of work of which this forms part. Aspects of this work have been funded by the SERC through grant GR/F62698 and through a DTI Nanotechnology LINK award in collaboration with Queensgate Instruments and Rank Taylor Hobson.

REFERENCES

1. Bonse, U. and Hart, M. (1965) An X-ray interferometer. *Appl. Phys. Lett.*, **6**, 155-6

2. Deslattes, R.D. (1969) Optical and X-ray interferometry of a silicon lattice spacing. *Appl. Phys. Lett.*, **15**, 386-8

3. Deslattes, R.D. and Henins, A. (1973) X-ray to visible wavelength ratios. *Phys. Rev. Lett.*, **31**, 972-5

4. Becker, P., Seyfried, P. and Siegert, H. (1982) The lattice parameter of highly pure silicon single crystals. *Z. Phys. B*, **48**, 17-21

5. Basile, G., Bergamin, A., Cavagnero, G., Mana, G. and Zosi, G. (1989) Progress at IMGC in the absolute determination of the silicon d(220) lattice spacing. *IEEE Trans. Instrum. Meas.*, **38**, 210-6

6. Hart, M. (1968) An angstrom ruler. *J. Phys. D: Appl. Phys.*, **1**, 1405-8

7. Chetwynd, D.G., Siddons, D.P. and Bowen, D.K. (1983) X-ray interferometer calibration of microdisplacement transducers. *J. Phys. E: Sci. Instrum.*, **16**, 871-4

8. Hart, M. (1981) Bragg angle measurement and mapping. *J. Cryst. Growth*, **55**, 409-27

9. Becker, P., Seyfried, P. and Siegert, H. (1987) Translation stage for a scanning X-ray optical interferometer. *Rev. Sci. Instrum.*, **58**, 207-11

10. Mewes, E-R, Lehrke, K., Rademacher, H-J, Seyfried, P. and Reim, G. (1974) Berichte uber Arbeiten am Rontgen-Verschiebinterferometer; PTB, Braunschweig, Report APh-8

11. Chetwynd, D.G., Cockerton, S.C., Smith, S.T. and Fung, W.W. (1991) The design and operation of monolithic X-ray interferometers for super-precision metrology. *Nanotechnology*, **2**, 1-10

12. Jones, R.V. and Young, I.R. (1956) Some parasitic deflexions in parallel spring movements. *J. Sci. Instrum.*, **33**, 11-15

13. Bowen, D.K., Chetwynd, D.G. and Schwarzenberger, D.R. (1990) Sub-nanometre displacements calibration using X-ray interferometry. *Meas. Sci. Technol.*, **1**, 107-19

14. Smith, S.T., Chetwynd, D.G. and Bowen, D.K. (1987) The design and assessment of monolithic spring systems. *J. Phys. E: Sci. Instrum.*, **20**, 977-83

15. Bergamin, A., Cavagnero, G. and Mana, G. (1989) Lattice bending in X-ray interferometers. *Z. Phys. B*, **76**, 25-31

16. Grievenkamp, J.E. (1984) Generalized data reduction for heterodyne interferometry. *Opt. Eng.*, **23**, 350-2

17. Schwarzenberger, D.R., Chetwynd, D.G. and Bowen, D.K. (1989) Phase measurement X-ray interferometry. *J. X-ray Sci. Technol.*, **1**, 134-42

18. Bowen, D.K., Chetwynd, D.G., Schwarzenberger, D.R. and Smith, S.T. (1990) Sub-nanometre transducer characterization by X-ray interferometry. *Precision Engineering*, **12**, 165-171

INDEX